# Calculus II

Tunc Geveci

Copyright © 2011 by Tunc Geveci. All rights reserved. No part of this publication may be reprinted, reproduced, transmitted, or utilized in any form or by any electronic, mechanical, or other means, now known or hereafter invented, including photocopying, microfilming, and recording, or in any information retrieval system without the written permission of University Readers, Inc.

First published in the United States of America in 2011 by Cognella, a division of University Readers, Inc.

Trademark Notice: Product or corporate names may be trademarks or registered trademarks, and are used only for identification and explanation without intent to infringe.

15 14 13 12 11        1 2 3 4 5

Printed in the United States of America

ISBN: 978-1-935551-44-7

www.cognella.com  800.200.3908

# Contents

**6 Techniques of Integration** — **1**
- 6.1 Integration by Parts .................................................. 1
- 6.2 Integrals of Rational Functions ..................................... 13
- 6.3 Integrals of Some Trigonometric and Hyperbolic Functions ........... 27
- 6.4 Trigonometric and Hyperbolic Substitutions ......................... 43
- 6.5 Numerical Integration ............................................... 55
- 6.6 Improper Integrals: Part 1 .......................................... 65
- 6.7 Improper Integrals: Part 2 .......................................... 79

**7 Applications of Integration** — **89**
- 7.1 Volumes by Slices or Cylindrical Shells ............................. 89
- 7.2 Length and Area ..................................................... 99
- 7.3 Some Physical Applications of the Integral ......................... 111
- 7.4 The Integral and Probability ....................................... 121

**8 Differential Equations** — **133**
- 8.1 First-Order Linear Differential Equations .......................... 133
- 8.2 Applications of First-Order Linear Differential Equations .......... 148
- 8.3 Separable Differential Equations ................................... 157
- 8.4 Applications of Separable Differential Equations ................... 171
- 8.5 Approximate Solutions and Slope Fields ............................. 179

**9 Infinite Series** — **187**
- 9.1 Taylor Polynomials: Part 1 ......................................... 187
- 9.2 Taylor Polynomials: Part 2 ......................................... 197
- 9.3 The Concept of an Infinite Series .................................. 207
- 9.4 The Ratio Test and the Root Test ................................... 217
- 9.5 Power Series: Part 1 ............................................... 228
- 9.6 Power Series: Part 2 ............................................... 241
- 9.7 The Integral Test and Comparison Tests ............................. 253
- 9.8 Conditional Convergence ............................................ 264
- 9.9 Fourier Series ..................................................... 272

**10 Parametrized Curves and Polar Coordinates** — **285**
- 10.1 Parametrized Curves ............................................... 285
- 10.2 Polar Coordinates ................................................. 294
- 10.3 Tangents and Area in Polar Coordinates ............................ 305
- 10.4 Arc Length of Parametrized Curves ................................. 310
- 10.5 Conic Sections .................................................... 316
- 10.6 Conic Sections in Polar Coordinates ............................... 325

| | | |
|---|---|---:|
| H | Taylor's Formula for the Remainder | **335** |
| I | Answers to Some Problems | **341** |
| J | Basic Differentiation and Integration formulas | **371** |

# Preface

This is the second volume of my calculus series, **Calculus I**, **Calculus II** and **Calculus III**. This series is designed for the usual three semester calculus sequence that the majority of science and engineering majors in the United States are required to take. Some majors may be required to take only the first two parts of the sequence.

**Calculus I** covers the usual topics of the first semester: **Limits, continuity, the derivative, the integral and special functions such exponential functions, logarithms, and inverse trigonometric functions.** Calculus II covers the material of the second semester: **Further techniques and applications of the integral, improper integrals, linear and separable first-order differential equations, infinite series, parametrized curves and polar coordinates.** Calculus III covers topics in **multivariable calculus: Vectors, vector-valued functions, directional derivatives, local linear approximations, multiple integrals, line integrals, surface integrals, and the theorems of Green, Gauss and Stokes.**

An important feature of my book is its **focus on the fundamental concepts, essential functions and formulas of calculus**. Students should not lose sight of the basic concepts and tools of calculus by being bombarded with functions and differentiation or antidifferentiation formulas that are not significant. I have written the examples and designed the exercises accordingly. I believe that "less is more". That approach enables one to demonstrate to the students the beauty and utility of calculus, without cluttering it with ugly expressions. Another important feature of my book is **the use of visualization as an integral part of the exposition**. I believe that the most significant contribution of technology to the teaching of a basic course such as calculus has been the effortless production of graphics of good quality. Numerical experiments are also helpful in explaining the basic ideas of calculus, and I have included such data.

**Remarks on some icons:** I have indicated the end of a proof by ■, the end of an example by □ and the end of a remark by ◊.

**Supplements:** An **instructors' solution manual** that contains the solutions of all the problems is available as a PDF file that can be sent to an instructor who has adopted the book. The student who purchases the book can access the **students' solutions manual** that contains the solutions of odd numbered problems via **www.cognella.com**.

**Acknowledgments: ScientificWorkPlace** enabled me to type the text and the mathematical formulas easily in a seamless manner. **Adobe Acrobat Pro** has enabled me to convert the LaTeX files to pdf files. **Mathematica** has enabled me to import high quality graphics to my documents. I am grateful to the producers and marketers of such software without which I would not have had the patience to write and rewrite the material in these volumes. I would also like to acknowledge my gratitude to two wonderful mathematicians who have influenced me most by demonstrating the beauty of Mathematics and teaching me to write clearly and precisely: **Errett Bishop and Stefan Warschawski**.

Last, but not the least, I am grateful to **Simla** for her encouragement and patience while I spent hours in front a computer screen.

**Tunc Geveci** (tgeveci@math.sdsu.edu)
San Diego, August 2010

# Chapter 6

# Techniques of Integration

In this chapter we introduce an important technique of integration that is referred to as **integration by parts**. We will focus on **the integration of rational functions** via partial fraction decompositions, **the integration of various trigonometric and hyperbolic functions**, and certain substitutions that are helpful in the integration of some expressions that involve the square-root. We will discuss **basic approximation schemes** for integrals. We will also discuss the meaning of the so-called **improper integrals** that involve unbounded intervals and/or functions with discontinuities.

## 6.1 Integration by Parts

Integration by parts is the rule for indefinite and definite integrals that corresponds to the product rule for differentiation, just as the substitution rule is the counterpart of the chain rule. The rule is helpful in the evaluation of certain integrals and leads to useful general relationships involving derivatives and integrals.

### Integration by Parts for Indefinite Integrals

Assume that $f$ and $g$ are differentiable in the interval $J$. By the product rule,

$$\frac{d}{dx}\left(f\left(x\right)g\left(x\right)\right) = \frac{df}{dx}g\left(x\right) + f\left(x\right)\frac{dg}{dx}$$

for each $x \in J$. This is equivalent to the statement that $f'g + fg'$ is an antiderivative of $fg$. Thus,

$$f(x)g(x) = \int \left(\frac{df}{dx}g(x) + f(x)\frac{dg}{dx}\right)dx$$

for each $x \in J$. By the linearity of indefinite integrals,

$$f(x)g(x) = \int \frac{df}{dx}g(x)dx + \int f(x)\frac{dg}{dx}dx$$

Therefore,

$$\int f(x)\frac{dg}{dx}dx = f(x)g(x) - \int g(x)\frac{df}{dx}dx$$

for each $x \in J$. This is the indefinite integral version of integration by parts:

**INTEGRATION BY PARTS FOR DEFINITE INTEGRALS** Assume that $f$ and $g$ are differentiable in the interval $J$. Then,

$$\int f(x)\frac{dg}{dx}dx = f(x)g(x) - \int g(x)\frac{df}{dx}dx$$

for each $x \in J$.

We can use the 'prime notation", of course:

$$\int f(x)\,g'(x)\,dx = f(x)\,g(x) - \int g(x)\,f'(x)\,dx$$

**Example 1**

a) Determine

$$\int xe^{-x}dx$$

b) Check that your response to part a) is valid by differentiation.

**Solution**

a) We will set $f(x) = x$ and $dg/dx = e^{-x}$, and apply integration by parts, as stated in Theorem 1. We have

$$\frac{df}{dx} = \frac{d}{dx}(x) = 1,$$

and

$$\frac{dg}{dx} = e^{-x} \Leftrightarrow g(x) = \int e^{-x}dx$$

The determination of $g(x)$ is itself an antidifferentiation problem. We set $u = -x$, so that $du/dx = -1$. By the substitution rule,

$$\int e^{-x}dx = -\int e^{-x}(-1)\,dx = -\int e^u \frac{du}{dx}dx = -\int e^u du = -e^u + C = -e^{-x} + C,$$

where $C$ is an arbitrary constant. In the implementation of integration by parts, any antiderivative will do. Let us set $g(x) = -e^{-x}$. Therefore,

$$\int xe^{-x}dx = \int f(x)g'(x)dx = f(x)g(x) - \int f'(x)g(x)dx$$

$$= x\left(-e^{-x}\right) - \int (1)\left(-e^{-x}\right)dx$$

$$= -xe^{-x} + \int e^{-x}dx = -xe^{-x} - e^{-x} + C,$$

where $C$ is an arbitrary constant.

b) The expression

$$\int xe^{-x}dx = -xe^{-x} - e^{-x} + C$$

is valid on the entire number line. Indeed, by the linearity of differentiation and the product rule,

## 6.1. INTEGRATION BY PARTS

$$\frac{d}{dx}\left(-xe^{-x} - e^{-x} + C\right) = -\frac{d}{dx}\left(xe^{-x}\right) - \frac{d}{dx}e^{-x} + \frac{d}{dx}(C)$$
$$= -\left(\frac{d}{dx}(x)\right)e^{-x} - x\left(\frac{d}{dx}e^{-x}\right) + e^{-x}$$
$$= -e^{-x} + xe^{-x} + e^{-x} = xe^{-x}$$

for each $x \in \mathbb{R}$. The use of the product rule is not surprising, since we derived the formula for integration by parts from the product rule. □

The symbolic expression

$$du = \frac{du}{dx}dx$$

is helpful in the implementation of the substitution rule. This symbolism is also helpful in the implementation of integration by parts. In the expression

$$\int f(x)\frac{dg}{dx}dx = f(x)g(x) - \int g(x)\frac{df}{dx}dx,$$

let us replace $f(x)$ by $u$ and $g(x)$ by $v$. Thus,

$$\int u\frac{dv}{dx}dx = uv - \int v\frac{du}{dx}dx.$$

Let us also replace

$$\frac{du}{dx}dx$$

by $du$, and

$$\frac{dv}{dx}dx$$

by $dv$. Therefore, we can express the formula for integration by parts as follows:

$$\int u\,dv = uv - \int v\,du.$$

Note that

$$v = \int \frac{dv}{dx}dx = \int dv.$$

**Example 2** Determine

$$\int x\sin(4x)\,dx.$$

**Solution**

We will apply the formula for integration by parts in the form

$$\int u\,dv = uv - \int v\,du,$$

with $u = x$ and $dv = \sin(4x)\,dx$. Therefore,

$$du = \frac{du}{dx}dx = dx,$$

and
$$v = \int dv = \int \sin(4x)dx = -\frac{1}{4}\cos(4x).$$

Therefore,
$$\int x\sin(4x)\,dx = \int u\,dv = uv - \int v\,du = x\left(-\frac{1}{4}\cos(4x)\right) - \int\left(-\frac{1}{4}\cos(4x)\right)dx$$
$$= -\frac{1}{4}x\cos(4x) + \frac{1}{4}\int\cos(4x)\,dx$$
$$= -\frac{1}{4}x\cos(4x) + \frac{1}{4}\left(\frac{1}{4}\sin(4x)\right) + C$$
$$= -\frac{1}{4}x\cos(4x) + \frac{1}{16}\sin(4x) + C,$$

where $C$ is an arbitrary constant. The above expression is valid on the entire number line. □

Indefinite integrals of the form
$$\int x^n \cos(ax)\,dx \text{ or } \int x^n \sin(ax)\,dx,$$

where $n$ is a positive integer and $a$ is a constant can be evaluated by applying integration by parts repeatedly, so that the power of $x$ is reduced at each step. The technique is to set $u = x^n$ and $dv = \cos(ax)\,dx$ or $dv = \sin(ax)\,dx$, so that
$$du = \frac{du}{dx}dx = nx^{n-1}dx$$

**Example 3** Determine
$$\int x^2 \cos(4x)\,dx.$$

**Solution**

We set $u = x^2$ and $dv = \cos(4x)\,dx$ so that
$$du = 2x\,dx \text{ and } v = \int \cos(4x)\,dx = \frac{1}{4}\sin(4x)$$

Therefore,
$$\int x^2 \cos(4x)\,dx = \int u\,dv = uv - \int v\,du$$
$$= x^2\left(\frac{1}{4}\sin(4x)\right) - \int\left(\frac{1}{4}\sin(4x)\right)2x\,dx$$
$$= \frac{1}{4}x^2\sin(4x) - \frac{1}{2}\int x\sin(4x)\,dx.$$

In Example 2 we evaluated the indefinite integral on the right-hand side via integration by parts:
$$\int x\sin(4x)\,dx = -\frac{1}{4}x\cos(4x) + \frac{1}{16}\sin(4x) + C$$

## 6.1. INTEGRATION BY PARTS

Thus,

$$\int x^2 \cos(4x)\,dx = \frac{1}{4}x^2 \sin(4x) - \frac{1}{2}\left(-\frac{1}{4}x\cos(4x) + \frac{1}{16}\sin(4x) + C\right)$$
$$= \frac{1}{4}x^2 \sin(4x) + \frac{1}{8}x\cos(4x) - \frac{1}{32}\sin(4x) - \frac{1}{2}C.$$

Since $C$ denotes an arbitrary constant, $-C/2$ is also an arbitrary constant. Therefore, we can replace $-C/2$ with $C$ and present the result as follows:

$$\int x^2 \cos(4x)\,dx = \frac{1}{4}x^2 \sin(4x) + \frac{1}{8}x\cos(4x) - \frac{1}{32}\sin(4x) + C.$$

□

In the following examples, we will not bother to indicate the existence of an arbitrary constant in an indefinite integrals that is encountered at an intermediate step.

Indefinite integrals of the form

$$\int x^n e^{ax}\,dx,$$

where $n$ is a positive integer and $a$ is a constant, can be evaluated by applying integration by parts repeatedly so that the power of $x$ is reduced at each step. The technique is to set $u = x^n$ and $dv = e^{ax}\,dx$.

**Example 4**

a) Determine
$$\int x^2 e^{-x}\,dx.$$

b) Compute
$$\int_0^4 x^2 e^{-x}\,dx.$$

**Solution**

a) We set $u = x^2$ and $dv = e^{-x}\,dx$, so that

$$du = \frac{du}{dx}dx = 2x\,dx \text{ and } v = \int e^{-x}\,dx = -e^{-x}.$$

Therefore,

$$\int x^2 e^{-x}\,dx = \int u\,dv = uv - \int v\,du = x^2\left(-e^{-x}\right) - \int \left(-e^{-x}\right) 2x\,dx = -x^2 e^{-x} + 2\int xe^{-x}\,dx.$$

In Example 1 we used the same technique to show that

$$\int xe^{-x}\,dx = -xe^{-x} - e^{-x}$$

Therefore

$$\int x^2 e^{-x}\,dx = -x^2 e^{-x} + 2\left(-xe^{-x} - e^{-x}\right) + C = -x^2 e^{-x} - 2xe^{-x} - 2e^{-x} + C$$

b) By the result of part a) and the Fundamental Theorem of Calculus,

$$\int_0^4 x^2 e^{-x} dx = e^{-x}\left(-x^2 - 2x - 2\right)\Big|_0^4 = 26e^{-4} + 2 \cong 2.47621$$

Thus, the area of the region $G$ between the graph of $y = x^2 e^{-x}$ and the interval $[0, 4]$ is $26e^{-4} + 2$. Figure 1 illustrates the region $G$. □

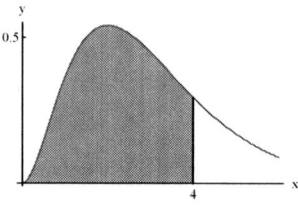

Figure 1

We can determine the indefinite integral of the natural logarithm via integration by parts:

**Example 5** Determine
$$\int \ln(x) dx.$$

**Solution**

We set $u = \ln(x)$ and $dv = dx$, so that

$$du = \frac{du}{dx} dx = \frac{1}{x} dx \text{ and } v = \int dx = x.$$

Therefore,

$$\int \ln(x) dx = \int u \, dv = uv - \int v \, du = \ln(x)x - \int x\left(\frac{1}{x}\right) dx$$
$$= x \ln(x) - \int dx$$
$$= x \ln(x) - x + C$$

Note that the above expression is valid if $x > 0$.

Indefinite integrals of the form

$$\int x^r \ln(x) \, dx,$$

where $r \neq -1$ can be determined with the technique that we used to antidifferentiate $\ln(x)$. We set $u = \ln(x)$ and $dv = x^r dx$ (if $r = -1$, the substitution $u = \ln(x)$ works).

**Example 6** Determine
$$\int \frac{\ln(x)}{\sqrt{x}} dx$$

## Solution

We have
$$\int \frac{\ln(x)}{\sqrt{x}}\,dx = \int x^{-1/2}\ln(x)\,dx = \int \ln(x)\,x^{-1/2}\,dx.$$

We set $u = \ln(x)$ and $dv = x^{-1/2}dx$, so that
$$du = \frac{1}{x}dx \text{ and } v = \int dv = \int x^{-1/2}dx = 2x^{1/2}$$

Thus,
$$\int \ln(x)\,x^{-1/2}dx = \int u\,dv = uv - \int v\,du = \ln(x)\left(2x^{1/2}\right) - \int 2x^{1/2}\left(\frac{1}{x}\right)dx$$
$$= 2\sqrt{x}\ln(x) - 2\int x^{-1/2}dx$$
$$= 2\sqrt{x}\ln(x) - 2\left(2x^{1/2}\right) + C$$
$$= 2\sqrt{x}\ln(x) - 4\sqrt{x} + C,$$

where $C$ is an arbitrary constant. □

The natural logarithm is the inverse of the natural exponential function. The technique that we used to determine the indefinite integral of the natural exponential function can be used to evaluate the indefinite integrals of other inverse functions such as arctangent, arcsine and arccosine.

**Example 7** Determine
$$\int \arctan(x)\,dx.$$

## Solution

We set $u = \arctan(x)$ and $dv = dx$, so that
$$du = \frac{du}{dx}dx = \frac{1}{x^2+1}dx \text{ and } v = x.$$

Thus,
$$\int \arctan(x)\,dx = \int u\,dv = uv - \int v\,du = \arctan(x)(x) - \int x\left(\frac{1}{x^2+1}\right)dx$$
$$= x\arctan(x) - \int \frac{x}{x^2+1}\,dx.$$

In order to evaluate the indefinite integral on the right-hand side, we set $w = x^2 + 1$, so that $dw = 2x\,dx$. By the substitution rule,
$$\int \frac{x}{x^2+1}\,dx = \frac{1}{2}\int \frac{1}{x^2+1}(2x)\,dx = \frac{1}{2}\int \frac{1}{w}\,dw$$
$$= \frac{1}{2}\ln(|w|) + C = \frac{1}{2}\ln(x^2+1) + C.$$

Therefore,
$$\int \arctan(x)\,dx = x\arctan(x) - \int \frac{x}{x^2+1}\,dx = x\arctan(x) - \frac{1}{2}\ln(x^2+1) + C$$

Indefinite integrals of the form

$$\int e^{ax} \sin(bx)\, dx \text{ or } \int e^{ax} \cos(bx)\, dx,$$

where $a$ and $b$ are constants, can be determined by a "cyclical" application of integration by parts as in the following example:

**Example 8**

a) Determine

$$\int e^{-x/2} \cos(x)\, dx.$$

b) Plot the graph of the region $G$ between the graph of $y = e^{-x/2} \cos(x)$ and the interval $[0, \pi]$ with the help of your graphing utility. Compute the area of $G$.

**Solution**

a) We set $u = e^{-x/2}$ and $dv = \cos(x)$ so that

$$du = -\frac{1}{2} e^{-x/2} dx \text{ and } v = \int \cos(x)\, dx = \sin(x).$$

Therefore,

$$\int e^{-x/2} \cos(x)\, dx = \int u\, dv = uv - \int v\, du$$

$$= e^{-x/2} \sin(x) - \int \sin(x) \left(-\frac{1}{2} e^{-x/2}\right) dx$$

$$= e^{-x/2} \sin(x) + \frac{1}{2} \int e^{-x/2} \sin(x)\, dx.$$

It appears that we have merely replaced $\cos(x)$ by $\sin(x)$ under the integral sine by applying integration by parts. We will persevere and apply integration by parts once more by setting $u = e^{-x/2}$ and $dv = \sin(x)$. Thus,

$$du = -\frac{1}{2} e^{-x/2} dx \text{ and } v = \int \sin(x)\, dx = -\cos(x),$$

and

$$\int e^{-x/2} \sin(x)\, dx = \int u\, dv = uv - \int v\, du$$

$$= e^{-x/2}(-\cos(x)) - \int (-\cos(x)) \left(-\frac{1}{2} e^{-x/2}\right) dx$$

$$= -e^{-x/2} \cos(x) - \frac{1}{2} \int e^{-x/2} \cos(x)\, dx.$$

Therefore,

$$\int e^{-x/2} \cos(x)\, dx = e^{-x/2} \sin(x) + \frac{1}{2} \int e^{-x/2} \sin(x)\, dx$$

$$= e^{-x/2} \sin(x) + \frac{1}{2} \left(-e^{-x/2} \cos(x) - \frac{1}{2} \int e^{-x/2} \cos(x)\, dx\right)$$

$$= e^{-x/2} \sin(x) - \frac{1}{2} e^{-x/2} \cos(x) - \frac{1}{4} \int e^{-x/2} \cos(x)\, dx$$

so that
$$\frac{5}{4}\int e^{-x/2}\cos(x)\,dx = e^{-x/2}\sin(x) - \frac{1}{2}e^{-x/2}\cos(x).$$
Therefore,
$$\int e^{-x/2}\cos(x)\,dx = \frac{4}{5}e^{-x/2}\sin(x) - \frac{2}{5}e^{-x/2}\cos(x) + C$$

b) Figure 2 illustrates the region $G$. Let $f(x) = e^{-x/2}\cos(x)$. The picture indicates that $f(\pi/2) = 0$, $f(x) > 0$ if $0 \le x < \pi/2$, and $f(x) < 0$ if $\pi/2 < x \le \pi$. Indeed,
$$e^{-x/2}\cos(x) = 0 \Leftrightarrow \cos(x) = 0,$$
since $e^{-x/2} > 0$ for each $x \in \mathbb{R}$. and the only zero of $\cos(x)$ in the interval $[0,\pi]$ is $x = \pi/2$. We have
$$\cos(x) > 0 \text{ if } 0 \le x < \pi/2 \text{ and } \cos(x) < 0 \text{ if } \pi/2 < x \le \pi.$$
Therefore, the area of $G$ is
$$\int_0^{\pi/2} e^{-x/2}\cos(x)\,dx - \int_{\pi/2}^{\pi} e^{-x/2}\cos(x)\,dx.$$

By part a),
$$\begin{aligned}\int_0^{\pi/2} e^{-x/2}\cos(x)\,dx &= \left.\frac{4}{5}e^{-x/2}\sin(x) - \frac{2}{5}e^{-x/2}\cos(x)\right|_0^{\pi/2} \\ &= \left(\frac{4}{5}e^{-\pi/4}\sin\left(\frac{\pi}{2}\right) - \frac{2}{5}e^{-\pi/4}\cos\left(\frac{\pi}{2}\right)\right) - \left(\frac{4}{5}e^0\sin(0) - \frac{2}{5}e^0\cos(0)\right) \\ &= \frac{4}{5}e^{-\pi/4} + \frac{2}{5},\end{aligned}$$
and
$$\begin{aligned}\int_{\pi/2}^{\pi} e^{-x/2}\cos(x)\,dx &= \left.\frac{4}{5}e^{-x/2}\sin(x) - \frac{2}{5}e^{-x/2}\cos(x)\right|_{\pi/2}^{\pi} \\ &= \left(\frac{4}{5}e^{-\pi/2}\sin(\pi) - \frac{2}{5}e^{-\pi/2}\cos(\pi)\right) - \left(\frac{4}{5}e^{-\pi/4}\sin\left(\frac{\pi}{2}\right) - \frac{2}{5}e^{-\pi/4}\cos\left(\frac{\pi}{2}\right)\right) \\ &= \frac{2}{5}e^{-\pi/2} - \frac{4}{5}e^{-\pi/4}.\end{aligned}$$
Therefore, the area of $G$ is
$$\left(\frac{4}{5}e^{-\pi/4} + \frac{2}{5}\right) - \left(\frac{2}{5}e^{-\pi/2} - \frac{4}{5}e^{-\pi/4}\right) = \frac{2}{5} - \frac{2}{5}e^{-\pi/2} + \frac{8}{5}e^{-\pi/4} \cong 1.046\,35$$

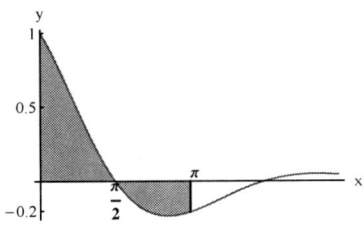

Figure 2

## Integration by Parts for Definite Integrals

We can make use of an indefinite integral that is obtained via integration by parts to evaluate a definite integral, as in Example 8. There is a version of integration by parts that is directly applicable to definite integrals:

**THE DEFINITE INTEGRAL VERSION OF INTEGRATION BY PARTS** Assume that $f'$ and $g'$ are continuous on $[a, b]$. Then,

$$\int_a^b f(x) \frac{dg}{dx} dx = [f(b)g(b) - f(a)g(a)] - \int_a^b g(x) \frac{df}{dx} dx$$

**Proof**

As in the proof of the indefinite integral version of integration by parts, the starting point is the product rule for differentiation:

$$\frac{d}{dx}(f(x)g(x)) = \frac{df}{dx}g(x) + f(x)\frac{dg}{dx}$$

Since $f'$ and $g'$ are continuous on $[a, b]$, $fg$, $f'g$ and $fg'$ are all continuous, hence integrable, on $[a, b]$. By the Fundamental Theorem of Calculus,

$$\int_a^b \frac{d}{dx}(f(x)g(x))\,dx = f(b)g(b) - f(a)g(a).$$

Therefore,

$$f(b)g(b) - f(a)g(a) = \int_a^b \left(\frac{df}{dx}g(x) + f(x)\frac{dg}{dx}\right) dx$$
$$= \int_a^b \frac{df}{dx}g(x)\,dx + \int_a^b f(x)\frac{dg}{dx}dx.$$

Thus,

$$\int_a^b f(x)\frac{dg}{dx}dx = [f(b)g(b) - f(a)g(a)] - \int_a^b g(x)\frac{df}{dx}dx$$

As in the case of indefinite integrals, we can express the definite integral version of integration by parts by using the symbolism

$$du = \frac{du}{dx}dx.$$

Indeed, if we replace $f(x)$ by $u$ and $g(x)$ by $v$, we have

$$\int_a^b u\,dv = [u(b)v(b) - u(a)v(a)] - \int_a^b v\,du$$
$$= uv\Big|_a^b - \int_a^b v\,du.$$

**Example 9** Compute

$$\int_0^{1/2} \arcsin(x)\,dx$$

by applying the definite integral version of integration by parts.

## 6.1. INTEGRATION BY PARTS

**Solution**

We set $u = \arcsin(x)$ and $dv = dx$, so that
$$du = \frac{du}{dx}dx = \frac{1}{\sqrt{1-x^2}}dx \text{ and } v = \int dx = x.$$
Therefore,
$$\int_0^{1/2} \arcsin(x)\,dx = \int_0^{1/2} u\,dv = uv\Big|_0^{1/2} - \int_0^{1/2} v\,du$$
$$= \arcsin(x)\,x\Big|_0^{1/2} - \int_0^{1/2} x\left(\frac{1}{\sqrt{1-x^2}}\right)dx$$
$$= \frac{1}{2}\arcsin\left(\frac{1}{2}\right) - \int_0^{1/2} \frac{x}{\sqrt{1-x^2}}dx$$
$$= \frac{\pi}{12} - \int_0^{1/2} \frac{x}{\sqrt{1-x^2}}dx$$

We can evaluate the integral on the right-hand side by a simple substitution: If we set $w = 1-x^2$, we have $dw = -2x\,dx$, so that
$$\int_0^{1/2} \frac{x}{\sqrt{1-x^2}}dx = -\frac{1}{2}\int_{w(0)}^{w(1/2)} \frac{1}{\sqrt{w}}dw = -\frac{1}{2}\int_1^{3/4} w^{-1/2}dw$$
$$= -\frac{1}{2}\left(2w^{1/2}\Big|_1^{3/4}\right) = -\frac{\sqrt{3}}{2} + 1.$$

Therefore,
$$\int_0^{1/2} \arcsin(x)\,dx = \frac{\pi}{12} - \int_0^{1/2} \frac{x}{\sqrt{1-x^2}}dx = \frac{\pi}{12} + \frac{\sqrt{3}}{2} - 1.$$

Thus, the area of the region between the graph of $y = \arcsin(x)$ and the interval $[0, 1/2]$ is
$$\frac{\pi}{12} + \frac{\sqrt{3}}{2} - 1 \cong 0.127\,825.$$

Figure 3 illustrates the region. $\square$

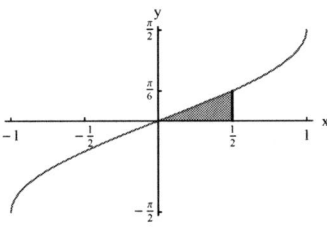

Figure 3

The definite integral version of integration by parts is indispensable in proving certain general facts. Here is an example:

**Example 10** Assume that $f''$ and $g''$ are continuous on the interval $[a, b]$ and that $f(a) = f(b) = g(a) = g(b) = 0$. Show that
$$\int_a^b f''(x)g(x)\,dx = \int_a^b f(x)g''(x)\,dx.$$

**Solution**
We apply integration by parts by setting $u = g(x)$ and $dv = f''(x)\,dx$ so that

$$du = g'(x)\,dx \text{ and } v = \int f''(x)\,dx = \int (f')'(x)\,dx = f'(x).$$

Thus,

$$\int_a^b f''(x)g(x)\,dx = \int_a^b g(x)(f')'(x)\,dx = \int_a^b u\,dv$$

$$= uv\Big|_a^b - \int_a^b v\,du$$

$$= (g(b)f'(b) - g(a)f'(a)) - \int_a^b f'(x)g'(x)\,dx$$

$$= -\int_a^b f'(x)g'(x)\,dx,$$

since $g(a) = g(b) = 0$. We apply integration by parts again, by setting $u = g'(x)$ and $dv = f'(x)\,dx$. Therefore,

$$du = g''(x)\,dx \text{ and } v = f(x).$$

Thus,

$$\int_a^b g'(x)f'(x)\,dx = \int_a^b u\,dv$$

$$= u(b)v(b) - u(a)v(a) - \int_a^b v\,du$$

$$= g'(b)f(b) - g'(a)f(a) - \int_a^b f(x)g''(x)\,dx$$

$$= -\int_a^b f(x)g''(x)\,dx,$$

since $f(a) = f(b) = 0$. Therefore,

$$\int_a^b f''(x)g(x)\,dx = -\int_a^b f'(x)g'(x)\,dx = -\left(-\int_a^b f(x)g''(x)\,dx\right) = \int_a^b f(x)g''(x)\,dx,$$

as claimed. $\square$

## Problems

In problems 1-13, determine the indefinite integral:

1.
$$\int xe^{2x}\,dx$$

2.
$$\int x^2 e^{-x/3}\,dx$$

3.
$$\int x\cos\left(\frac{1}{2}x\right)dx$$

4.
$$\int x^2 \sin\left(\frac{1}{4}x\right)dx$$

## 6.2. INTEGRALS OF RATIONAL FUNCTIONS

5. $\int x \sinh(x)\, dx$

6. $\int x^2 \cosh(x)\, dx$

7. $\int \arccos(x)\, dx$

8. $\int \arctan(x/2)\, dx$

9. $\int x^2 \ln(x)\, dx$

10. $\int \dfrac{\ln(x)}{x^{1/3}}\, dx$

11. $\int e^{-x} \sin(x)\, dx$

12. $\int e^{x/4} \sin(4x)\, dx$

13. $\int e^{-x} \cos\left(\dfrac{x}{2}\right)\, dx$

In problems 14 and 15 evaluate the integral (simplify as much as possible):

14. $\displaystyle\int_{\pi}^{2\pi} x \cos(3x)\, dx$

15. $\displaystyle\int_{e}^{e^2} x^2 \ln^2(x)\, dx$

In problems 16 and 17, evaluate the integral by applying the definite integral version of integration by parts:

16. $\displaystyle\int_{1/2}^{\sqrt{3}/2} \arccos(x)\, dx$

17. $\displaystyle\int_{1/4}^{1/2} x \sin(\pi x)\, dx$

18. Assume that
$$\int_1^3 f(x) g'(x)\, dx = -4,\ f(1) = -4,\ f(3) = 2,\ g(1) = 4,\ g(3) = -2.$$
Evaluate
$$\int_1^3 f'(x) g(x)\, dx.$$

19. Assume that $f''$ and $g''$ are continuous on the interval $[a, b]$, $f(a) = g(a) = 0$, and $f'(b) = g'(b) = 0$. Show that
$$\int_a^b f''(x) g(x)\, dx = \int_a^b f(x) g''(x)\, dx.$$

20. Assume that $f''$ is continuous on the interval $[a, b]$ and $f(a) = f(b) = 0$. Show that
$$\int_a^b f(x) f''(x)\, dx = -\int_a^b (f'(x))^2\, dx.$$

## 6.2 Integrals of Rational Functions

In this section we will be able to express the indefinite integrals of rational functions as combinations of rational functions, the natural logarithm and arctangent. This will be accomplished with the help of the algebraic device that is referred to as **the partial fraction decomposition** of a rational function

## Special Cases

We have already seen the following indefinite integrals that involve rational functions:

$$\int \frac{1}{x^n} dx = \int x^{-n} dx = \frac{x^{-n+1}}{-n+1} + C, \ n \neq 1,$$

$$\int \frac{1}{x} dx = \ln(|x|) + C$$

$$\int \frac{1}{x^2+1} dx = \arctan(x) + C$$

(as usual, $C$ denotes an arbitrary constant).

We were able to expand the scope of the above list of formulas by making use of the substitution rule, as in the following examples.

**Example 1** Determine

$$\int \frac{x}{(x^2+9)^2} dx.$$

**Solution**

We set $u = x^2 + 9$, so that $du = 2x dx$. Thus,

$$\int \frac{x}{(x^2+9)^2} dx = \frac{1}{2} \int \frac{1}{(x^2+9)^2} (2x) \, dx = \frac{1}{2} \int \frac{1}{u^2} du$$

$$= \frac{1}{2} \int u^{-2} du$$

$$= \frac{1}{2} \left(\frac{u^{-1}}{-1}\right) + C = -\frac{1}{2u} + C = -\frac{1}{2(x^2+9)} + C.$$

□

**Example 2**

a) Determine the indefinite integral

$$\int \frac{x^2}{x^3+1} dx.$$

b) Compute the integral

$$\int_{-4}^{-2} \frac{x^2}{x^3+1} dx.$$

**Solution**

a) We set $u = x^3 + 1$ so that $du = 3x^2 dx$. Thus,

$$\int \frac{x^2}{x^3+1} dx = \frac{1}{3} \int \frac{1}{x^3+1} (3x^2) \, dx$$

$$= \frac{1}{3} \int \frac{1}{u} du$$

$$= \frac{1}{3} \ln(|u|) + C = \frac{1}{3} \ln(|x^3+1|) + C.$$

Note that the above expression for the indefinite integral is valid on any interval that is contained in $(-\infty, -1)$ or $(-1, +\infty)$.

## 6.2. INTEGRALS OF RATIONAL FUNCTIONS

b) By part a) and the Fundamental Theorem of Calculus,

$$\int_{-4}^{-2} \frac{x^2}{x^3+1} dx = \frac{1}{3} \ln\left(\left|x^3+1\right|\right)\Big|_{-4}^{-2}$$

$$= \frac{1}{3} \ln\left(|-7|\right) - \frac{1}{3} \ln\left(|-63|\right)$$

$$= \frac{1}{3} \ln(7) - \frac{1}{3} \ln(63) = -\frac{1}{3} \ln\left(\frac{63}{7}\right) = -\frac{1}{3} \ln(9) \cong -0.732\,408.$$

Figure 1 illustrates the region between the graph of

$$y = \frac{x^2}{x^3+1}$$

and the interval $[-4, -2]$. The signed area of the region is $-\ln(9)/3$, and the area of $G$ is $\ln(9)/3$. □

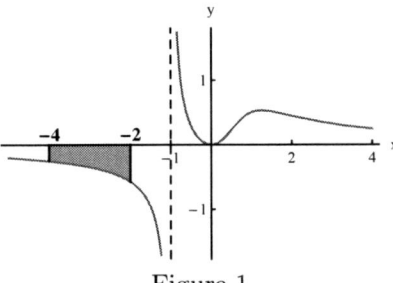

Figure 1

We say that the quadratic expression $ax^2 + bx + c$ is **irreducible** if it cannot be expressed as a product of linear factors whose coefficients are real numbers. As reviewed in Section A1 of Appendix A, this is the case if and only if the discriminant $b^2 - 4ac$ is negative. If $ax^2 + bx + c$ is irreducible, the determination of the indefinite integral

$$\int \frac{1}{ax^2+bx+c} dx$$

can be reduced to the basic formula

$$\int \frac{1}{u^2+1} du = \arctan(u) + C$$

by the completion of the square, as in the following example.

**Example 3** Determine

$$\int \frac{1}{4x^2-8x+13} dx$$

**Solution**

The expression $4x^2 - 8x + 13$ is irreducible since its discriminant is negative:

$$(-8)^2 - 4(4)(13) = -144$$

We complete the square:

$$4x^2 - 8x + 13 = 4\left(x^2 - 2x\right) + 13 = 4(x-1)^2 - 4 + 13 = 4(x-1)^2 + 9$$

Therefore,

$$\int \frac{1}{4x^2 - 8x + 13}\,dx = \int \frac{1}{4(x-1)^2 + 9}\,dx = \int \frac{1}{9\left(\frac{4(x-1)^2}{9} + 1\right)}\,dx$$

$$= \frac{1}{9}\int \frac{1}{\left(\frac{2(x-1)}{3}\right)^2 + 1}\,dx.$$

The final form of the integrand suggests the substitution

$$u = \frac{2(x-1)}{3}.$$

Then, $du/dx = \frac{2}{3}$. Thus,

$$\frac{1}{9}\int \frac{1}{\left(\frac{2(x-1)}{3}\right)^2 + 1}\,dx = \frac{1}{9}\int \frac{1}{u^2 + 1}\left(\frac{3}{2}\right)\frac{du}{dx}\,dx = \frac{1}{6}\int \frac{1}{u^2 + 1}\,du$$

$$= \frac{1}{6}\arctan(u) + C$$

$$= \frac{1}{6}\arctan\left(\frac{2(x-1)}{3}\right) + C,$$

where $C$ is an arbitrary constant. $\square$

If $ax^2 + bx + c$ is irreducible and $n$ is an integer greater than 1, an indefinite integral of the form

$$\int \frac{1}{(ax^2 + bx + c)^n}\,dx$$

can be reduced to the case

$$\int \frac{1}{(x^2 + 1)^n}\,dx$$

by the completion of the square and an appropriate substitution. In such cases the following **reduction formula** is helpful:

**Proposition 1** If $n$ is an integer and $n > 1$, then

$$\int \frac{1}{(x^2 + 1)^n}\,dx = \left(\frac{2n-3}{2(n-1)}\right)\int \frac{1}{(x^2 + 1)^{n-1}}\,dx + \frac{x}{2(n-1)(x^2+1)^{n-1}}$$

The above expression is referred to as a reduction formula since an indefinite integral that involves the positive integer $n$ is expressed in terms of the same type of indefinite integral where $n$ is replaced by $n - 1$.

### The Proof of Proposition 1

The proof will make use of integration by parts. Our starting point is the following identity:

$$\frac{1}{(x^2+1)^n} = \frac{(x^2+1) - x^2}{(x^2+1)^n} = \frac{1}{(x^2+1)^{n-1}} - \frac{x^2}{(x^2+1)^n}.$$

## 6.2. INTEGRALS OF RATIONAL FUNCTIONS

Thus,
$$\int \frac{1}{(x^2+1)^n} dx = \int \frac{1}{(x^2+1)^{n-1}} dx - \int \frac{x^2}{(x^2+1)^n} dx.$$

Let's apply integration by parts to the last indefinite integral with
$$u = x \text{ and } dv = \frac{x}{(x^2+1)^n} dx$$

Then,
$$du = dx \text{ and } v = \int \frac{x}{(x^2+1)^n} dx$$

We can determine $v$ via the substitution $w = x^2 + 1$. Then, $dw/dx = 2x$, so that

$$v = \int \frac{x}{(x^2+1)^n} dx = \frac{1}{2} \int \frac{1}{w^n} \frac{dw}{dx} dx = \frac{1}{2} \int w^{-n} dw$$
$$= \frac{1}{2} \left( \frac{w^{-n+1}}{-n+1} \right)$$
$$= -\frac{1}{2(n-1)w^{n-1}} = -\frac{1}{2(n-1)(x^2+1)^{n-1}}$$

Therefore,
$$\int \frac{x^2}{(x^2+1)^n} dx = \int u\,dv = uv - \int v\,du$$
$$= -\frac{x}{2(n-1)(x^2+1)^{n-1}} + \int \frac{1}{2(n-1)(x^2+1)^{n-1}} dx$$
$$= -\frac{x}{2(n-1)(x^2+1)^{n-1}} + \frac{1}{2(n-1)} \int \frac{1}{(x^2+1)^{n-1}} dx.$$

Thus,
$$\int \frac{1}{(x^2+1)^n} dx = \int \frac{1}{(x^2+1)^{n-1}} dx - \int \frac{x^2}{(x^2+1)^n} dx$$
$$= \int \frac{1}{(x^2+1)^{n-1}} dx + \frac{x}{2(n-1)(x^2+1)^{n-1}} + \frac{1}{2(n-1)} \int \frac{1}{(x^2+1)^{n-1}} dx$$
$$= \left( \frac{2n-3}{2(n-1)} \right) \int \frac{1}{(x^2+1)^{n-1}} dx + \frac{x}{2(n-1)(x^2+1)^{n-1}}$$

∎

The following example illustrates the use of the above reduction formula.

**Example 4** Determine
$$\int \frac{1}{(x^2+1)^3} dx$$

**Solution**

In the reduction formula we set $n = 3$:
$$\int \frac{1}{(x^2+1)^3} dx = \frac{x}{4(x^2+1)^2} + \frac{3}{4} \int \frac{1}{(x^2+1)^2} dx.$$

Now we apply the reduction formula with $n = 2$ to the indefinite integral on the right-hand side:
$$\int \frac{1}{(x^2+1)^2} dx = \frac{1}{2} \int \frac{1}{x^2+1} dx + \frac{x}{2(x^2+1)}.$$

Since
$$\int \frac{1}{x^2+1} dx = \arctan(x),$$

we have
$$\int \frac{1}{(x^2+1)^2} dx = \frac{1}{2} \arctan(x) + \frac{x}{2(x^2+1)}.$$

Therefore,
$$\int \frac{1}{(x^2+1)^3} dx = \frac{x}{4(x^2+1)^2} + \frac{3}{4} \int \frac{1}{(x^2+1)^2} dx$$
$$= \frac{x}{4(x^2+1)^2} + \frac{3}{4} \left( \frac{1}{2} \arctan(x) + \frac{x}{2(x^2+1)} \right)$$
$$= \frac{x}{4(x^2+1)^2} + \frac{3x}{8(x^2+1)} + \frac{3}{8} \arctan(x)$$

(we can add an arbitrary constant, of course). □

## Arbitrary Rational Functions

The antidifferentiation of an arbitrary rational function can be reduced to the antidifferentiation of the special cases that we considered with the help of an algebraic device called **partial fraction decomposition**. If $f$ is a rational function, we have

$$f(x) = \frac{P(x)}{Q(x)},$$

where $P(x)$ and $Q(x)$ are polynomials. If the degree of the numerator $P(x)$ is greater than or equal to the degree of denominator $Q(x)$ we can divide, and express $f(x)$ as

$$f(x) = p(x) + \frac{r(x)}{Q(x)},$$

where $p(x)$ and $r(x)$ are polynomials, and the degree of $r(x)$ is less than the degree of $Q(x)$. The partial fraction decomposition of $f(x)$ is obtained by expressing $r(x)/Q(x)$ as a sum of expressions of the form

$$\frac{A}{(x-d)^k} \quad \text{and} \quad \frac{Bx+C}{(ax^2+bx+c)^l},$$

where $(x-d)^k$ and $(ax^2+bx+c)^l$ are factors of the denominator $Q(x)$, and $ax^2+bx+c$ is irreducible over the real numbers. The partial fraction contains a sum of the form

$$\frac{A_1}{x-d} + \frac{A_2}{(x-d)^2} + \cdots + \frac{A_m}{(x-d)^m}$$

if $m$ is the highest power of $(x-d)$ in the factorization of the denominator $Q(x)$. If $n$ is the highest power of the irreducible factor $(ax^2+bx+c)$ of $Q(x)$, the partial fraction decomposition contains a sum of the form

$$\frac{B_1 x + C_1}{ax^2+bx+c} + \frac{B_2 x + C_2}{(ax^2+bx+c)^2} + \cdots + \frac{B_n x + C_n}{(ax^2+bx+c)^n}.$$

Let's look at some examples.

## 6.2. INTEGRALS OF RATIONAL FUNCTIONS

**Example 5** Let
$$f(x) = \frac{x^3 - x - 5}{x^2 - x - 2}$$
a) Determine the partial fraction decomposition of $f(x)$.
b) Determine
$$\int f(x)\,dx$$

**Solution**

a) Since the degree of the numerator is greater than the degree of the denominator, we divide (long division will do):
$$f(x) = x + 1 + \frac{2x - 3}{x^2 - x - 2}$$
The next step is to express the denominator as a product of linear factors, if possible. This can be done by solving the quadratic equation $x^2 - x - 2 = 0$ with the help of the quadratic formula. Since
$$x^2 - x - 2 = 0 \Leftrightarrow x = 2 \text{ or } x = -1,$$
we have $x^2 - x - 2 = (x - 2)(x + 1)$ (this factorization could have been obtained by the ad hoc procedures that are familiar from precalculus courses in the present case, but the technique that is based on the quadratic formula works even if the factors involve irrational numbers).
Now we have to determine the numbers $A$ and $B$ such that
$$\frac{2x - 3}{(x - 2)(x + 1)} = \frac{A}{x - 2} + \frac{B}{x + 1}$$
for each $x$ such that $x \neq 2$ and $x \neq -1$ (so that the expressions are defined). This is the case if
$$\frac{2x - 3}{(x - 2)(x + 1)} = \frac{A(x + 1) + B(x - 2)}{(x - 2)(x + 1)},$$
i.e., if
$$2x - 3 = A(x + 1) + B(x - 2)$$
for each $x$ such that $x \neq 2$ and $x \neq -1$. The above equality must be valid if $x = 2$ or $x = -1$, as well. Indeed, by the continuity of the functions defined by the expressions on either side of the equality, we have
$$\lim_{x \to 2}(2x - 3) = \lim_{x \to 2}(A(x + 1) + B(x - 2)),$$
so that
$$2x - 3|_{x=2} = A(x + 1) + B(x - 2)|_{x=2}$$
Similarly,
$$2x - 3|_{x=-1} = A(x + 1) + B(x - 2)|_{x=-1}$$
(we will not repeat the justification of such an evaluation in the following examples).
Thus, we have
$$1 = 3A \text{ and } -5 = -3B \Rightarrow A = \frac{1}{3} \text{ and } B = \frac{5}{3}.$$
Therefore,
$$\frac{2x - 3}{(x - 2)(x + 1)} = \frac{A}{x - 2} + \frac{B}{x + 1} = \frac{1}{3(x - 2)} + \frac{5}{3(x + 1)}$$
(if $x \neq 2$ and $x \neq -1$). Thus,
$$f(x) = x + 1 + \frac{2x - 3}{x^2 - x - 2} = x + 1 + \frac{1}{3(x - 2)} + \frac{5}{3(x + 1)}$$

b) By the result of part a), and the linearity of antidifferentiation,

$$\int f(x)\,dx = \int \left( x + 1 + \frac{1}{3(x-2)} + \frac{5}{3(x+1)} \right)$$
$$= \int x\,dx + \int 1\,dx + \frac{1}{3}\int \frac{1}{x-2}\,dx + \frac{5}{3}\int \frac{1}{x+1}\,dx$$
$$= \frac{1}{2}x^2 + x + \frac{1}{3}\ln(|x-2|) + \frac{5}{3}\ln(|x+1|) + C,$$

where $C$ is an arbitrary constant. $\square$

**Example 6** Let
$$f(x) = \frac{2x^2 + 6x - 26}{(x-2)(x+1)(x-4)}.$$

a) Determine the partial fraction decomposition of $f(x)$.
b) Determine an antiderivative of $f$.

**Solution**

a) Since the degree of the numerator is 2 and the degree of the denominator is 3, there is no need to divide. We seek $A, B$ and $C$ so that

$$\frac{2x^2 + 6x - 26}{(x-2)(x+1)(x-4)} = \frac{A}{x-2} + \frac{B}{x+1} + \frac{C}{x-4}.$$

This is the case if

$$2x^2 + 6x - 26 \equiv A(x+1)(x-4) + B(x-2)(x-4) + C(x-2)(x+1)$$

for each $x$ such that $x \neq 2$, $x \neq -1$ and $x \neq 4$. As in Example 5 we can replace $x$ by $2, -1$ or $4$, since the functions defined by the expressions on either side of the above equality are continuous. Therefore,

$$-6 = -6A, \quad -30 = 15B \text{ and } 30 = 10C$$

so that
$$A = 1, \; B = -2 \text{ and } C = 3.$$

Therefore,
$$f(x) = \frac{2x^2 + 6x - 26}{(x-2)(x+1)(x-4)} \equiv \frac{1}{x-2} - \frac{2}{x+1} + \frac{3}{x-4}.$$

b) By part a),

$$\int \frac{2x^2 + 6x - 26}{(x-2)(x+1)(x-4)}\,dx = \int \left( \frac{1}{x-2} - \frac{2}{x+1} + \frac{3}{x-4} \right) dx$$
$$= \int \frac{1}{x-2}\,dx - 2\int \frac{1}{x+1}\,dx + 3\int \frac{1}{x-4}\,dx$$
$$= \ln(|x-2|) - 2\ln(|x+1|) + 3\ln(|x-4|) + C$$

where $C$ is an arbitrary constant. $\square$

**Remark** (**Gaussian Elimination**) With reference to Example 6, we can follow a different route in order to determine $A, B$ and $C$ so that

$$2x^2 + 6x - 26 = A(x+1)(x-4) + B(x-2)(x-4) + C(x-2)(x+1).$$

We expand and collect the terms on the right-hand side so that it is expressed in descending powers of $x$:

$$2x^2 + 6x - 26 = (A + B + C)x^2 + (-3A - 6B - C)x + (-4A + 8B - 2C)$$

for each $x \in R$. This is the case if and only if the coefficients of like powers of $x$ are the same. Therefore,

$$A + B + C = 2$$
$$-3A - 6B - C = 6$$
$$-4A + 8B - 2C = -26$$

This is a system of equations in the unknowns $A$, $B$ and $C$. The system is referred to as **a linear system of equations**, since the highest power of each unknown is 1. Even though you must have solved such systems by ad hoc eliminations and substitutions in precalculus courses, there is a systematic way of solving such equations that goes back to the famous mathematician **Gauss**. The idea is simple: We can interchange the order of the equations or add a multiple of one equation to another equation without changing the solutions. We will perform such operations systematically in order to transform the given system to a system that is easily solvable. The procedure is referred to as **Gaussian elimination.** Let us illustrate Gaussian elimination in the case of the above system. We begin by looking at the first equation. If that equation had not involved the first unknown $A$, we would have found one that did, and interchanged it with the first equation. In the present case, the first equation involves $A$, so that there is no need to carry out such an equation interchange. We eliminate $A$ from the second equation and the third equation by adding 3 times the first equation to the second equation and 4 times the the first equation to the third equation. The resulting system is shown below:

$$A + B + C = 2$$
$$-3B + 2C = 12$$
$$12B + 2C = -18$$

Let us multiply the second equation by $-1/3$ so that the coefficient of $B$ is 1 (that will simplify the subsequent arithmetic). Thus, the system is transformed to the following form:

$$A + B + C = 2$$
$$B - \frac{2}{3}C = -4$$
$$12B + 2C = -18$$

Now we eliminate the second unknown $B$ from the third equation. We can do this by subtracting 12 times the second equation from the third equation. The resulting system is the following:

$$A + B + C = 2,$$
$$B - \frac{2}{3}C = -4,$$
$$10C = 30.$$

This is the end of the **elimination phase.** Then we determine the unknowns one by one, starting with the last unknown in the last equation, and working our way backwards to the first equation. This phase of the method is referred to as **back substitution**. Thus, the last equation yields the value of $C$:

$$10C = 30 \Rightarrow C = \frac{30}{10} = 3.$$

The second equation determines $B$ when we substitute the value of $C$:

$$B - \frac{2}{3}C = -4 \Rightarrow B - 2 = -4 \Rightarrow B = -2.$$

The first equation determines $A$ when we substitute the values of $B$ and $C$:

$$A - 2 + 3 = 2 \Rightarrow A = 1.$$

◊

**Example 7** Let

$$f(x) = \frac{4x^2 - 7x + 1}{(x-1)^2(x-2)}.$$

a) Determine the partial fraction decomposition of $f(x)$.
b) Determine $\int f(x)\,dx$.
c) Compute $\int_3^4 f(x)\,dx$.

**Solution**

a) We must determine $A, B$ and $C$ so that

$$\frac{4x^2 - 7x + 1}{(x-1)^2(x-2)} = \frac{A}{x-1} + \frac{B}{(x-1)^2} + \frac{C}{x-2},$$

i.e.,

$$4x^2 - 7x + 1 = A(x-1)(x-2) + B(x-2) + C(x-1)^2.$$

**Method 1** (the substitution of suitable values for $x$)
If we set $x = 1$, we obtain

$$-2 = -B,$$

so that $B = 2$. If we set $x = 2$, we obtain

$$3 = C.$$

We ran out of the obvious choices for $x$, but we must still determine $A$. We can replace $x$ by any value other than the previous choices in order to determine $A$. The number 0 seems to be a reasonable choice. If we set $x = 0$, we obtain the equation,

$$1 = 2A - 2B + C.$$

Since $B = 2$ and $C = 3$, we have

$$1 = 2A - 2(2) + 3 = 2A - 1.$$

Therefore, $A = 1$.
Thus,

$$\frac{4x^2 - 7x + 1}{(x-1)^2(x-2)} = \frac{1}{x-1} + \frac{2}{(x-1)^2} + \frac{3}{x-2}.$$

**Method 2 (Gaussian elimination)**
We have

$$4x^2 - 7x + 1 = A(x-1)(x-2) + B(x-2) + C(x-1)^2$$
$$= (A+C)x^2 + (-3A + B - 2C)x + (2A - 2B + C)$$

## 6.2. INTEGRALS OF RATIONAL FUNCTIONS

We equate the coefficients of like powers of $x$:
$$A + C = 4$$
$$-3A + B - 2C = -7$$
$$2A - 2B + C = 1$$

We eliminate $A$ from the second and third equations by adding 3 times the first equation to the second equation and $-2$ times the first equation to the third equation:
$$A + C = 4$$
$$B + C = 5$$
$$-2B - C = -7$$

We eliminate $B$ from the third equation by adding 2 times the second equation to the third equation:
$$A + C = 4$$
$$B + C = 5$$
$$C = 3$$

Now we implement back-substitution:
$$C = 3,$$
$$B = 5 - C = 2,$$
$$A = 4 - C = 1,$$

as before.

b) By part a),
$$\int \frac{4x^2 - 7x + 1}{(x-1)^2(x-2)} dx = \int \left( \frac{1}{x-1} + \frac{2}{(x-1)^2} + \frac{3}{x-2} \right) dx$$
$$= \int \frac{1}{x-1} dx + 2\int \frac{1}{(x-1)^2} dx + 3\int \frac{1}{x-2}$$
$$= \ln(|x-1|) - \frac{2}{x-1} + 3\ln(|x-2|) + C$$

c)
$$\int_3^4 \frac{4x^2 - 7x + 1}{(x-1)^2(x-2)} dx = \left. \ln(|x-1|) - \frac{2}{x-1} + 3\ln(|x-2|) \right|_3^4$$
$$= \ln(3) - \frac{2}{3} + 3\ln(2) - \ln(2) + 1$$
$$= \ln(3) + \frac{1}{3} + 2\ln(2) \cong 2.81824$$

Thus, the area of the shaded region in Figure 2 is approximately 2.8. □

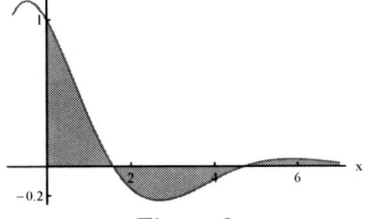

Figure 2

**Example 8** Let
$$f(x) = \frac{3x^2 - 7x + 7}{(x-3)(x^2+4)}.$$

a) Determine the partial fraction decomposition of $f(x)$.
b) Determine $\int f(x)\,dx$.
c) Compute $\int_0^2 f(x)\,dx$.

**Solution**

a) Since the factor $x^2 + 4$ is irreducible, the partial fraction decomposition of $f(x)$ is in the form
$$\frac{A}{x-3} + \frac{Bx+C}{x^2+4}.$$

Therefore, we need to have
$$\frac{3x^2 - 7x + 7}{(x-3)(x^2+4)} = \frac{A}{x-3} + \frac{Bx+C}{x^2+4},$$

i.e.,
$$3x^2 - 7x + 7 = A(x^2+4) + (Bx+C)(x-3).$$

There is only one obvious choice for the value of $x$ (i.e., 3). We will choose to set up the system of equations that will determine $A$, $B$ and $C$ and solve the system by Gaussian elimination. Thus,
$$\begin{aligned}3x^2 - 7x + 7 &= A(x^2+4) + (Bx+C)(x-3) \\ &= (A+B)x^2 + (-3B+C)x + (4A - 3C)\end{aligned}$$

This is the case if
$$A + B = 3$$
$$-3B + C = -7$$
$$4A - 3C = 7$$

The first unknown $A$ does not appear in the second equation. We will eliminate $A$ from the third equation by adding $-4$ times the first equation to the third equation:
$$A + B = 3$$
$$-3B + C = -7$$
$$-4B - 3C = -5$$

We will multiply the second equation by $-1/3$ so that the coefficient of $B$ is 1:
$$A + B = 3$$
$$B - \frac{1}{3}C = \frac{7}{3}$$
$$-4B - 3C = -5$$

## 6.2. INTEGRALS OF RATIONAL FUNCTIONS

We will eliminate $B$ from the third equation by multiplying the second equation by 4 and adding to the third equation:

$$A + B = 3$$
$$B - \frac{1}{3}C = \frac{7}{3}$$
$$-\frac{13}{3}C = \frac{13}{3}.$$

Now we implement back-substitution:

$$C = -1,$$
$$B = \frac{1}{3}C + \frac{7}{3} = -\frac{1}{3} + \frac{7}{3} = 2,$$
$$A = 3 - B = 1$$

Therefore,

$$f(x) = \frac{A}{x-3} + \frac{Bx+C}{x^2+4} = \frac{1}{x-3} + \frac{2x-1}{x^2+4}.$$

b) By part a),

$$\int f(x)\,dx = \int \frac{1}{x-3}\,dx + \int \frac{2x-1}{x^2+4}\,dx = \ln(|x-3|) + \int \frac{2x-1}{x^2+4}\,dx.$$

Now,

$$\int \frac{2x-1}{x^2+4}\,dx = 2\int \frac{x}{x^2+4}\,dx - \int \frac{1}{x^2+4}\,dx.$$

If we set $u = x^2 + 4$, we have $du/dx = 2x$, so that

$$\int \frac{x}{x^2+4}\,dx = \frac{1}{2}\int \frac{1}{x^2+4}(2x)\,dx = \frac{1}{2}\int \frac{1}{u}\frac{du}{dx}\,dx = \frac{1}{2}\int \frac{1}{u}\,du = \frac{1}{2}\ln(|u|) = \frac{1}{2}\ln(x^2+4).$$

As for the second antiderivative,

$$\int \frac{1}{x^2+4}\,dx = \int \frac{1}{4\left(\frac{x^2}{4}+1\right)}\,dx = \frac{1}{4}\int \frac{1}{\left(\frac{x}{2}\right)^2+1}\,dx.$$

Therefore, if we set $u = x/2$, we have $du/dx = 1/2$, so that

$$\frac{1}{4}\int \frac{1}{\left(\frac{x}{2}\right)^2+1}\,dx = \frac{1}{2}\int \frac{1}{\left(\frac{x}{2}\right)^2+1}\left(\frac{1}{2}\right)\,dx$$
$$= \frac{1}{2}\int \frac{1}{u^2+1}\frac{du}{dx}\,dx = \frac{1}{2}\int \frac{1}{u^2+1}\,du = \frac{1}{2}\arctan(u) = \frac{1}{2}\arctan\left(\frac{x}{2}\right).$$

Therefore,

$$\int \frac{2x-1}{x^2+4}\,dx = 2\int \frac{x}{x^2+4}\,dx - \int \frac{1}{x^2+4}\,dx = \ln(x^2+4) - \frac{1}{2}\arctan\left(\frac{x}{2}\right).$$

Finally,

$$\int f(x)\,dx = \int \frac{1}{x-3}\,dx + \int \frac{2x-1}{x^2+4}\,dx = \ln(|x-3|) + \int \frac{2x-1}{x^2+4}\,dx$$
$$= \ln(|x-3|) + \ln(x^2+4) - \frac{1}{2}\arctan\left(\frac{x}{2}\right) + C,$$

where $C$ is an arbitrary constant.

c)

$$\int_0^2 \frac{3x^2 - 7x + 7}{(x-3)(x^2+4)} dx = \ln(|x-3|) + \ln(x^2+4) - \frac{1}{2}\arctan\left(\frac{x}{2}\right)\Big|_0^2$$
$$= \ln(1) + \ln(8) - \frac{1}{2}\arctan(1) - \ln(3) - \ln(4) + \frac{1}{2}\arctan(0)$$
$$= \ln(2) - \ln(3) - \frac{\pi}{8} \cong -0.798\,164$$

Thus, the area of the shaded region in Figure 3 is approximately 0.8. □

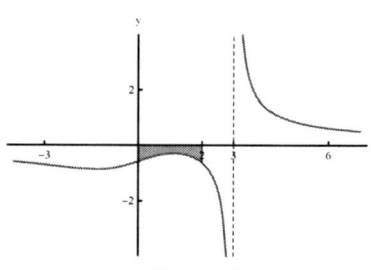

Figure 3

## Problems

In problems 1-4 determine the required antiderivative.

1. $$\int \frac{1}{4x^2 + 8x + 5} dx$$

2. $$\int \frac{x}{x^2 + 2x + 5} dx$$

3. $$\int \frac{1}{16x^2 - 96x + 153} dx$$

4. $$\int \frac{2x - 1}{x^2 - 8x + 17} dx$$

In problems 5-16,
a) compute the partial fraction decomposition of the integrand,
b) compute the required antiderivative:

5. $$\int \frac{-3x + 16}{x^2 + x - 12} dx$$

6. $$\int \frac{x + 7}{x^2 - 7x + 10} dx$$

7. $$\int \frac{x + 5}{x^2 + 5x + 6} dx$$

8. $$\int \frac{x - 17}{x^2 - 6x + 5} dx$$

9. $$\int \frac{2x^3 - 5x^2 - 24x + 13}{x^2 - x - 12} dx$$

10. $$\int \frac{x^3 + x^2 - 2x + 3}{x^2 + x - 2} dx$$

11. $$\int \frac{x - 5}{(x - 2)^2} dx$$

12. $$\int \frac{-2x^2 + 7x + 13}{(x - 1)(x + 2)^2} dx$$

**13.**
$$\int \frac{x-1}{x^2+4}dx$$

**14.**
$$\int \frac{x-4}{4x^2+1}dx$$

**15.**
$$\int \frac{4x^2+3x+5}{(x+2)(x^2+1)}dx$$

**16.**
$$\int \frac{1}{(x-1)(x^2+1)^2}dx$$

Hint: Make use of the reduction formula:
$$\int \frac{1}{(x^2+1)^n}dx = \left(\frac{2n-3}{2(n-1)}\right)\int \frac{1}{(x^2+1)^{n-1}}dx + \frac{x}{2(n-1)(x^2+1)^{n-1}}.$$

In problems 17 and 18,
a) compute the partial fraction decomposition of the integrand,
b) compute the required antiderivative,
c) compute the given integral:

**17.**
$$\int \frac{6}{(x+4)(x-2)}dx\;;\;\int_{-3}^{1}\frac{6}{(x+4)(x-2)}dx$$

**18.**
$$\int \frac{1}{x^2+6x+13}dx\;;\;\int_{-3}^{-1}\frac{1}{x^2+6x+13}dx$$

**19.** [CAS] Determine
$$\int \frac{1}{x^4+1}dx.$$
with the help of a CAS.

**20.** [CAS] Determine
$$\int \frac{1}{x^6+1}dx.$$
with the help of a CAS.

## 6.3 Integrals of Some Trigonometric and Hyperbolic Functions

In this section we will discuss integrals that involve the products of powers of $\sin(x)$ and $\cos(x)$, or $\sinh(x)$ and $\cosh(x)$. We will also consider the quotients of such expressions.

### Products of sin(x) and cos(x) or sinh(x) and cosh(x)

We will consider the determination of indefinite integrals of the form
$$\int \sin^m(x)\cos^n(x)\,dx \text{ or } \int \sinh^m(x)\cosh^n(x)\,dx,$$
where $m$ and $n$ are nonnegative integers.

Let's begin the cases where $m$ or $n$ is odd. In such a case the indefinite integral can be determined without too much effort, as illustrated by the following examples.

**Example 1** Determine
$$\int \sin^5(x) \cos^2(x)\, dx.$$

**Solution**

The power of $\sin(x)$ is odd. We express $\sin^5(x)$ as $\sin^4(x)\sin(x)$, and make use of the identity $\sin^2(x) + \cos^2(x) = 1$ in order to express $\sin^4(x)$ in terms of $\cos^2(x)$:

$$\begin{aligned}\int \sin^5(x)\cos^2(x)\, dx &= \int \sin^4(x)\cos^2(x)\sin(x)\, dx \\ &= \int \left(\sin^2(x)\right)^2 \cos^2(x)\sin(x)\, dx \\ &= \int \left(1 - \cos^2(x)\right)^2 \cos^2(x)\sin(x)\, dx.\end{aligned}$$

The above expression suggests the substitution $u = \cos(x)$. Then $du/dx = -\sin(x)$ so that

$$\begin{aligned}\int \sin^5(x)\cos^2(x)\, dx &= \int \left(1 - \cos^2(x)\right)^2 \cos^2(x)\sin(x)\, dx \\ &= -\int \left(1 - u^2\right)^2 u^2 \frac{du}{dx}\, dx \\ &= -\int \left(1 - u^2\right)^2 u^2\, du \\ &= -\int \left(u^2 - 2u^4 + u^6\right) du \\ &= -\frac{1}{3}u^3 + \frac{2}{5}u^5 - \frac{1}{7}u^7 \\ &= -\frac{1}{3}\cos^3(x) + \frac{2}{5}\cos^5(x) - \frac{1}{7}\cos^7(x) + C\end{aligned}$$

□

**Example 2** Determine
$$\int \sin^2(x) \cos^3(x)\, dx$$

**Solution**

The power of $\cos(x)$ is odd. We express $\cos^3(x)$ as $\cos^2(x)\cos(x)$, and make use of the identity $\sin^2(x) + \cos^2(x) = 1$ in order to express $\cos^2(x)$ in terms of $\sin^2(x)$:

$$\begin{aligned}\int \sin^2(x)\cos^3(x)\, dx &= \int \sin^2(x)\cos^2(x)\cos(x)\, dx \\ &= \int \sin^2(x)\left(1 - \sin^2(x)\right)\cos(x)\, dx.\end{aligned}$$

Then we set $u = \sin(x)$ so that $du/dx = \cos(x)$. Therefore,

$$\begin{aligned}\int \sin^2(x)\left(1 - \sin^2(x)\right)\cos(x)\, dx &= \int u^2\left(1 - u^2\right)\frac{du}{dx}\, dx \\ &= \int \left(u^2 - u^4\right) du \\ &= \frac{1}{3}u^3 - \frac{1}{5}u^5 + C \\ &= \frac{1}{3}\sin^3(x) - \frac{1}{5}\sin^5(x) + C,\end{aligned}$$

## 6.3. INTEGRALS OF SOME TRIGONOMETRIC AND HYPERBOLIC FUNCTIONS

where $C$ is an arbitrary constant. □

Indefinite integrals of the form

$$\int \sinh^m(x) \cosh^n(x)\, dx$$

where $m$ or $n$ is odd is handled in a similar manner. The identity $\cosh^2(x) - \sinh^2(x) = 1$ replaces the identity $\cos^2(x) + \sin^2(x) = 1$.

**Example 3** Determine

$$\int \sinh^4(x) \cosh^5(x)\, dx.$$

**Solution**

The power of $\cosh(x)$ is odd. We express $\cosh^3(x)$ as $\cosh^2(x) \cosh(x)$, and make use of the identity $\cosh^2(x) - \sinh^2(x) = 1$ in order to express $\cosh^2(x)$ in terms of $\sinh^2(x)$:

$$\int \sinh^4(x) \cosh^5(x)\, dx = \int \sinh^4(x) \cosh^4(x) \cosh(x)\, dx$$

$$= \int \sinh^4(x) \left(1 + \sinh^2(x)\right)^2 \cosh(x)\, dx.$$

The above expression suggests the substitution $u = \sinh(x)$. Then $du/dx = \cosh(x)$ so that

$$\int \sinh^4(x) \cosh^5(x)\, dx = \int \sinh^4(x) \left(1 + \sinh^2(x)\right)^2 \cosh(x)\, dx$$

$$= \int u^4 \left(1 + u^2\right)^2 \frac{du}{dx}\, dx$$

$$= \int u^4 \left(1 + u^2\right)^2 du$$

$$= \int u^4 \left(1 + 2u^2 + u^4\right) du$$

$$= \int \left(u^4 + 2u^6 + u^8\right) du$$

$$= \frac{1}{5} u^5 + \frac{2}{7} u^7 + \frac{1}{9} u^9$$

$$= \frac{1}{5} \sinh^5(x) + \frac{2}{7} \sinh^7(x) + \frac{1}{9} \sinh^9(x) + C,$$

where $C$ is an arbitrary constant. □

We have to try a different tact in order to evaluate an indefinite integral of the form

$$\int \sin^m(x) \cos^n(x)\, dx \quad \text{or} \quad \int \sinh^m(x) \cosh^n(x)\, dx,$$

where both $m$ and $n$ are nonnegative even integers. Let's begin with the simplest cases.

**Example 4** Evaluate

$$\int \sin^2(x)\, dx.$$

## Solution

We will make use of the identity

$$\sin^2(x) = \frac{1-\cos(2x)}{2}.$$

Thus,

$$\int \sin^2(x)\, dx = \int \left(\frac{1-\cos(2x)}{2}\right) dx = \frac{1}{2}\int dx - \frac{1}{2}\int \cos(2x)\, dx$$

$$= \frac{x}{2} - \frac{1}{2}\left(\frac{1}{2}\sin(2x)\right) = \frac{x}{2} - \frac{1}{4}\sin(2x)$$

We can add an arbitrary constant $C$, of course:

$$\int \sin^2(x)\, dx = \frac{x}{2} - \frac{1}{4}\sin(2x) + C.$$

☐

We can evaluate

$$\int \cos^2(x)\, dx$$

in a similar fashion by using the identity

$$\cos^2(x) = \frac{1+\cos(2x)}{2}.$$

The result is that

$$\int \cos^2(x)\, dx = \frac{x}{2} + \frac{1}{4}\sin(2x) + C$$

(exercise).

The identities

$$\sin^2(x) = \frac{1-\cos(2x)}{2} \quad \text{and} \quad \cos^2(x) = \frac{1+\cos(2x)}{2}$$

can be used to evaluate antiderivatives of the form

$$\int \sin^n(x)\, dx \quad \text{and} \quad \int \cos^n(x)\, dx,$$

where $n$ is any even positive integer, as in the following example:

**Example 5** Determine

$$\int \cos^4(x)\, dx.$$

**Solution**

We have

$$\int \cos^4(x)\, dx = \int \left(\cos^2(x)\right)^2 dx = \int \left(\frac{1+\cos(2x)}{2}\right)^2 dx$$

$$= \int \frac{1}{4}dx + \frac{1}{2}\int \cos(2x)\, dx + \frac{1}{4}\int \cos^2(2x)\, dx$$

$$= \frac{x}{4} + \frac{1}{4}\sin(2x) + \frac{1}{4}\int \cos^2(2x)\, dx.$$

## 6.3. INTEGRALS OF SOME TRIGONOMETRIC AND HYPERBOLIC FUNCTIONS

If we set $u = 2x$ then $du = 2dx$ and

$$\int \cos^2(2x)\,dx = \frac{1}{2}\int \cos^2(u)\,du = \frac{1}{2}\left(\frac{u}{2} + \frac{1}{4}\sin(2u)\right)$$
$$= \frac{u}{4} + \frac{1}{8}\sin(2u) = \frac{x}{2} + \frac{1}{8}\sin(4x).$$

Therefore,

$$\int \cos^4(x)\,dx = \frac{x}{4} + \frac{1}{4}\sin(2x) + \frac{1}{4}\int \cos^2(2x)\,dx$$
$$= \frac{x}{4} + \frac{1}{4}\sin(2x) + \frac{1}{4}\left(\frac{x}{2} + \frac{1}{8}\sin(4x)\right)$$
$$= \frac{3}{8}x + \frac{1}{4}\sin(2x) + \frac{1}{32}\sin(4x)$$

(we can add an arbitrary constant, of course). □

We can transform the antidifferentiation of a product of even powers of $\sin(x)$ and $\cos(x)$ to the antidifferentiation of a power of either $\sin(x)$ or $\cos(x)$ by making use of the identity $\cos^2(x) + \sin^2(x) = 1$.

**Example 6** Determine

$$\int \sin^2(x)\cos^2(x)\,dx.$$

**Solution**

We will choose to express the integrand in powers of $\cos(x)$:

$$\int \sin^2(x)\cos^2(x)\,dx = \int \left(1 - \cos^2(x)\right)\cos^2(x)\,dx$$
$$= \int \cos^2(x) - \int \cos^4(x)\,dx.$$

We have remarked that

$$\int \cos^2(x) = \frac{x}{2} + \frac{1}{4}\sin(2x).$$

In Example 5 we derived the expression

$$\int \cos^4(x)\,dx = \frac{3}{8}x + \frac{1}{4}\sin(2x) + \frac{1}{32}\sin(4x).$$

Therefore,

$$\int \sin^2(x)\cos^2(x)\,dx = \int \cos^2(x) - \int \cos^4(x)\,dx$$
$$= \left(\frac{x}{2} + \frac{1}{4}\sin(2x)\right) - \left(\frac{3}{8}x + \frac{1}{4}\sin(2x) + \frac{1}{32}\sin(4x)\right)$$
$$= \frac{1}{8}x - \frac{1}{32}\sin(4x).$$

□

Sometimes it is more convenient to express such indefinite integrals in terms of powers of $\cos(x)$ and $\sin(x)$. For example,

$$\int \sin^2(x)\,dx = \frac{x}{2} - \frac{1}{4}\sin(2x) = \frac{x}{2} - \frac{1}{4}(2\sin(x)\cos(x)) = \frac{x}{2} - \frac{1}{2}\sin(x)\cos(x).$$

Similarly,
$$\int \cos^2(x)\,dx = \frac{x}{2} + \frac{1}{2}\sin(x)\cos(x).$$

As for the result of Example 6 we have

$$\begin{aligned}
\int \sin^2(x)\cos^2(x)\,dx &= \frac{1}{8}x - \frac{1}{32}\sin(4x) \\
&= \frac{1}{8}x - \frac{2}{32}\sin(2x)\cos(2x) \\
&= \frac{1}{8}x - \frac{1}{16}\left(2\sin(x)\cos(x)\right)\left(\cos^2(x) - \sin^2(x)\right) \\
&= \frac{1}{8}x - \frac{1}{8}\sin(x)\cos^3(x) + \frac{1}{8}\sin^3(x)\cos(x).
\end{aligned}$$

Such indefinite integrals can be computed directly in terms of powers of $\sin(x)$ and $\cos(x)$ with the help of **reduction formulas**:

For any integer $k \geq 2$, we have

$$\int \sin^k(x)\,dx = -\frac{1}{k}\sin^{k-1}(x)\cos(x) + \frac{k-1}{k}\int \sin^{k-2}(x)\,dx,$$

$$\int \cos^k(x)\,dx = \frac{1}{k}\cos^{k-1}(x)\sin(x) + \frac{k-1}{k}\int \cos^{k-2}(x)\,dx.$$

As in Section 6.2, such formulas are referred to as reduction formulas since the implementation of each formula reduces the power of $\sin(x)$ or $\cos(x)$ that appears under the integral sign.

We will derive the reduction formula that involves powers of $\sin(x)$. The derivation of the formula that involves powers of $\cos(x)$ is similar and left as an exercise.

The derivation is based on integration by parts. We set $u = \sin^{k-1}(x)$ and $dv = \sin(x)\,dx$. Thus,

$$du = \frac{du}{dx}dx = (k-1)\sin^{k-2}(x)\cos(x)\,dx, \text{ and } v = \int \sin(x)\,dx = -\cos(x).$$

Therefore,

$$\begin{aligned}
\int \sin^k(x)\,dx &= \int \sin^{k-1}(x)\sin(x)\,dx \\
&= \int u\,dv \\
&= uv - \int v\,du \\
&= \sin^{k-1}(x)(-\cos(x)) - \int (-\cos(x))(k-1)\sin^{k-2}(x)\cos(x)\,dx \\
&= -\sin^{k-1}(x)\cos(x) + (k-1)\int \sin^{k-2}(x)\cos^2(x)\,dx \\
&= -\sin^{k-1}(x)\cos(x) + (k-1)\int \sin^{k-2}(x)\left(1 - \sin^2(x)\right)\,dx \\
&= -\sin^{k-1}(x)\cos(x) + (k-1)\int \sin^{k-2}(x)\,dx - (k-1)\int \sin^k(x)\,dx.
\end{aligned}$$

## 6.3. INTEGRALS OF SOME TRIGONOMETRIC AND HYPERBOLIC FUNCTIONS

Thus,
$$k \int \sin^k(x)dx = -\cos(x) \sin^{k-1}(x) + (k-1) \int \sin^{k-2}(x)dx,$$

so that
$$\int \sin^k(x)dx = -\frac{1}{k} \sin^{k-1}(x) \cos(x) + \frac{k-1}{k} \int \sin^{k-2}(x)dx.$$

■

**Example 7** Use the reduction formula to determine
$$\int \sin^4(x)dx.$$

**Solution**

We apply the reduction formula for powers of $\sin(x)$ with $k = 4$:
$$\int \sin^4(x)dx = -\frac{1}{4} \sin^3(x) \cos(x) + \frac{3}{4} \int \sin^2(x)dx.$$

We have already seen that
$$\int \sin^2(x)dx = -\frac{1}{2} \sin(x) \cos(x) + \frac{1}{2}x.$$

The above expression follows from the reduction formula with $k = 2$, as well. Indeed,
$$\int \sin^2(x) dx = -\frac{1}{2} \sin(x) \cos(x) + \frac{1}{2} \int \sin^0(x)dx = -\frac{1}{2} \sin(x) \cos(x) + \frac{1}{2} \int 1 dx$$
$$= \frac{1}{2} \sin(x) \cos(x) + \frac{1}{2}x.$$

Therefore,
$$\int \sin^4(x)dx = -\frac{1}{4} \sin^3(x) \cos(x) + \frac{3}{4} \int \sin^2(x)dx$$
$$= -\frac{1}{4} \sin^3(x) \cos(x) + \frac{3}{4} \left( -\frac{1}{2} \sin(x) \cos(x) + \frac{1}{2}x \right)$$
$$= -\frac{1}{4} \sin^3(x) \cos(x) - \frac{3}{8} \sin(x) \cos(x) + \frac{3}{8}x$$

(we can add an arbitrary constant, of course). □

The determination of indefinite integrals of the form
$$\int \sinh^m(x) \cosh^n(x) dx,$$
where both $m$ and $n$ are are nonnegative even integers, is similar to the corresponding cases that involve $\sin(x)$ and $\cos(x)$. We will discuss the technique of reduction formulas only, even though it is also feasible to make use of the counterparts of "the double angle formulas".

**The Reduction Formulas for** $\sinh(x)$ **and** $\cosh(x)$:
$$\int \sinh^k(x)dx = \frac{1}{k} \sinh^{k-1}(x) \cosh(x) - \frac{k-1}{k} \int \sinh^{k-2}(x)dx,$$
$$\int \cosh^k(x)dx = \frac{1}{k} \cosh^{k-1}(x) \sinh(x) + \frac{k-1}{k} \int \cosh^{k-2}(x)dx$$

for any integer $k \geq 2$.

We will derive the formula that involves $\sinh(x)$. The derivation of the other formula is similar. As in the case of $\sin(x)$, the derivation is based on integration by parts. We will make use of the identity
$$\cosh^2(x) - \sinh^2(x) = 1,$$
instead of $\cos^2(x) + \sin^2(x) = 1$.

We set $u = \sinh^{k-1}(x)$ and $dv = \sinh(x)\,dx$. Thus,
$$du = \frac{du}{dx}dx = (k-1)\sinh^{k-2}(x)\cosh(x)\,dx, \text{ and } v = \int \sinh(x)\,dx = \cosh(x).$$

Therefore,
$$\begin{aligned}\int \sinh^k(x)\,dx &= \int \sinh^{k-1}(x)\sinh(x)\,dx \\ &= \int u\,dv \\ &= uv - \int v\,du \\ &= \sinh^{k-1}(x)\cosh(x) - \int \cosh(x)(k-1)\sinh^{k-2}(x)\cosh(x)\,dx \\ &= \sin^{k-1}(x)\cosh(x) - (k-1)\int \sinh^{k-2}(x)\cosh^2(x)\,dx.\end{aligned}$$

Since $\cosh^2(x) - \sinh^2(x) = 1$, we have $\cosh^2(x) = \sinh^2(x) + 1$. Thus,
$$\begin{aligned}\int \sinh^k(x)\,dx &= \sin^{k-1}(x)\cosh(x) - (k-1)\int \sinh^{k-2}(x)\left(\sinh^2(x)+1\right)dx \\ &= \sin^{k-1}(x)\cosh(x) - (k-1)\int \sinh^k(x)\,dx - (k-1)\int \sinh^{k-2}(x)\,dx.\end{aligned}$$

Therefore,
$$k\int \sinh^k(x)\,dx = \sin^{k-1}(x)\cosh(x) - (k-1)\int \sinh^{k-2}(x)\,dx$$

so that
$$\int \sinh^k(x)\,dx = \frac{1}{k}\sin^{k-1}(x)\cosh(x) - \frac{(k-1)}{k}\int \sinh^{k-2}(x)\,dx.$$

∎

**Example 8** Determine
$$\int \sinh^2(x)\,dx.$$

**Solution**

We implement the reduction formula that involves $\sinh(x)$, with $k=2$:
$$\begin{aligned}\int \sinh^k(x) &= \frac{1}{2}\sinh(x)\cosh(x) - \frac{1}{2}\int \sinh^0(x)\,dx \\ &= \frac{1}{2}\sinh(x)\cosh(x) - \frac{1}{2}\int 1\,dx \\ &= \frac{1}{2}\sinh(x)\cosh(x) - \frac{1}{2}x + C.\end{aligned}$$

□

## 6.3. INTEGRALS OF SOME TRIGONOMETRIC AND HYPERBOLIC FUNCTIONS

**Example 9** Determine
$$\int \cosh^2(x)\, dx.$$

**Solution**

We implement the reduction formula that involves $\cosh(x)$, with $k = 2$:

$$\begin{aligned}
\int \cosh^2(x)\,dx &= \frac{1}{2}\cosh(x)\sinh(x) + \frac{1}{2}\int \cosh^0(x)\,dx \\
&= \frac{1}{2}\cosh(x)\sinh(x) + \frac{1}{2}\int 1\,dx \\
&= \frac{1}{2}\cosh(x)\sinh(x) + \frac{1}{2}x + C
\end{aligned}$$

□

**Example 10** Determine
$$\int \cosh^4(x)\,dx.$$

**Solution**

We apply the reduction formula for $\cosh(x)$ with $k = 4$:

$$\int \cosh^4(x)\,dx = \frac{1}{4}\cosh^3(x)\sinh(x) + \frac{3}{4}\int \cosh^2(x)\,dx.$$

In Example 9 we derived the formula

$$\int \cosh^2(x)\,dx = \frac{1}{2}\sinh(x)\cosh(x) + \frac{1}{2}x.$$

Thus,

$$\begin{aligned}
\int \cosh^4(x)\,dx &= \frac{1}{4}\cosh^3(x)\sinh(x) + \frac{3}{4}\left(\frac{1}{2}\sinh(x)\cosh(x) + \frac{1}{2}x\right) \\
&= \frac{1}{4}\cosh^3(x)\sinh(x) + \frac{3}{8}\sinh(x)\cosh(x) + \frac{3}{8}x
\end{aligned}$$

(we can add an arbitrary constant, of course). □

### Special Integrals that involve $\sin(mx)$ and $\cos(nx)$

Certain integrals that involve $\sin(mx)$ and $\cos(nx)$, where $m$ and $n$ are nonnegative integers, have special significance, as you will see in Section 9.9 where we will discuss Fourier series. We will need the following trigonometric identities for their evaluation:

$$\sin(A)\cos(B) = \frac{1}{2}\sin(A+B) + \frac{1}{2}\sin(A-B)$$

$$\sin(A)\sin(B) = \frac{1}{2}\cos(A-B) - \frac{1}{2}\cos(A+B)$$

$$\cos(A)\cos(B) = \frac{1}{2}\cos(A-B) + \frac{1}{2}\cos(A+B)$$

The above identities follow from the sum and difference formulas for sine and cosine, as you can confirm.

**Proposition 1** Assume that $m$ and $n$ are nonnegative integers. Then,

a)
$$\int_{-\pi}^{\pi} \sin(mx) \cos(nx) \, dx = 0,$$

b)
$$\int_{-\pi}^{\pi} \sin(mx) \sin(nx) \, dx = 0 \text{ if } m \neq n,$$

c)
$$\int_{-\pi}^{\pi} \cos(mx) \cos(nx) \, dx = 0 \text{ if } m \neq n,$$

d)
$$\int_{-\pi}^{\pi} \sin^2(nx) \, dx = \pi \text{ and } \int_{-\pi}^{\pi} \cos^2(nx) \, dx = \pi$$

if $n \geq 1$.

**Proof**

a) If $m = 0$, $\sin(mx) = \sin(0) = 0$, so that the integral is 0. Therefore, we can assume that $m \geq 1$. We have

$$\sin(mx) \cos(nx) = \frac{1}{2} \sin(mx + nx) + \frac{1}{2} \sin(mx - nx)$$

Therefore,

$$\int_{-\pi}^{\pi} \sin(mx) \cos(nx) \, dx = \frac{1}{2} \int_{-\pi}^{\pi} \sin((m+n)x) \, dx + \frac{1}{2} \int_{-\pi}^{\pi} \sin((m-n)x) \, dx.$$

If we set $u = (m+n)x$, we have $du/dx = m+n$. Therefore,

$$\int_{-\pi}^{\pi} \sin((m+n)x) \, dx = \frac{1}{m+n} \int_{-(m+n)\pi}^{(m+n)\pi} \sin(u) \, du$$
$$= \frac{1}{m+n} \left( -\cos((m+n)\pi) + \cos(-(m+n)\pi) \right)$$
$$= \frac{1}{m+n} \left( -\cos((m+n)\pi) + \cos((m+n)\pi) \right)$$
$$= 0$$

(we have used the fact that cosine is an even function).
Similarly,
$$\int_{-\pi}^{\pi} \sin((m-n)x) \, dx = 0.$$

if $m \neq n$. Therefore,
$$\int_{-\pi}^{\pi} \sin(mx) \cos(nx) \, dx$$

if $m \neq n$.

If $m = n$,
$$\int_{-\pi}^{\pi} \sin(mx)\cos(nx)\,dx = \int_{-\pi}^{\pi} \sin(nx)\cos(nx)\,dx$$
$$= \frac{1}{2}\int_{-\pi}^{\pi} \sin(2nx)\,dx$$
$$= \frac{1}{4n}\int_{-2n\pi}^{2n\pi} \sin(u)\,du = 0$$
$$= \frac{1}{4n}\left(-\cos(u)\Big|_{-2n\pi}^{2n\pi}\right) = 0.$$

b) and c): Assume that $m \neq n$. Then,
$$\int_{-\pi}^{\pi} \sin(mx)\sin(nx)\,dx = \frac{1}{2}\int_{-\pi}^{\pi} \cos((m-n)x)\,dx - \frac{1}{2}\int_{-\pi}^{\pi} \cos((m+n)x)\,dx.$$
You can confirm that each integral on the right-hand side is 0, as in the proof of part a). Similarly,
$$\int_{-\pi}^{\pi} \cos(mx)\cos(nx)\,dx = \frac{1}{2}\int_{-\pi}^{\pi} \cos((m-n)x)\,dx + \frac{1}{2}\int_{-\pi}^{\pi} \cos((m+n)x)\,dx = 0.$$

d) If $n = 1, 2, 3, \ldots$,
$$\int_{-\pi}^{\pi} \sin^2(nx)\,dx = \frac{1}{n}\int_{-n\pi}^{n\pi} \sin^2(u)\,du$$
$$= \frac{1}{n}\left(\frac{1}{2}u - \frac{1}{2}\cos(u)\sin(u)\Big|_{-n\pi}^{n\pi}\right)$$
$$= \frac{1}{n}\left(\frac{n\pi}{2} + \frac{n\pi}{2}\right) = \pi.$$

Similarly,
$$\int_{-\pi}^{\pi} \cos^2(nx)\,dx = \pi$$
if $n = 1, 2, 3 \cdots$. ■

**Example 11** Let
$$f(x) = \sin(2x)\sin(3x).$$
By Proposition 1,
$$\int_{-\pi}^{\pi} f(x)\,dx = \int_{-\pi}^{\pi} \sin(2x)\sin(3x)\,dx = 0.$$
Figure 1 shows the graph of $f$ on the interval $[-\pi, \pi]$. □

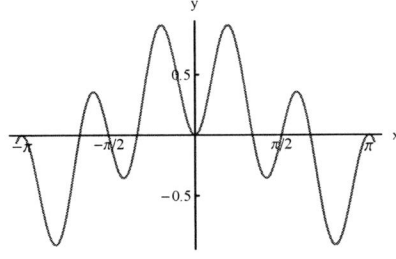

Figure 1

## Rational Functions of sin(x) and cos(x) or sinh(x) and cosh(x) (Optional)

Let $p(x)$ and $q(x)$ be linear combinations of expressions of the form $\sin^m(x)\cos^n(x)$, where $m$ and $n$ are nonnegative integers. If

$$f(x) = \frac{p(x)}{q(x)},$$

we will refer to $f(x)$ as **a rational function of sin(x) and and cos(x)**. Similarly, if $p(x)$ and $q(x)$ are linear combinations of expressions of the form $\sinh^m(x)\cosh^n(x)$, we will refer to $f(x)$ as **a rational function of sinh(x) and cosh(x)**. For example,

$$\frac{1}{\sin(x) + \cos(x)}$$

is a rational function of $\sin(x)$ and $\cos(x)$, and

$$\frac{1}{\sinh(x) + \cosh(x)}$$

is a rational function $\sinh(x)$ and $\cosh(x)$. We will discuss substitutions that transform the integration of such functions to the integration of ordinary rational functions.

**Proposition 2** Let

$$t = \tan\left(\frac{x}{2}\right).$$

Then

$$\sin(x) = \frac{2t}{1+t^2}, \ \cos(x) = \frac{1-t^2}{1+t^2} \text{ and } dx = \frac{2}{1+t^2}dt.$$

**Proof**

Note that

$$t = \tan\left(\frac{x}{2}\right) \Leftrightarrow \frac{x}{2} = \arctan(t) \pm n\pi \Leftrightarrow x = 2\arctan(t) \pm 2n\pi,$$

where $n = 0, 1, 2, \ldots$, since tangent is periodic with period $\pi$. Therefore,

$$dx = \frac{dx}{dt}dt = \left(2\frac{d}{dt}\arctan(t)\right)dt = \frac{2}{1+t^2}dt.$$

In order to express $\sin(x)$ and $\cos(x)$ in terms of $t$, we will begin with the identity

$$\cos^2\left(\frac{x}{2}\right) + \sin^2\left(\frac{x}{2}\right) = 1.$$

We divide by $\cos^2(x/2)$, so that

$$1 + \tan^2\left(\frac{x}{2}\right) = \frac{1}{\cos^2\left(\frac{x}{2}\right)} \Rightarrow \cos^2\left(\frac{x}{2}\right) = \frac{1}{1+t^2}.$$

Therefore,

$$\sin^2\left(\frac{x}{2}\right) = 1 - \cos^2\left(\frac{x}{2}\right) = 1 - \frac{1}{1+t^2} = \frac{t^2}{1+t^2}.$$

## 6.3. INTEGRALS OF SOME TRIGONOMETRIC AND HYPERBOLIC FUNCTIONS

Thus,

$$\sin(x) = 2\sin\left(\frac{x}{2}\right)\cos\left(\frac{x}{2}\right) = 2\frac{\sin\left(\frac{x}{2}\right)}{\cos\left(\frac{x}{2}\right)}\cos^2\left(\frac{x}{2}\right)$$
$$= 2\tan\left(\frac{x}{2}\right)\cos^2\left(\frac{x}{2}\right) = 2t\left(\frac{1}{1+t^2}\right) = \frac{2t}{1+t^2}.$$

We also have

$$\cos(x) = \cos^2\left(\frac{x}{2}\right) - \sin^2\left(\frac{x}{2}\right) = \frac{1}{1+t^2} - \frac{t^2}{1+t^2} = \frac{1-t^2}{1+t^2}.$$

∎

The substitution that is described by Proposition 2 enables us to transform the integration of a rational function of $\sin(x)$ and $\cos(x)$ to the integration of an ordinary rational function.

**Example 12** Evaluate

$$\int \sec(x)\,dx.$$

**Solution**

The secant function is a rational function of sine and cosine since

$$\sec(x) = \frac{1}{\cos(x)}.$$

We will implement the substitution

$$t = \tan\left(\frac{x}{2}\right),$$

as described by Proposition 2. Thus,

$$\int \frac{1}{\cos(x)}\,dx = \int \frac{1+t^2}{1-t^2}\left(\frac{2}{1+t^2}\right)dt = \int \frac{2}{1-t^2}\,dt.$$

We have

$$\frac{2}{1-t^2} = -\frac{1}{t-1} + \frac{1}{t+1}$$

(check). Therefore,

$$\int \frac{1}{\cos(x)}\,dx = \int \frac{2}{1-t^2}\,dt = -\int \frac{1}{t-1}\,dt + \int \frac{1}{t+1}\,dt$$
$$= -\ln(|t-1|) + \ln(|t+1|)$$
$$= \ln\left(\left|\frac{t+1}{t-1}\right|\right)$$
$$= \ln\left(\left|\frac{\tan(x/2)+1}{\tan(x/2)-1}\right|\right)$$

(we can add an arbitrary constant, of course). □

**Remark** Since two antiderivatives of the same function can differ by a constant, different computer algebra systems may express the indefinite integral of a given function in different ways. For example, the statements

$$\int \frac{1}{\cos(x)}\,dx = \ln(|\sec(x) + \tan(x)|),$$

$$\int \frac{1}{\cos(x)}\,dx = -\ln(|\cos(x/2) - \sin(x/2)|) + \ln(|\cos(x/2) + \sin(x/2)|)$$

are the responses of two computer algebra systems (the absolute value signs have been added, since the computer algebra systems in question do not bother to include the absolute value sign in indefinite integrals that lead to the natural logarithm). You can check that each response is valid by differentiation. ◊

**Example 13** Determine
$$\int \frac{1}{\cos(x) + \sin(x)} dx.$$

**Solution**

We will use the substitution $t = \tan(x/2)$ as described by Proposition 2. Thus,
$$\int \frac{1}{\cos(x) + \sin(x)} dx = \int \frac{1}{\frac{1-t^2}{1+t^2} + \frac{2t}{1+t^2}} \frac{2}{1+t^2} dt = -2 \int \frac{1}{t^2 - 2t - 1} dt.$$

Since
$$t^2 - 2t - 1 = 0 \Leftrightarrow t = 1 \pm \sqrt{2},$$

we have
$$t^2 - 2t - 1 = \left(t - \left(1 + \sqrt{2}\right)\right)\left(t - \left(1 - \sqrt{2}\right)\right).$$

Therefore, the partial fraction decomposition of
$$\frac{1}{t^2 - 2t - 1}$$

is in the form
$$\frac{A}{t - \left(1 + \sqrt{2}\right)} + \frac{B}{t - \left(1 - \sqrt{2}\right)}.$$

You should check that
$$\frac{1}{t^2 - 2t - 1} = \frac{\sqrt{2}}{4\left(t - \left(1 + \sqrt{2}\right)\right)} - \frac{\sqrt{2}}{4\left(t - \left(1 - \sqrt{2}\right)\right)}.$$

Therefore,
$$-2 \int \frac{1}{t^2 - 2t - 1} dt = -2 \int \frac{\sqrt{2}}{4\left(t - \left(1 + \sqrt{2}\right)\right)} dt + 2 \int \frac{\sqrt{2}}{4\left(t - \left(1 - \sqrt{2}\right)\right)} dt$$
$$= -\frac{\sqrt{2}}{2} \ln\left(\left|t - 1 - \sqrt{2}\right|\right) + \frac{\sqrt{2}}{2} \ln\left(\left|t - 1 + \sqrt{2}\right|\right).$$

Thus,
$$\int \frac{1}{\cos(x) + \sin(x)} dx = -2 \int \frac{1}{t^2 - 2t - 1} dt$$
$$= -\frac{\sqrt{2}}{2} \ln\left(\left|t - 1 - \sqrt{2}\right|\right) + \frac{\sqrt{2}}{2} \ln\left(\left|t - 1 + \sqrt{2}\right|\right)$$
$$= -\frac{\sqrt{2}}{2} \ln\left(\left|\tan\left(\frac{x}{2}\right) - 1 - \sqrt{2}\right|\right) + \frac{\sqrt{2}}{2} \ln\left(\left|\tan\left(\frac{x}{2}\right) - 1 + \sqrt{2}\right|\right)$$

(an arbitrary constant can be added) .□

As illustrated by the above examples, the evaluation of the definite integral of a rational function of $\sin(x)$ and $\cos(x)$ can be rather tedious, so that the help provided by a computer algebra

## 6.3. INTEGRALS OF SOME TRIGONOMETRIC AND HYPERBOLIC FUNCTIONS

systems is appreciated. Nevertheless, the fact that such a task can be transformed to the evaluation of the indefinite integral of an ordinary rational function should be comforting.

We can transform the integration of a rational function of $\sinh(x)$ and $\cosh(x)$ to the integration of an ordinary rational function by the substitution $u = e^x$, as described by the following proposition:

**Proposition 3** Let $u = e^x$. Then,

$$\sinh(x) = \frac{1}{2}u - \frac{1}{2u}, \quad \cosh(x) = \frac{1}{2}u + \frac{1}{2u} \quad \text{and} \quad dx = \frac{1}{u}du.$$

**Proof**

By the definitions of hyperbolic sine and hyperbolic cosine,

$$\sinh(x) = \frac{e^x - e^{-x}}{2} = \frac{1}{2}e^x - \frac{1}{2e^x} = \frac{1}{2}u - \frac{1}{2u},$$

and

$$\cosh(x) = \frac{e^x + e^{-x}}{2} = \frac{1}{2}u + \frac{1}{2u}.$$

For any $u > 0$,

$$u = e^x \Leftrightarrow x = \ln(u).$$

Therefore,

$$dx = \frac{dx}{du}du = \frac{1}{u}du.$$

∎

**Example 14** Determine

$$\int \frac{1}{\sinh(x) + \cosh(x)}dx$$

**Solution**

We will implement the substitution $u = e^x$, as described by Proposition 3. Thus,

$$\int \frac{1}{\sinh(x) + \cosh(x)}dx = \int \frac{1}{\left(\frac{1}{2}u - \frac{1}{2u}\right) + \left(\frac{1}{2}u + \frac{1}{2u}\right)}\frac{1}{u}du$$

$$= \int \frac{1}{u^2}du = -\frac{1}{u} + C = -\frac{1}{e^x} + C = -e^{-x} + C.$$

□

## Problems

In problems 1 - 4 determine the given indefinite integral:

1. $$\int \cos^3(x)\sin^2(x)\,dx$$

2. $$\int \sin^3(x)\cos^4(x)\,dx$$

3. $$\int \cosh^3(x)\sinh^2(x)\,dx$$

4. $$\int \sinh^3(x)\cosh^4(x)\,dx$$

**5.** Use the double-angle formulas to determine

$$\int \cos^2(x)\,dx.$$

**6.** Use the double-angle formulas to determine

$$\int \sin^4(x)\,dx$$

**7.** Use integration by parts to determine

$$\int \cos^2(x)\,dx$$

**8.** Use integration by parts to determine

$$\int \cosh^2(x)\,dx$$

**9.** Use integration by parts to determine

$$\int \sinh^2(x)\,dx$$

**10.** Make use of the appropriate reduction formula to determine

$$\int \cos^4(x)\,dx$$

In problems 11 and 12 make use of integration by parts to determine the missing expressions so that the equation is valid.

**11.**
$$\int \sin^6(x)\,dx = \ldots + \ldots \int \sin^4(x)\,dx$$

**12.**
$$\int \cosh^4(x)\,dx = \ldots + \ldots \int \cosh^2(x)\,dx$$

**13.** The function cosecant is defined as the reciprocal of sine:

$$\csc(x) = \frac{1}{\sin(x)}.$$

a) Evaluate

$$\int \csc(x)\,dx$$

by making use of $t = \tan(x/2)$ substitution.

b) [CAS] Evaluate

$$\int \csc(x)\,dx$$

with the help of a CAS.

**14.**
a) Evaluate

$$\int \frac{1}{1+\sin(x)}\,dx$$

by making use of $t = \tan(x/2)$ substitution.
b) [CAS] Evaluate
$$\int \frac{1}{1 + \sin(x)} dx$$
with the help of a CAS.

**15.** The function hyperbolic secant is defined as the reciprocal of hyperbolic cosine:
$$\operatorname{sech}(x) = \frac{1}{\cosh(x)}.$$

a) Evaluate
$$\int \operatorname{sech}(x) \, dx$$
by making use of $u = e^x$ substitution.
b) [CAS] Evaluate
$$\int \operatorname{sech}(x) \, dx$$
with the help of a CAS.

**16.**
a) Evaluate
$$\int \frac{1}{2 + \sinh(x)} dx$$
by making use of $u = e^x$ substitution.
b) [CAS] Evaluate
$$\int \frac{1}{2 + \sinh(x)} dx$$
with the help of a CAS.

## 6.4 Trigonometric and Hyperbolic Substitutions

In this section we will discuss the integration of some functions that involve expressions of the form $\sqrt{a^2 - x^2}$, $\sqrt{x^2 + a^2}$ or $\sqrt{x^2 - a^2}$, where $a$ is a positive constant. We will make use of trigonometric or hyperbolic substitutions in order to remove the radical sign. The resulting expressions are rational functions of $\sin(x)$ and $\cos(x)$, or $\sinh(x)$ and $\cosh(x)$, and can be antidifferentiated by the techniques of Section 6.3.

### The Substitution $x = a \sin(u)$

The substitution $x = a \sin(u)$ is useful in some cases that involve the expression $\sqrt{a^2 - x^2}$, where $a$ is a positive constant. The following proposition states the facts that are needed in the implementation of this substitution.

**Proposition 1** Let $a$ be a positive constant. We set
$$x = a \sin(u) \Leftrightarrow u = \arcsin\left(\frac{x}{a}\right)$$
(so that $-a \leq x \leq a$ and $-\pi/2 \leq u \leq \pi/2$). Then,
$$dx = a \cos(u) \, du \text{ and } \sqrt{a^2 - x^2} = a \cos(u).$$

**Proof**

We have
$$dx = \frac{dx}{du}du = \left(\frac{d}{du}(a\sin(u))\right)du = a\cos(u)\,du.$$

We also have
$$\sqrt{a^2 - x^2} = a\sqrt{1 - \left(\frac{x}{a}\right)^2} = a\sqrt{1 - \sin^2(u)} = a\sqrt{\cos^2(u)} = a\left|\cos(u)\right|.$$

Since $u = \arcsin(x/a)$, we have $-\pi/2 \leq u \leq \pi/2$. Therefore $\cos(u) \geq 0$, so that
$$\sqrt{a^2 - x^2} = a\left|\cos(u)\right| = a\cos(u).$$

∎

**Example 1** Determine
$$\int \sqrt{a - x^2}\,dx,$$
where $a$ is a positive constant.

**Solution**

We will implement the substitution $x = a\sin(u)$, as summarized by Proposition 1. Thus,
$$\int \sqrt{a^2 - x^2}\,dx = \int a\cos(u)\,a\cos(u)\,du = a^2\int \cos^2(u)\,du.$$

In Section 6.3 we stated that
$$\int \cos^2(u)\,du = \frac{1}{2}\cos(u)\sin(u) + \frac{1}{2}u.$$

Therefore,
$$\int \sqrt{a^2 - x^2}\,dx = a^2 \int \cos^2(u)\,du = a^2\left(\frac{1}{2}\cos(u)\sin(u) + \frac{1}{2}u\right)$$
$$= \frac{a^2}{2}\cos(u)\sin(u) + \frac{a^2}{2}u$$
$$= \frac{a^2}{2}\left(\frac{\sqrt{a^2 - x^2}}{a}\right)\left(\frac{x}{a}\right) + \frac{a^2}{2}\arcsin\left(\frac{x}{a}\right)$$
$$= \frac{x\sqrt{a^2 - x^2}}{2} + \frac{a^2}{2}\arcsin\left(\frac{x}{a}\right)$$

(an arbitrary constant may be added). □

**Example 2** Show that the area of a sector of disk with radius $r$ and central angle $\theta$ is
$$\frac{1}{2}r^2\theta.$$

## 6.4. TRIGONOMETRIC AND HYPERBOLIC SUBSTITUTIONS

**Solution**

Let's place the center of the disk at the origin of the $xy$-coordinate plane so that its boundary is the circle
$$x^2 + y^2 = r^2.$$

We will assume that $0 < \theta < \pi/2$ (the other cases can be dealt with similarly). Let's place the sector so that one edge is on the $x$-axis, as in Figure 1. In particular, the semicircle is the graph of the equation
$$y = \sqrt{r^2 - x^2}.$$

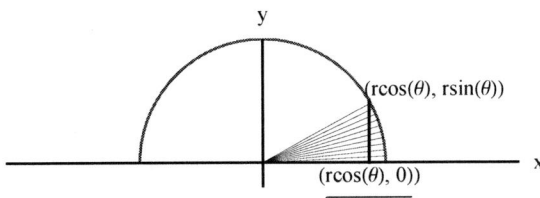

Figure 1: $y = \sqrt{r^2 - x^2}$

Part of the sector is the triangular region with vertices $(0,0)$, $(r\cos(\theta), 0)$ and $(r\cos(\theta), r\sin(\theta))$ with area
$$\frac{1}{2}(r\cos(\theta))(r\sin(\theta)) = \frac{1}{2}r^2 \cos(\theta)\sin(\theta).$$

The remaining part of the sector has area
$$\int_{r\cos(\theta)}^{r} \sqrt{r^2 - x^2}\,dx.$$

In Example 1 we determined an antiderivative of $\sqrt{r^2 - x^2}$:
$$\int \sqrt{r^2 - x^2} = \frac{x\sqrt{r^2 - x^2}}{2} + \frac{r^2}{2}\arcsin\left(\frac{x}{r}\right).$$

Therefore,
$$\int_{r\cos(\theta)}^{r} \sqrt{r^2 - x^2}\,dx = \left. \frac{x}{2}\sqrt{r^2 - x^2} + \frac{r^2}{2}\arcsin\left(\frac{x}{r}\right) \right|_{r\cos(\theta)}^{r}$$
$$= \frac{r^2}{2}\arcsin(1) - \frac{r\cos(\theta)}{2}\sqrt{r^2 - r^2\cos^2(\theta)} - \frac{r^2}{2}\arcsin(\cos(\theta))$$
$$= \frac{r^2}{2}\left(\frac{\pi}{2}\right) - \frac{r^2}{2}\cos(\theta)\sin(\theta) - \frac{r^2}{2}\arcsin(\cos(\theta))$$

If we set
$$\beta = \arcsin(\cos(\theta))$$

we have
$$\sin(\beta) = \cos(\theta) \text{ and } -\frac{\pi}{2} \leq \beta \leq \frac{\pi}{2}.$$

Since $0 < \theta \leq \pi/2$ we have
$$\beta = \frac{\pi}{2} - \theta$$

(confirm with the help of a triangle). Thus,

$$\int_{r\cos(\theta)}^{r} \sqrt{r^2 - x^2}\,dx = \frac{r^2}{2}\left(\frac{\pi}{2}\right) - \frac{r^2}{2}\cos(\theta)\sin(\theta) - \frac{r^2}{2}\arcsin(\cos(\theta))$$

$$= \frac{r^2}{2}\left(\frac{\pi}{2} - \cos(\theta)\sin(\theta) - \left(\frac{\pi}{2} - \theta\right)\right)$$

$$= \frac{r^2}{2}\left(\theta - \cos(\theta\sin(\theta)))\right)$$

Therefore the area of the sector is

$$\frac{1}{2}r^2\cos(\theta)\sin(\theta) + \frac{1}{2}r^2\left(\theta - \cos(\theta\sin(\theta))\right) = \frac{1}{2}r^2\theta,$$

as claimed. $\square$

**Example 3** Compute

$$\int_0^2 x^2\sqrt{4 - x^2}\,dx.$$

**Solution**

We will make use of the substitution

$$x = 2\sin(u) \Leftrightarrow u = \arcsin\left(\frac{x}{2}\right)$$

so that

$$-2 \le x \le 2 \text{ and } \frac{\pi}{2} \le u \le \frac{\pi}{2}.$$

As in the proof of Proposition 1,

$$\sqrt{4 - x^2} = 2\cos(u) \text{ and } dx = 2\cos(u)\,du.$$

We will make use of the definite integral version of the substitution rule. Thus,

$$\int_0^2 x^2\sqrt{4 - x^2}\,dx = \int_{u(0)}^{u(2)} (2\sin(u))^2 (2\cos(u))\, 2\cos(u)\,du$$

$$= 16\int_{\arcsin(0)}^{\arcsin(1)} \sin^2(u)\cos^2(u)\,du$$

$$= 16\int_0^{\pi/2} \sin^2(u)\cos^2(u)\,du$$

In Example 6 of Section 6.3 we derived the expression

$$\int \sin^2(u)\cos^2(u)\,du = \frac{1}{8}u - \frac{1}{8}\sin(u)\cos^3(u) + \frac{1}{8}\sin^3(u)\cos(u).$$

Therefore,

$$\int_0^2 x^2\sqrt{4 - x^2}\,dx = 16\int_0^{\pi/2} \sin^2(u)\cos^2(u)\,du$$

$$= 16\left(\frac{1}{8}u - \frac{1}{8}\sin(u)\cos^3(u) + \frac{1}{8}\sin^3(u)\cos(u)\Big|_0^{\pi/2}\right) = \pi$$

## 6.4. TRIGONOMETRIC AND HYPERBOLIC SUBSTITUTIONS

Thus, the area between the graph of $y = x^2\sqrt{4-x^2}$ and the interval $[0,2]$ is $\pi$. $\square$

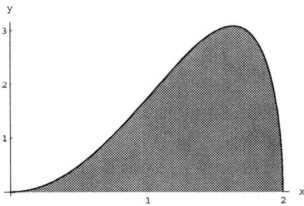

Figure 2: $y = x^2\sqrt{4-x^2}$

### The Substitution $x = a\sinh(u)$

The substitution $x = a\sinh(u)$ is useful in some cases that involve the expression $\sqrt{x^2 + a^2}$, where $a$ is a positive constant. The following proposition states the facts that are needed in the implementation of this substitution.

**Proposition 2** Let $a$ be a positive constant. We set

$$x = a\sinh(u) \Leftrightarrow u = \operatorname{arcsinh}\left(\frac{x}{a}\right)$$

Then,

$$dx = a\cosh(u)\,du \text{ and } \sqrt{x^2 + a^2} = a\cosh(u).$$

**Proof**

We have

$$dx = \frac{dx}{du}du = \left(\frac{d}{du}(a\sinh(u))\right)du = a\cosh(u)\,du.$$

We also have

$$x^2 + a^2 = a^2\sinh^2(u) + a^2 = a^2\left(\sinh^2(u) + 1\right) = a^2\cosh^2(u),$$

since $\cosh^2(u) - \sinh^2(u) = 1$. Thus,

$$\sqrt{x^2 + a^2} = \sqrt{a^2\cosh^2(u)} = a\,|\cosh(u)| = |a|\,|\cosh(u)| = a\cosh(u),$$

since $a > 0$ and $\cosh(u) > 0$ for each $u \in \mathbb{R}$. ∎

**Example 4** Determine

$$\int \sqrt{x^2 + a^2}\,dx,$$

where $a$ is a positive constant.

**Solution**

We will implement the substitution $x = a\sinh(u)$, as summarized by Proposition 2. Thus,

$$\int \sqrt{x^2 + a^2}\,dx = \int a\cosh(u)\,a\cosh(u)\,du = a^2\int \cosh^2(u)\,du.$$

As in Example 9 of Section 6.3,
$$\int \cosh^2(u)\,dx = \frac{1}{2}\sinh(u)\cosh(u) + \frac{1}{2}u.$$
Therefore,
$$\begin{aligned}\int \sqrt{x^2+a^2}\,dx &= a^2\left(\frac{1}{2}\sinh(u)\cosh(u) + \frac{1}{2}u\right)\\ &= \frac{a^2}{2}\sinh(u)\cosh(u) + \frac{a^2}{2}u\\ &= \frac{a^2}{2}\left(\frac{\sqrt{x^2+a^2}}{a}\right)\left(\frac{x}{a}\right) + \frac{a^2}{2}\operatorname{arcsinh}\left(\frac{x}{a}\right)\\ &= \frac{x}{2}\sqrt{x^2+a^2} + \frac{a^2}{2}\operatorname{arcsinh}\left(\frac{x}{a}\right)\end{aligned}$$
(an arbitrary constant may be added). □

The indefinite integral of Example 4 can be expressed in terms of the natural logarithm, since
$$\operatorname{arcsinh}(x) = \ln\left(x + \sqrt{x^2+a^2}\right).$$
Thus,
$$\begin{aligned}\int \sqrt{x^2+a^2}\,dx &= \frac{x}{2}\sqrt{x^2+a^2} + \frac{a^2}{2}\operatorname{arcsinh}\left(\frac{x}{a}\right)\\ &= \frac{x}{2}\sqrt{x^2+a^2} + \frac{a^2}{2}\ln\left(\frac{x}{a} + \sqrt{\left(\frac{x}{a}\right)^2+1}\right)\\ &= \frac{x}{2}\sqrt{x^2+a^2} + \frac{a^2}{2}\ln\left(\frac{x+\sqrt{x^2+a^2}}{a}\right)\\ &= \frac{x}{2}\sqrt{x^2+a^2} + \frac{a^2}{2}\ln\left(x+\sqrt{x^2+a^2}\right) - \frac{a^2}{2}\ln(a).\end{aligned}$$
Since the last term is a constant, we can express the indefinite integral as
$$\int \sqrt{x^2+a^2}\,dx = \frac{x}{2}\sqrt{x^2+a^2} + \frac{a^2}{2}\ln\left(x+\sqrt{x^2+a^2}\right) + C.$$
Many computer algebra systems yield the above expression.

**Example 5** Determine
$$\int \frac{x^2}{\sqrt{x^2+1}}\,dx$$

**Solution**

We set
$$x = \sinh(u) \Leftrightarrow u = \operatorname{arcsinh}(x),$$
so that
$$\sqrt{x^2+1} = \cosh(u) \text{ and } dx = \cosh(u)\,du,$$
as in Proposition 2. Therefore,
$$\int \frac{x^2}{\sqrt{x^2+1}}\,dx = \int \frac{\sinh^2(x)}{\cosh(x)}\cosh(x)\,dx = \int \sinh^2(x)\,dx.$$

## 6.4. TRIGONOMETRIC AND HYPERBOLIC SUBSTITUTIONS

In Example 8 of Section 6.3 we derived the expression
$$\int \sinh^2(u)\,du = \frac{1}{2}\sinh(u)\cosh(u) - \frac{1}{2}u.$$

Therefore,
$$\int \frac{x^2}{\sqrt{x^2+1}}\,dx = \frac{1}{2}\sinh(u)\cosh(u) - \frac{1}{2}u$$
$$= \frac{1}{2}x\sqrt{x^2+1} - \frac{1}{2}\operatorname{arcsinh}(x).$$

We can express the antiderivative in terms of the natural logarithm:
$$\int \frac{x^2}{\sqrt{x^2+1}}\,dx = \frac{1}{2}x\sqrt{x^2+1} - \frac{1}{2}\operatorname{arcsinh}(x) = \frac{1}{2}x\sqrt{x^2+1} - \frac{1}{2}\ln\left(x+\sqrt{x^2+1}\right).$$

□

**Example 6** Determine
$$\int x^2\sqrt{a^2+x^2}\,dx,$$
where $a$ is a positive constant.

**Solution**

We implement the substitution $x = a\sinh(u)$, as summarized by Proposition 2. Thus,
$$\int x^2\sqrt{a^2+x^2}\,dx = \int a^2\sinh^2(u)(a\cosh(u))\,a\cosh(u)\,du$$
$$= a^4\int \sinh^2(u)\cosh^2(u)\,du.$$

Since
$$\cosh^2(u) - \sinh^2(u) = 1 \Rightarrow \sinh^2(u) = \cosh^2(u) - 1,$$
we have
$$\int \sinh^2(u)\cosh^2(u)\,du = \int \left(\cosh^2(u) - 1\right)\cosh^2(u)\,du$$
$$= \int \cosh^4(u)\,du - \int \cosh^2(u)\,du.$$

As in Example 10 of Section 6.3,
$$\int \cosh^4(u)\,du = \frac{1}{4}\cosh^3(u)\sinh(u) + \frac{3}{8}\sinh(u)\cosh(u) + \frac{3}{8}u.$$

As in Example 9 of Section 6.3,
$$\int \cosh^2(u)\,du = \frac{1}{2}\cosh(u)\sinh(u) + \frac{1}{2}u.$$

Therefore,
$$\int \sinh^2(u)\cosh^2(u)\,du = \int \cosh^4(u)\,du - \int \cosh^2(u)\,du$$
$$= \left(\frac{1}{4}\cosh^3(u)\sinh(u) + \frac{3}{8}\sinh(u)\cosh(u) + \frac{3}{8}u\right)$$
$$- \left(\frac{1}{2}\cosh(u)\sinh(u) + \frac{1}{2}u\right)$$
$$= \frac{1}{4}\cosh^3(u)\sinh(u) - \frac{1}{8}\sinh(u)\cosh(u) - \frac{1}{8}u.$$

Thus,

$$\int x^2\sqrt{a^2+x^2}dx = a^4 \int \sinh^2(u)\cosh^2(u)\,du$$

$$= \frac{a^4}{4}\cosh^3(u)\sinh(u) - \frac{a^4}{8}\sinh(u)\cosh(u) - \frac{a^4}{8}u$$

$$= \frac{a^4}{4}\left(\frac{\sqrt{x^2+a^2}}{a}\right)^3\left(\frac{x}{a}\right) - \frac{a^4}{8}\left(\frac{x}{a}\right)\left(\frac{\sqrt{x^2+a^2}}{a}\right) - \frac{a^4}{8}\operatorname{arcsinh}\left(\frac{x}{a}\right)$$

$$= \frac{1}{4}\left(x^2+a^2\right)^{\frac{3}{2}}x - \frac{1}{8}a^2 x\sqrt{x^2+a^2} - \frac{a^4}{8}\operatorname{arcsinh}\left(\frac{x}{a}\right).$$

Many computer algebra systems express the indefinite integral in terms of the natural logarithm. Indeed,

$$\int x^2\sqrt{a^2+x^2}dx = \frac{1}{4}\left(x^2+a^2\right)^{\frac{3}{2}}x - \frac{1}{8}a^2 x\sqrt{x^2+a^2} - \frac{a^4}{8}\ln\left(\frac{x}{a} + \sqrt{\left(\frac{x}{a}\right)^2+1}\right)$$

$$= \frac{1}{4}\left(x^2+a^2\right)^{\frac{3}{2}}x - \frac{1}{8}a^2 x\sqrt{x^2+a^2} - \frac{a^4}{8}\ln\left(x+\sqrt{x^2+a^2}\right) - \frac{a^4}{8}\ln(a).$$

The last term is a constant, and does not appear in the expressions obtained from the computer algebra systems. □

The lengthy calculations of Example 6 should make us appreciate the work of the people who design computer algebra systems.

## The Substitution $x = a\cosh(u)$

The substitution $x = a\cosh(u)$ is useful in some cases that involve the expression $\sqrt{x^2-a^2}$, where $a$ is a positive constant. The following proposition states the facts that are needed in the implementation of this substitution.

**Proposition 3** Let $a$ be a positive constant, and $x \geq a$. We set

$$x = a\cosh(u) \Leftrightarrow u = \operatorname{arcosh}\left(\frac{x}{a}\right).$$

Then,
$$dx = a\sinh(u)\,du \text{ and } \sqrt{x^2-a^2} = a\sinh(u).$$

**Proof**

We have
$$x = \frac{dx}{du}du = \left(\frac{d}{du}(a\cosh(u))\right)du = a\sinh(u)\,du.$$

We also have
$$\sqrt{x^2-a^2} = \sqrt{a^2\cosh^2(u)-a^2} = a\sqrt{\cosh^2(u)-1} = a\sqrt{\sinh^2(u)},$$

since $a > 0$ and $\cosh^2(u) - \sinh^2(u) = 1$. Since $u = \operatorname{arcosh}(x/a) \geq 0$ we have $\sinh(u) \geq 0$. Therefore,

$$\sqrt{x^2-a^2} = a\sinh(u).$$

■

## 6.4. TRIGONOMETRIC AND HYPERBOLIC SUBSTITUTIONS

**Remark** The expression $\sqrt{x^2 - a^2}$ is a real number if $x \leq -a$ as well. Even though Proposition 3 covers the case $x \geq a$, in case $x \leq -a$ we can introduce the intermediate variable $v = -x$. Then $v \geq a$ and we can make use of the substitution

$$v = a \cosh(u) \Leftrightarrow u = \operatorname{arcosh}\left(\frac{v}{a}\right).$$

◊

**Example 7** Determine

$$\int \sqrt{x^2 - a^2}\, dx,$$

where $a$ is a positive constant and $x \geq a$.

**Solution**

We implement the substitution $x = a \cosh(u)$, as summarized by Proposition 3. Thus,

$$\int \sqrt{x^2 - a^2}\, dx = \int a \sinh(u) a \sinh(u)\, du = a^2 \int \sinh^2(u)\, du.$$

As in Example 8 of Section 6.3,

$$\int \sinh^2(u)\, du = \frac{1}{2} \cosh(u) \sinh(u) - \frac{1}{2} u.$$

Therefore

$$\int \sqrt{x^2 - a^2}\, dx = \frac{a^2}{2} \cosh(u) \sinh(u) - \frac{a^2}{2} u$$

$$= \frac{a^2}{2} \left(\frac{x}{a}\right) \left(\frac{\sqrt{x^2 - a^2}}{a}\right) - \frac{a^2}{2} \operatorname{arcosh}\left(\frac{x}{a}\right)$$

$$= \frac{x}{2} \sqrt{x^2 - a^2} - \frac{a^2}{2} \operatorname{arcosh}\left(\frac{x}{a}\right).$$

We can express the antiderivative in terms of the natural logarithm:

$$\int \sqrt{x^2 - a^2}\, dx = \frac{x}{2} \sqrt{x^2 - a^2} - \frac{a^2}{2} \operatorname{arcosh}\left(\frac{x}{a}\right)$$

$$= \frac{x}{2} \sqrt{x^2 - a^2} - \frac{a^2}{2} \ln\left(\frac{x}{a} + \sqrt{\frac{x^2}{a^2} - 1}\right)$$

$$= \frac{x}{2} \sqrt{x^2 - a^2} - \frac{a^2}{2} \ln\left(x + \sqrt{x^2 - a^2}\right) - \frac{a^2}{2} \ln(a).$$

We can drop the constant $-a^2 \ln(a)/2$, and write

$$\int \sqrt{x^2 - a^2}\, dx = \frac{x}{2} \sqrt{x^2 - a^2} - \frac{1}{2} \ln\left(x + \sqrt{x^2 - a^2}\right), \quad x \geq a$$

(as usual, an arbitrary constant may be added). □

**Example 8** Let $G$ be the region in the first quadrant that is bounded by the $x$-axis, the hyperbola $x^2 - y^2 = 1$ and the line $x = 2$.

a) Sketch $G$.
b) Compute the area of $G$.

**Solution**

a) Figure 3 shows $G$.

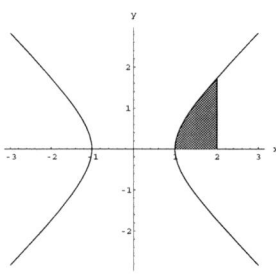

Figure 3

b) A point $(x, y)$ is on the given hyperbola if and only if
$$y^2 = x^2 - 1 \Leftrightarrow y = \pm\sqrt{x^2 - 1}.$$

Therefore, the area of $G$ is
$$\int_1^2 \sqrt{x^2 - 1}\, dx.$$

We make use of the indefinite integral in Example 7 with $a = 1$. Thus,
$$\int_1^2 \sqrt{x^2 - 1}\, dx = \frac{x}{2}\sqrt{x^2 - 1} - \frac{1}{2}\operatorname{arcosh}(x)\Big|_1^2 = \sqrt{3} - \frac{1}{2}\operatorname{arcosh}(2) + \frac{1}{2}\operatorname{arcosh}(1).$$

Since
$$\operatorname{arccosh}(x) = \ln\left(x + \sqrt{x^2 - 1}\right)$$

we can express the result in terms of the natural logarithm:
$$\int_1^2 \sqrt{x^2 - 1}\, dx = \sqrt{3} - \frac{1}{2}\ln\left(2 + \sqrt{3}\right) \cong 1.07357$$

(confirm). □

## Problems

**1.**
a) Evaluate
$$\int \sqrt{9 - 4x^2}\, dx$$

Hint:
$$\int \cos^2(u)\, du = \frac{u}{2} + \frac{1}{2}\sin(u)\cos(u)$$

b) Evaluate
$$\int_0^{3/4} \sqrt{9 - 4x^2}\, dx$$

## 2.
a) Express the integral
$$\int_{2}^{2\sqrt{2}} \sqrt{16 - x^2}\,dx$$
as an integral in terms of powers of sin $(u)$ and cos$(u)$.
b) Evaluate the integral that you obtained in part a) in terms of powers of sin $(u)$ and cos$(u)$.
Hint:
$$\int \cos^2(u)\,du = \frac{u}{2} + \frac{1}{2}\sin(u)\cos(u)$$

## 3. Determine the area of the region that is bounded by the ellipse
$$\frac{x^2}{4} + \frac{y^2}{9} = 1.$$

Hint:
$$\int \cos^2(u)\,du = \frac{u}{2} + \frac{1}{2}\sin(u)\cos(u)$$

## 4.
a)
$$\int_{-\sqrt{2}}^{\sqrt{3}} x^2\sqrt{4 - x^2}\,dx$$
as an integral in terms of powers of sin $(u)$ and cos$(u)$.
b) Evaluate the integral.
Hint:
$$\int \sin^2(x)\cos^2(x)\,dx = \frac{1}{8}x - \frac{1}{8}\sin(x)\cos^3(x) + \frac{1}{8}\sin^3(x)\cos(x)$$

## 5.
a) Evaluate
$$\int \sqrt{9x^2 + 16}\,dx$$
by making use of the appropriate hyperbolic substitution.
Hint:
$$\int \cosh^2(u)\,du = \frac{1}{2}\cosh(u)\sinh(u) + \frac{1}{2}u$$

b) [CAS] Compare your response with the response of a CAS.

## 6.
a) Evaluate
$$\int \sqrt{4x^2 - 1}\,dx$$
by making use of the appropriate hyperbolic substitution. Assume that $x > 1/2$.
Hint:
$$\int \sinh^2(u)\,du = \frac{1}{2}\cosh(u)\sinh(u) - \frac{1}{2}u.$$

b) [CAS] Compare your response with the response of a CAS.

## 7.
a) Express the integral
$$\int_{2\sinh(2)}^{2\sinh(3)} \sqrt{x^2 + 4}\,dx$$

as an integral in terms of powers of $\sinh(u)$ and $\cosh(u)$.
b) Evaluate the integral that you obtained in part a) in terms of powers of $\sinh(u)$ and $\cosh(u)$. Hint:
$$\int \cosh^2(u)\, du = \frac{1}{2}\cosh(u)\sinh(u) + \frac{1}{2}u.$$

c) [C] Compare your result with the result of the numerical integrator of your computational utility.

**8.**
a) Express the integral
$$\int_{3\cosh(1)}^{3\cosh(2)} \sqrt{x^2 - 9}\, dx$$
as an integral in terms of powers of $\sinh(u)$ and $\cosh(u)$..
b) Evaluate the integral that you obtained in part a) in terms of powers of $\sinh(u)$ and $\cosh(u)$.
Hint:
$$\int \sinh^2(u)\, du = \frac{1}{2}\cosh(u)\sinh(u) - \frac{1}{2}u.$$

c) [C] Compare your result with the result of the numerical integrator of your computational utility.

**9.** Evaluate
$$\int \sqrt{-x^2 + 2x + 3}\, dx$$
Hint: Complete the square and make use of the formula
$$\int \cos^2(u)\, du = \frac{u}{2} + \frac{1}{2}\sin(u)\cos(u)$$

**10.**
a) Evaluate
$$\int \sqrt{x^2 - 2x + 5}\, dx$$
by making use of the appropriate hyperbolic substitution.
Hint: Complete the square and make use of the formula
$$\int \cosh^2(u) = \frac{1}{2}\cosh(u)\sinh(u) + \frac{1}{2}u.$$

b) [CAS] Compare your response with the response of a CAS.

**11.**
a) Determine the area of the bounded region bounded by the hyperbola $x^2 - y^2 = 1$ and the line $x = 2$.
Hint:
$$\int \sinh^2(u)\, du = \frac{1}{2}\cosh(u)\sinh(u) - \frac{1}{2}u.$$

b) [C] Compare your result with the result of the numerical integrator of your computational utility.

**12.**
a) Determine the area of the region bounded by the hyperbola $y^2 - x^2 + 2x = 2$ and the lines $x = 1$ and $x = 3$.

Hint complete the square and use the formula

$$\int \cosh^2(u) = \frac{1}{2}\cosh(u)\sinh(u) + \frac{1}{2}u.$$

b) [C] Compare your result with the result of the numerical integrator of your computational utility.

## 6.5 Numerical Integration

It may not be possible to find an antiderivative of a given function as a combination of familiar special functions, even with the help of a computer algebra system. Therefore we may have to rely on our computational utility for the approximate calculation of many integrals. In this section we will discuss basic approximation procedures that form the basis of professional numerical integration schemes.

### Riemann Sums

Let's begin by reviewing the special Riemann sums that we have used to approximate integrals. We partition the given interval $[a, b]$ to $n$ subintervals of length

$$\Delta x = \frac{b-a}{n},$$

and set

$$x_k = a + k\Delta x, \ k = 0, 1, 2, \ldots, n.$$

Thus, $x_{k-1}$ is the left endpoint of the $k$th subinterval $[x_{k-1}, x_k]$ and $x_k$ is the right endpoint of $[x_{k-1}, x_k]$.

A **left-endpoint sum** for a given function $f$ is

$$l(f, a, b, n) = \sum_{k=1}^{n} f(x_{k-1}) \Delta x.$$

A **right-endpoint sum** for $f$ is

$$r(f, a, b, n) = \sum_{k=1}^{n} f(x_k) \Delta x.$$

A **midpoint sum** for $f$ is

$$m(f, a, b, n) = \sum_{k=1}^{n} f(c_k) \Delta x,$$

where

$$c_k = \frac{x_{k-1} + x_k}{2}$$

is the midpoint of the $k$th subinterval $[x_{k-1}, x_k]$. The technique of approximating integrals by midpoint sums is referred to as **the midpoint rule**. More generally, the act of approximating integrals is referred to as **numerical integration,** and an approximation scheme for integration is referred to as a **numerical integration rule**.

A midpoint sum approximates a given integral more accurately than a left-endpoint sum and a right-endpoint sum with the same number of partition points, provided that $f$ is sufficiently smooth:

**Proposition 1** Assume that $f'$ is continuous on the interval $[a,b]$. Set $C_1 = \max(|f'(x)| : x \in [a,b])$, and $\Delta x = (b-a)/n$. Then

$$\left| \sum_{k=1}^{n} f(x_{k-1}) \Delta x - \int_{a}^{b} f(x)dx \right| \leq C_1 (b-a) \Delta x,$$

$$\left| \sum_{k=1}^{n} f(x_k) \Delta x - \int_{a}^{b} f(x)dx \right| \leq C_1 (b-a) \Delta x.$$

If, $f''$ is continuous on $[a,b]$ and $C_2 = \max(|f''(x)| : x \in [a,b])$, then

$$\left| \sum_{k=1}^{n} f(c_k) \Delta x - \int_{a}^{b} f(x)dx \right| \leq \frac{C_2(b-a)}{24} (\Delta x)^2.$$

The proof of Proposition ?? is left to a course in numerical analysis.

**Example 1** Let $f(x) = \sin(x)$. Approximate $\int_{0}^{\pi/2} f(x)\, dx$ by left-endpoint sums, right-endpoint sums and midpoint sums that correspond to the partitioning of the interval $[0, \pi/2]$ to 16, 32, 64 and 128 subintervals of equal length. Calculate the absolute errors of the approximations and compare their accuracy.

**Solution**

We have
$$\int_{0}^{\pi/2} f(x)\,dx = \int_{0}^{\pi/2} \sin(x)\,dx = -\cos(x)\Big|_{0}^{\pi/2} = \cos(0) = 1.$$

Set
$$\Delta x = \frac{\pi}{2n}.$$

Table 1 displays $l(f, 0, \pi/2, n)$ and $|l(f, 0, \pi/2, n) - 1|$ for $n = 16, 32, 64, 128$:

| $n$ | $l(f, 0, \pi/2, n)$ | $|l(f, 0, \pi/2, n) - 1|$ |
|-----|---------------------|----------------------------|
| 16  | 0.950 109           | $5.0 \times 10^{-2}$       |
| 32  | 0.975 256           | $2.5 \times 10^{-2}$       |
| 64  | 0.987 678           | $1.2 \times 10^{-2}$       |
| 128 | 0.993 852           | $6.1 \times 10^{-3}$       |

Table 1

Table 2 displays the corresponding data for right-endpoint sums:

| $n$ | $r(f, 0, \pi/2, n)$ | $|r(f, 0, \pi/2, n) - 1|$ |
|-----|---------------------|----------------------------|
| 16  | 1.048 28            | $4.8 \times 10^{-2}$       |
| 32  | 1.024 34            | $2.4 \times 10^{-2}$       |
| 64  | 1.012 22            | $1.2 \times 10^{-2}$       |
| 128 | 1.006 12            | $6.1 \times 10^{-3}$       |

Table 2

Table 3 displays the corresponding data for midpoint sums:

| $n$ | $m(f, 0, \pi/2, n)$ | $|m(f, 0, \pi/2, n) - 1|$ |
|-----|---------------------|----------------------------|
| 16  | 1.000 4             | $4.0 \times 10^{-4}$       |
| 32  | 1.000 1             | $1.0 \times 10^{-4}$       |
| 64  | 1.000 03            | $2.5 \times 10^{-5}$       |
| 128 | 1.000 01            | $6.3 \times 10^{-6}$       |

## 6.5. NUMERICAL INTEGRATION

Table 3

In all cases the absolute error gets smaller as the number of subintervals increases so that the length of the subintervals decreases. The rate at which the absolute error decreases is almost the same when we make use of left-endpoint sums and right-endpoint sums. This is consistent with Proposition 1 since

$$\left| \sum_{k=1}^{n} f(x_{k-1}) \Delta x - \int_{a}^{b} f(x)dx \right| \leq C_1 (b-a) \Delta x,$$

$$\left| \sum_{k=1}^{n} f(x_k) \Delta x - \int_{a}^{b} f(x)dx \right| \leq C_1 (b-a) \Delta x.$$

On the other hand, the error decreases at a much faster rate when we use midpoint sums. This is also consistent with Proposition 1, since

$$\left| \sum_{k=1}^{n} f(c_k) \Delta x - \int_{a}^{b} f(x)dx \right| \leq \frac{C_2 (b-a)}{24} (\Delta x)^2,$$

and $(\Delta x)^2$ is much smaller than $\Delta x$ if $\Delta x$ is small. □

We have discussed left-endpoint sums, right-endpoint sums and midpoint sums based on the partitioning of a given interval to subintervals of equal length, for the sake of the simplicity of the calculations. The subintervals can be of different lengths, of course. Indeed, it is more efficient to place more points in the parts of the interval where the values of the function vary more rapidly.

### The Trapezoid Rule

Now we will discuss an approximation scheme for integrals that is not based on Riemann sums. The **trapezoid rule** is based on the approximation of a function by **piecewise linear functions**, as illustrated in Figure 1.

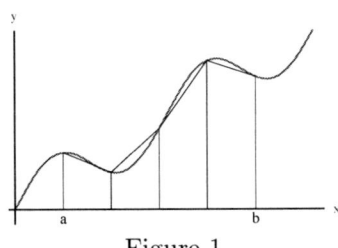

Figure 1

Assume that
$$P = \{x_0, x_1, x_2, \ldots, x_n\}$$
is a partition of $[a, b]$ so that
$$a = x_0 < x_1 < x_2 < \cdots < x_{k-1} < x_k < \cdots < x_n = b.$$

Set $\Delta x_k = x_k - x_{k-1}$. The trapezoid rule replaces $f$ on the $k$th subinterval $[x_{k-1}, x_k]$ by the linear function that has the same values as $f$ at $x_{k-1}$ and $x_k$.
Thus, the graph of the linear function is the line segment that passes through the points $(x_{k-1}, f(x_{k-1}))$ and $(x_k, f(x_k))$, as illustrated in Figure 2.

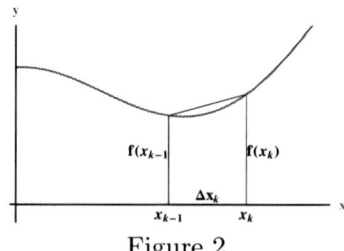

Figure 2

The area of the trapezoid that is formed by the points $(x_{k-1}, 0)$, $(x_{k-1}, f(x_{k-1}))$, $(x_k, 0)$ and $(x_k, f(x_k))$ is

$$\left(\frac{f(x_{k-1}) + f(x_k)}{2}\right)(x_k - x_{k-1}) = \left(\frac{f(x_{k-1}) + f(x_k)}{2}\right)\Delta x_k.$$

Indeed, the linear function $l_k$ that has the same values as $f$ at $x_{k-1}$ and $x_k$ is defined by the expression

$$l_k(x) = f(x_{k-1}) + \frac{f(x_k) - f(x_{k-1})}{x_k - x_{k-1}}(x - x_{k-1}) = f(x_{k-1}) + \frac{f(x_k) - f(x_{k-1})}{\Delta x_k}(x - x_{k-1}).$$

Therefore,

$$\int_{x_{k-1}}^{x_k} l_k(x) = \int_{x_{k-1}}^{x_k} \left(f(x_{k-1}) + \frac{f(x_k) - f(x_{k-1})}{\Delta x_k}(x - x_{k-1})\right) dx$$

$$= f(x_{k-1})\Delta x_k + \frac{f(x_k) - f(x_{k-1})}{\Delta x_k}\left(\frac{(x - x_{k-1})^2}{2}\bigg|_{x_{k-1}}^{x_k}\right)$$

$$= f(x_{k-1})\Delta x_k + \frac{f(x_k) - f(x_{k-1})}{\Delta x_k}\frac{(x_k - x_{k-1})^2}{2}$$

$$= f(x_{k-1})\Delta x_k + \frac{f(x_k) - f(x_{k-1})}{\Delta x_k}\frac{(\Delta x_k)^2}{2}$$

$$= f(x_{k-1})\Delta x_k + \frac{1}{2}(f(x_k) - f(x_{k-1}))\Delta x_k$$

$$= \frac{f(x_{k-1}) + f(x_k)}{2}\Delta x_k.$$

The above result is valid for a function of variable sign.
Thus,

$$\int_a^b f(x)\,dx = \sum_{k=1}^n \int_{x_{k-1}}^{x_k} f(x)\,dx \cong \sum_{k=1}^n \int_{x_{k-1}}^{x_k} l_k(x)\,dx = \sum_{k=1}^n \frac{f(x_{k-1}) + f(x_k)}{2}\Delta x_k.$$

Let's summarize: **The trapezoid rule approximates** $\int_a^b f(x)dx$ by the sum

$$\sum_{k=1}^n \frac{f(x_{k-1}) + f(x_k)}{2}\Delta x_k.$$

If each subinterval has the same length

$$\Delta x = \frac{b-a}{n},$$

## 6.5. NUMERICAL INTEGRATION

then the sum can be expressed as

$$T(f,a,b,n) = \sum_{k=1}^{n} \frac{f(x_{k-1} + f(x_k))}{2} \Delta x = \Delta x \sum_{k=1}^{n} \frac{f(x_{k-1}) + f(x_k)}{2},$$

where $x_k = a + k\Delta x$ for $k = 0, 1, 2, \ldots, n$. Note the difference between the trapezoid rule and the midpoint rule. Since

$$m(f,a,b,n) = \Delta x \sum_{k=1}^{n} f\left(\frac{x_{k-1} + x_k}{2}\right),$$

the midpoint rule is a special Riemann sum that samples the value of the integrand at the midpoint of each subinterval, whereas the trapezoid rule uses the average of the values of the integrand at the endpoints of each subinterval.

The accuracy of the trapezoid rule is comparable to the accuracy of the midpoint rule:

**Proposition 2** If $f''$ is continuous on $[a,b]$, and $C_2 = \max(|f''(x)| : x \in [a,b])$, then

$$\left| T(f,a,b,n) - \int_a^b f(c_k)\, \Delta x \right| \leq \frac{C_2(b-a)}{12} (\Delta x)^2,$$

where $\Delta x = (b-a)/n$.

We leave the proof of the above error estimate to a course in numerical analysis.

**Example 2** Let

$$f(x) = \frac{1}{x^2 + 1}.$$

Approximate $\int_0^1 f(x)\, dx$ by the trapezoid rule based on the partitioning of the interval $[0,1]$ to $n$ intervals of equal length, with $n = 10, 20, 40$ and $80$. Calculate the absolute errors in the approximations.

**Solution**

We have

$$\int_0^1 f(x)\, dx = \int_0^1 \frac{1}{x^2 + 1}\, dx = \arctan(x)\Big|_0^1 = \arctan(1) = \frac{\pi}{4} \cong 0.785\,398.$$

With the notation preceding Proposition 2,

$$T(f,0,1,n) = \Delta x \sum_{k=1}^{n} \frac{f(x_{k-1}) + f(x_k)}{2} = \frac{1}{n} \sum_{k=1}^{n} \frac{f(x_{k-1}) + f(x_k)}{2},$$

where $x_k = k\Delta x = k/n$, $k = 0, 1, 2, \ldots, n$. Table 4 displays $T(f,0,1,n)$ and $|T(f,0,1,n) - \pi/4|$ for $n = 10, 20, 40, 80$. As an exercise, calculate the corresponding midpoint sums and compare the accuracy of the two approximation schemes. □

| $n$ | $T(f,0,1,n)$ | $|T(f,0,1,n) - \pi/4|$ |
|---|---|---|
| 10 | 0.784 981 | $4.2 \times 10^{-4}$ |
| 20 | 0.785 294 | $1.0 \times 10^{-4}$ |
| 40 | 0.785 372 | $2.6 \times 10^{-5}$ |
| 80 | 0.785 392 | $6.5 \times 10^{-6}$ |

Table 4

## Simpson's Rule

**Simpson's rule** is an approximation scheme for integrals that is based on the approximation of a function by a piecewise quadratic function. Let

$$a = x_0 < x_1 < x_2 < \cdots < x_{2k} < x_{2k+1} < x_{2k+2} < \cdots < x_{2n-2} < x_{2n-1} < x_{2n} = b,$$

where $x_{2k+1}$ is the midpoint of the interval $[x_{2k}, x_{2k+2}]$ for $k = 0, 1, 2, \ldots, n-1$. We set

$$\Delta x_k = x_{2k+1} - x_{2k} = x_{2k+2} - x_{2k+1}.$$

We will replace

$$\int_{x_{2k}}^{x_{2k+2}} f(x)\, dx$$

by

$$\int_{x_{2k}}^{x_{2k+2}} q_k(x)\, dx,$$

where $q$ is the quadratic function such that

$$q_k(x_{2k}) = f(x_{2k}),\ q_k(x_{2k+1}) = f(x_{2k=1})\ \text{and}\ q_k(x_{2k+2}) = f(x_{2k+2}),$$

as illustrated in Figure 3.

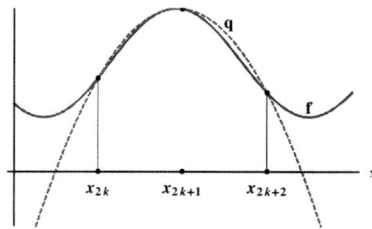

Figure 3

We have

$$\int_{x_{2k}}^{x_{2k+2}} q_k(x)\, dx = (f(x_{2k}) + 4f(x_{2k+1}) + f(x_{2k+2}))\frac{\Delta x_k}{3}$$

(you can find the derivation of the above expression at the end of this section). Therefore,

$$\int_a^b f(x)\, dx \cong \sum_{k=0}^{n-1} \int_{x_{2k}}^{x_{2k+2}} q_k(x)\, dx$$

$$= \int_0^{x_2} q_0(x)\, dx + \int_{x_2}^{x_4} q_1(x)\, dx + \cdots + \int_{x_{2n-2}}^{x_{2n}} q_{n-1}(x)\, dx$$

$$= (f(x_0) + 4f(x_1) + f(x_2))\frac{\Delta x_0}{3} + (f(x_2) + 4f(x_3) + f(x_4))\frac{\Delta x_1}{3} + \cdots$$

$$+ (f(x_{2n-2}) + 4f(x_{2n-1}) + f(x_{2n}))\frac{\Delta x_{n-1}}{3}$$

$$= \sum_{k=0}^{n-1} \left(\frac{f(x_{2k}) + 4f(x_{2k+1}) + f(x_{2k+2})}{3}\right) \Delta x_k.$$

Thus, Simpson's rule approximates the integral of $f$ on $[a, b]$ by a sum of the form

$$\sum_{k=0}^{n-1} \left(\frac{f(x_{2k}) + 4f(x_{2k+1}) + f(x_{2k+2})}{3}\right) \Delta x_k$$

## 6.5. NUMERICAL INTEGRATION

If we have subintervals of equal length

$$\Delta x = \frac{b-a}{2n},$$

and $x_k = k\Delta x$, $k = 0, 1, 2, \ldots, 2n$, we set

$$\begin{aligned} S(f,a,b,n) &= \sum_{k=0}^{n-1} \left( \frac{f(x_{2k}) + 4f(x_{2k+1}) + f(x_{2k+2})}{3} \right) \Delta x \\ &= \frac{b-a}{2n} \sum_{k=0}^{n-1} \left( \frac{f(x_{2k}) + 4f(x_{2k+1}) + f(x_{2k+2})}{3} \right). \end{aligned}$$

**Proposition 3** Assume that the fourth derivative of $f$ is continuous on $[a,b]$. Set

$$C_4 = \max \left\{ \left| \frac{d^4}{dx^4} f(x) \right| : x \in [a,b] \right\}.$$

Then,

$$\left| S(f,a,b,n) - \int_a^b f(x)dx \right| \leq \frac{C_4(b-a)}{180} (\Delta x)^4,$$

where

$$\Delta x = \frac{b-a}{2n}.$$

We leave the proof of Proposition 3 to a course in numerical analysis. Since $(\Delta x)^4$ is much smaller than $(\Delta x)^2$ if $\Delta x$ is small, Simpson's rule is even more accurate than the midpoint rule and the trapezoid rule, provided that the integrand is sufficiently smooth.

**Example 3** Approximate

$$\int_0^1 \frac{1}{\sqrt{4x^2+1}} dx$$

by $S(f,0,1,n)$, where $n = 1, 2$ and $4$. Calculate the absolute errors in the approximations.

**Solution**

We set $u = 2x$, so that $du = 2dx$ and

$$\int_0^1 \frac{1}{\sqrt{4x^2+1}} dx = \int_0^1 \frac{1}{\sqrt{(2x)^2+1}} dx = \frac{1}{2} \int_0^2 \frac{1}{\sqrt{u^2+1}} du$$

$$= \frac{1}{2} \left( \operatorname{arcsinh}(u) \big|_0^2 \right)$$

$$= \frac{1}{2} \operatorname{arcsinh}(2) = \frac{1}{2} \ln\left(2 + \sqrt{5}\right) \cong 0.721\,818.$$

Let's set

$$f(x) = \frac{1}{\sqrt{4x^2+1}}.$$

If

$$\Delta x = \frac{1}{2n} \text{ and } x_k = k\Delta x = \frac{k}{2n}, \ k = 0, 1, 2, \ldots, 2n,$$

we have

$$S(f,0,1,n) = \sum_{k=0}^{n-1} \left( \frac{f(x_{2k}) + 4f(x_{2k+1}) + f(x_{2k+2})}{3} \right) \Delta x$$

$$= \frac{\Delta x}{3} \sum_{k=0}^{n-1} (f(x_{2k}) + 4f(x_{2k+1}) + f(x_{2k+2}))$$

$$= \frac{1}{6n} \sum_{k=0}^{n-1} (f(x_{2k}) + 4f(x_{2k+1}) + f(x_{2k+2})).$$

Table 5 displays the required data. We see that the absolute error is very small even with a few subintervals. As an exercise, calculate the corresponding midpoint sums and the trapezoid rule approximations, and compare the accuracy of the three approximation schemes. $\Box$

| $n$ | $S(f,0,1,n)$ | $\left\lvert S(f,0,1,n) - \int_0^1 f(x)\,dx \right\rvert$ |
|---|---|---|
| 1 | 0.712 607 | $9.2 \times 10^{-3}$ |
| 2 | 0.721 495 | $3.2 \times 10^{-4}$ |
| 4 | 0.721 816 | $1.6 \times 10^{-6}$ |

Table 5

In this section you have acquired some idea about the midpoint rule, the trapezoid rule and Simpson's rule. These basic rules form the basis of many professional numerical integrators. A professional code, such as the one on your computational utility, is much more sophisticated than our examples that are based on the subdivision of an interval to subintervals of equal length. As we remarked earlier, the intermediate points are placed closer to each other in parts of the interval where the function varies more rapidly.

## The Derivation of Simpson's Rule

Let

$$a = x_0 < x_1 < x_2 < \cdots < x_{2k} < x_{2k+1} < x_{2k+2} < \cdots < x_{2n-2} < x_{2n-1} < x_{2n} = b,$$

where $x_{2k+1}$ is the midpoint of the interval $[x_{2k}, x_{2k+2}]$ for $k = 0, 1, 2, \ldots, n-1$. We set

$$\Delta x_k = x_{2k+1} - x_{2k} = x_{2k} - x_{2k+1}.$$

The quadratic function $q(x)$ such that

$$q_k(x_{2k}) = f(x_{2k}), \ q_k(x_{2k+1}) = f(x_{2k+1}) \text{ and } q_k(x_{2k+2}) = f(x_{2k+2}), \ k = 0, 1, \ldots, n-1,$$

can be expressed as

$$q_k(x) = f(x_{2k}) + \frac{f(x_{2k+1}) - f(x_{2k})}{\Delta x_k}(x - x_{2k})$$
$$+ \frac{f(x_{2k+2}) - 2f(x_{2k+1}) + f(x_{2k})}{(\Delta x_k)^2} \left( \frac{(x - x_{2k})(x - x_{2k+1})}{2} \right)$$

Indeed,

$$q(x_{2k}) = f(x_{2k}),$$
$$q(x_{2k+1}) = f(x_{2k}) + (f(x_{2k+1}) - f(x_{2k})) = f(x_{k+1}),$$
$$q(x_{2k+2}) = f(x_{2k}) + 2(f(x_{2k+1}) - f(x_{2k})) + (f(x_{2k+2}) - 2f(x_{2k+1}) + f(x_{2k}))$$
$$= f(x_{2k+2}).$$

## 6.5. NUMERICAL INTEGRATION

Therefore,

$$\int_{x_{2k}}^{x_{2k+2}} q_k(x)\,dx = f(x_{2k}) \int_{x_{2k}}^{x_{2k+2}} 1\,dx + \frac{f(x_{2k+1}) - f(x_{2k})}{\Delta x_k} \int_{x_{2k}}^{x_{2k+2}} (x - x_{2k})\,dx$$

$$+ \frac{f(x_{2k+2}) - 2f(x_{2k+1}) + f(x_{2k})}{(\Delta x_k)^2} \int_{x_{2k}}^{x_{2k+2}} \frac{(x - x_{2k})(x - x_{2k+1})}{2}\,dx$$

$$= 2\Delta x_k f(x_{2k}) + 2\Delta x_k \left( f(x_{2k+1}) - f(x_{2k}) \right)$$

$$+ \left( \frac{f(x_{2k+2}) - 2f(x_{2k+1}) + f(x_{2k})}{(\Delta x_k)^2} \right) \left( \frac{1}{3}(\Delta x_k)^3 \right)$$

$$= \left( \frac{f(x_{2k+2}) + 4f(x_{2k+1}) + f(x_{2k})}{3} \right) \Delta x_k.$$

Thus,

$$\int_a^b f(x)\,dx \cong \sum_{k=0}^{n-1} \int_{x_{2k}}^{x_{2k+2}} q_k(x)\,dx = \sum_{k=0}^{n-1} \left( \frac{f(x_{2k+2}) + 4f(x_{2k+1}) + f(x_{2k})}{3} \right) \Delta x_k.$$

∎

## Problems

**1** [C] Compute the left-endpoint sums $l_n = l(f, a, b, n)$ and the absolute errors

$$\left| l(f, a, b, n) - \int_a^b f(x)\,dx \right|$$

where

$$f(x) = \cos^2(x), \quad a = 0, \quad b = \pi/2, \quad n = 100, 200, 300, 400.$$

We have

$$\int_0^{\pi/2} \cos^2(x)\,dx = \frac{\pi}{4} \cong 0.785\,398$$

Notice the necessity to subdivide the interval by many small subintervals in order to achieve modest accuracy.

**2** [C] Compute the right-endpoint sums $r(f, a, b, n)$ and the absolute errors

$$\left| r(f, a, b, n) - \int_a^b f(x)\,dx \right|.$$

where

$$f(x) = \frac{1}{9 + 4x^2}, \quad a = 0, \quad b = 2, \quad n = 100, 200, 300, 400.$$

We have

$$\int_0^2 \frac{1}{9 + 4x^2}\,dx = \frac{1}{6} \arctan\left( \frac{4}{3} \right) \cong 0.154\,549$$

Note the necessity to subdivide the interval by many small subintervals in order to achieve modest accuracy.

In problems 3 and 4 Compute the midpoint sums $m(f, a, b, n)$ and the absolute errors

$$\left| m(f, a, b, n) - \int_a^b f(x)\,dx \right|.$$

Note the achievement of good accuracy via a modest number of subintervals.

**3.**
$$f(x) = e^{-x}, \ a = -1, \ b = 2, \ n = 4, 8, 16, 32.$$
We have
$$\int_1^2 e^{-x} dx = e^{-1} - e^{-2} \cong 0.232\,544$$

**4.**
$$f(x) = \frac{\ln(x)}{x}, \ a = 1, \ b = e^2, \ n = 4, 8, 16, 32$$
We have
$$\int_1^{e^2} \frac{\ln(x)}{x} dx = 2.$$

In problems 5 and 6 compute the trapezoid rule sums $T(f, a, b, n)$ and the absolute errors
$$\left| T(f, a, b, n) - \int_a^b f(x)\,dx \right|.$$

Note the achievement of good accuracy via a modest number of subintervals.

**5.**
$$f(x) = \frac{1}{\sqrt{4 - x^2}}, \ a = 0, \ b = 1, \ n = 4, 8, 16, 32$$
We have
$$\int_0^1 \frac{1}{\sqrt{4 - x^2}} dx = \frac{\pi}{6} \cong 0.523\,599.$$

**6.**
$$f(x) = \frac{3x - 14}{x^2 - x - 6}, \ a = -1, \ b = 2, \ n = 4, 8, 16, 32.$$
We have
$$\int_{-1}^2 f(x)\,dx = 10\ln(2) \cong 6.931\,47$$

In problems 7 and 8, compute the Simpson's rule sums $S(f, a, b, n)$ and the absolute errors
$$\left| S(f, a, b, n) - \int_a^b f(x)\,dx \right|.$$

Note the achievement of excellent accuracy via a small number of subintervals.

**7.**
$$f(x) = \sqrt{1 + x^2}, \ a = 2, \ b = 4, \ n = 2 \text{ and } 4.$$
We have
$$\int_2^4 \sqrt{1 + x^2}\,dx = \frac{1}{2}\ln\left(\sqrt{17} + 4\right) - \frac{1}{2}\ln\left(\sqrt{5} + 2\right) - \sqrt{5} + 2\sqrt{17} \cong 6.335\,68$$

**8.**
$$f(x) = xe^{-x/2}, \ a = 1, \ b = 4, \ n = 2 \text{ and } 4.$$
We have
$$\int_1^4 f(x)\,dx = 6e^{-1/2} - 12e^{-2} \cong 2.01457$$

## 6.6 Improper Integrals: Part 1

In this section we will expand the scope of the integral to cover some cases that involve unbounded intervals or functions with discontinuities.

### Improper integrals on unbounded intervals

Let's begin with a specific case:

**Example 1** Let $f(x) = 1/x^2$. We can compute the integral of $f$ on any interval of the form $[1, b]$:

$$\int_1^b f(x)\, dx = \int_1^b \frac{1}{x^2}\, dx = -\frac{1}{x}\Big|_1^b = -\frac{1}{b} + 1.$$

Therefore,

$$\lim_{b \to +\infty} \int_1^b f(x)\, dx = \lim_{b \to +\infty} \left(-\frac{1}{b} + 1\right) = 1.$$

Since we can interpret the integral of $f$ from 1 to $b$ as the area between the graph of $f$ and the interval $[1, b]$, it is reasonable to say that the area of the region $G$ between the graph of $f$ and the interval $[1, +\infty)$ is 1. The region $G$ is illustrated in Figure 1. □

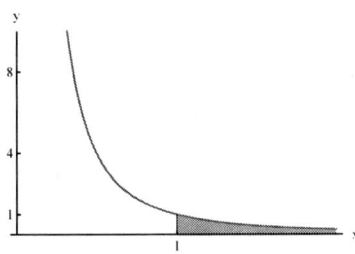

Figure 1: $y = 1/x^2$

Example 1 illustrates an "improper integral" on an unbounded interval:

**Definition 1** Assume that $f$ is continuous on any interval of the form $[a, b]$ where $b > a$. We say that **the improper integral $\int_a^\infty f(x)\, dx$ converges** if

$$\lim_{b \to +\infty} \int_a^b f(x)\, dx$$

exists (as a finite limit). In this case we define **the value of the improper integral** as the above limit, and denote it by the same symbol. Thus,

$$\int_a^\infty f(x)\, dx = \lim_{b \to +\infty} \int_a^b f(x)\, dx.$$

We say that the improper integral $\int_a^\infty f(x)\, dx$ **diverges** if

$$\lim_{b \to +\infty} \int_a^b f(x)\, dx$$

does not exist.

In the case of convergence we may interpret the value of an improper integral as the signed area of the region between the graph of $f$ and the interval $[a, +\infty)$, just as in the case of a bounded interval.

With reference to Example 1, if $f(x) = 1/x^2$, the improper integral of $f$ on the interval $[1, +\infty)$ converges and has the value 1:

$$\int_1^\infty \frac{1}{x^2} dx = \lim_{b \to +\infty} \int_1^b \frac{1}{x^2} dx = 1.$$

**Example 2** Determine whether the improper integral

$$\int_1^\infty \frac{1}{x} dx$$

converges or diverges, and its value if it converges.

**Solution**

For any $b > 1$,

$$\int_1^b \frac{1}{x} dx = \ln(b)$$

Therefore,

$$\lim_{b \to +\infty} \int_1^b \frac{1}{x} dx = \lim_{b \to +\infty} \ln(b) = +\infty.$$

This is merely shorthand for the statement that $\ln(b)$ is arbitrarily large if $b$ is large enough. Thus, the improper integral

$$\int_1^\infty \frac{1}{x} dx$$

diverges. □

Examples 1 and 2 involve improper integrals of the type $\int_a^\infty 1/x^p dx$. Let's record the general case for future reference:

**Proposition 1** *Assume that $p$ is an arbitrary real number and that $a > 0$. The improper integral $\int_a^\infty 1/x^p dx$ converges if $p > 1$ and diverges if $p \leq 1$.*

**Proof**

Let $p = 1$. We have

$$\int_a^b \frac{1}{x} dx = \ln(b) - \ln(a),$$

so that

$$\lim_{b \to +\infty} \int_a^b \frac{1}{x} dx = \lim_{b \to +\infty} (\ln(b) - \ln(a)) = +\infty.$$

Therefore the improper integral $\int_a^\infty 1/x \, dx$ diverges.

Now assume that $p \neq 1$. We have

$$\int_a^b \frac{1}{x^p} dx = \int_a^b x^{-p} dx = \left. \frac{x^{-p+1}}{-p+1} \right|_a^b = \frac{b^{-p+1}}{-p+1} - \frac{a^{-p+1}}{-p+1}.$$

If $p > 1$,

$$\lim_{b \to +\infty} \int_a^b \frac{1}{x^p} dx = \lim_{b \to +\infty} \left( \frac{b^{-p+1}}{-p+1} - \frac{a^{-p+1}}{-p+1} \right) = \lim_{b \to +\infty} \left( -\frac{1}{(p-1)b^{p-1}} + \frac{1}{(p-1)a^{p-1}} \right)$$

$$= \frac{1}{(p-1)a^{p-1}},$$

## 6.6. IMPROPER INTEGRALS: PART 1

since $\lim_{b \to +\infty} 1/b^{p-1} = 0$. Therefore, the improper integral $\int_a^\infty 1/x^p \, dx$ converges and

$$\int_a^\infty \frac{1}{x^p} dx = \frac{1}{(p-1) a^{p-1}}.$$

Now assume that $p < 1$. We have

$$\int_a^b \frac{1}{x^p} dx = \frac{b^{-p+1}}{-p+1} - \frac{a^{-p+1}}{-p+1} = \frac{b^{1-p}}{1-p} - \frac{a^{1-p}}{1-p}.$$

Since $1 - p > 0$, we have $\lim_{b \to +\infty} b^{1-p} = +\infty$. Therefore,

$$\lim_{b \to +\infty} \int_a^b \frac{1}{x^p} dx = +\infty,$$

so that the improper integral $\int_a^\infty 1/x^p \, dx$ diverges. ∎

**Example 3** Determine whether the improper integral

$$\int_0^\infty \sin(x) \, dx$$

converges or diverges. Compute its value if it converges.

**Solution**

Since

$$\int \sin(x) \, dx = -\cos(x),$$

we have

$$\int_0^b \sin(x) \, dx = -\cos(x)\Big|_0^b = -\cos(b) + 1$$

for any $b$. The limit of $-\cos(b) + 1$ as $x$ tends to $+\infty$ does not exists. Indeed,

$$-\cos\left((2n+1)\frac{\pi}{2}\right) + 1 = 1$$

and

$$-\cos(2n\pi) + 1 = 0$$

for $n = 1, 2, 3, \ldots$. Therefore, the improper integral

$$\int_0^\infty \sin(x) \, dx$$

diverges. Note that

$$0 \leq \int_0^b \sin(x) \, dx \leq 2$$

for each $b > 0$, since $0 \leq -\cos(b) + 1 \leq 2$. This example shows that an improper integral

$$\int_a^\infty f(x) \, dx$$

may diverge, even though

$$\int_a^b f(x) \, dx$$

does not diverge to infinity as $b$ tends to infinity. □

**Example 4** Determine whether the improper integral

$$\int_0^\infty e^{-x/2} \cos(x)\, dx$$

converges or diverges. Compute its value if it converges.

**Solution**

In Example 8 of Section 6.1 we derived the expression

$$\int e^{-x/2} \cos(x)\, dx = \frac{4}{5} e^{-x/2} \sin(x) - \frac{2}{5} e^{-x/2} \cos(x)$$

via integration by parts. Therefore, for any $b > 0$,

$$\int_0^b e^{-x/2} \cos(x)\, dx = \left. \frac{4}{5} e^{-x/2} \sin(x) - \frac{2}{5} e^{-x/2} \cos(x) \right|_0^b$$
$$= \frac{4}{5} e^{-b/2} \sin(b) - \frac{2}{5} e^{-b/2} \cos(b) + \frac{2}{5}.$$

Note that $\lim_{b \to +\infty} e^{-b/2} \sin(b) = 0$ and $\lim_{b \to +\infty} e^{-b/2} \cos(b) = 0$. Indeed,

$$\left| e^{-b/2} \sin(b) \right| = e^{-b/2} |\sin(b)| \le e^{-b/2} \text{ and } \left| e^{-b} \cos(b) \right| \le e^{-b/2},$$

since $|\sin(b)| \le 1$ and $|\cos(b)| \le 1$ for any real number $b$, and $\lim_{b \to +\infty} e^{-b/2} = 0$. Therefore,

$$\lim_{b \to +\infty} \int_0^b e^{-x/2} \cos(x)\, dx = \lim_{b \to +\infty} \left( \frac{4}{5} e^{-b/2} \sin(b) - \frac{2}{5} e^{-b/2} \cos(b) + \frac{2}{5} \right) = \frac{2}{5}.$$

Thus, the given improper integral converges and its value is $2/5$:

$$\int_0^\infty e^{-x/2} \cos(x)\, dx = \lim_{b \to +\infty} \int_0^b e^{-x/2} \cos(x)\, dx = \frac{2}{5}.$$

Figure 2 shows the graph of $f$, where $f(x) = e^{-x/2} \cos(x)$. The function $f$ has varying sign. We can interpret the above result as the signed area of the region between the graph of $f$ and the interval $[0, +\infty)$. □

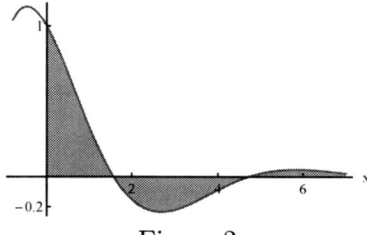

Figure 2

We can also consider improper integrals of the form $\int_{-\infty}^a f(x)\, dx$:

**Definition 2** Assume that $f$ is continuous on the interval $(-\infty, a]$. We say that the **improper integral** $\int_{-\infty}^a f(x)\, dx$ converges if

$$\lim_{b \to -\infty} \int_b^a f(x)\, dx$$

## 6.6. IMPROPER INTEGRALS: PART 1

exists (as a finite limit). In this case we define **the value of the improper integral** as the limit and denote it by the same symbol. Thus,

$$\int_{-\infty}^{a} f(x)\,dx = \lim_{b \to -\infty} \int_{b}^{a} f(x)\,dx.$$

If the above limit does not exist, we say that the improper integral $\int_{-\infty}^{a} f(x)\,dx$ **diverges**.

**Example 5** Determine whether the improper integral

$$\int_{-\infty}^{-1} \frac{1}{x^2+1}\,dx$$

converges or diverges, and its value if it converges.

**Solution**

We have

$$\int_{b}^{-1} \frac{1}{x^2+1}\,dx = \arctan(x)\Big|_{b}^{-1} = \arctan(-1) - \arctan(b) = -\frac{\pi}{4} - \arctan(b).$$

Therefore,

$$\lim_{b \to -\infty} \int_{b}^{-1} \frac{1}{x^2+1}\,dx = \lim_{b \to -\infty}\left(-\frac{\pi}{4} - \arctan(b)\right) = -\frac{\pi}{4} - \lim_{b \to -\infty} \arctan(b)$$
$$= -\frac{\pi}{4} - \left(-\frac{\pi}{2}\right) = \frac{\pi}{4}.$$

Thus, the given improper integral converges, and we have

$$\int_{-\infty}^{-1} \frac{1}{x^2+1}\,dx = \lim_{b \to -\infty} \int_{b}^{-1} \frac{1}{x^2+1}\,dx = \frac{\pi}{4}.$$

Figure 3 shows the graph of $f$, where $f(x) = 1/(x^2+1)$. The area between the graph $f$ and the interval $(-\infty, -1]$ is $\pi/4$. □

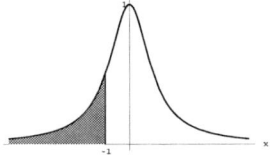

Figure 3

We may also consider **"two-sided"** improper integrals:

**Definition 3** Assume that $f$ is continuous on $\mathbb{R} = (-\infty, +\infty)$. We say that **the improper integral** $\int_{-\infty}^{+\infty} f(x)\,dx$ **converges** if there exists an $a \in \mathbb{R}$ such that the improper integrals $\int_{-\infty}^{a} f(x)\,dx$ and $\int_{a}^{\infty} f(x)\,dx$ converge. If this is the case, we define the value of the improper integral to be the sum of the values of the "one-sided" improper integrals and set

$$\int_{-\infty}^{+\infty} f(x)\,dx = \int_{-\infty}^{a} f(x)\,dx + \int_{a}^{\infty} f(x)\,dx.$$

The improper integral $\int_{-\infty}^{+\infty} f(x)\,dx$ is said to **diverge** if $\int_{-\infty}^{a} f(x)\,dx$ or $\int_{a}^{\infty} f(x)\,dx$ diverges.

**Remark** The convergence or divergence of the improper integral $\int_{-\infty}^{+\infty} f(x)\,dx$, and its value in the case of convergence, do not depend on "the intermediate point" $a$. This assertion should be plausible due to the geometric interpretation of the integral. You can prove this fact by using the additivity of the integral with respect to intervals (exercise). ◊

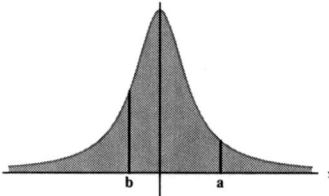

Figure 4: The area under the graph of $f$ does not depend on the choice of $a$ or $b$ as the intermediate point

**Example 6** Determine whether the improper integral

$$\int \frac{1}{4x^2 - 8x + 13}\,dx$$

converges or diverges and its value if it converges.

**Solution**

In Example 3 of Section 6.2 we showed that

$$\int \frac{1}{4x^2 - 8x + 13}\,dx = \frac{1}{6}\arctan\left(\frac{2(x-1)}{3}\right) + C.$$

The form of the antiderivative suggests that 1 is a convenient intermediate point. We have

$$\int_1^b \frac{1}{4x^2 - 8x + 13}\,dx = \frac{1}{6}\arctan\left(\frac{2(x-1)}{3}\right)\Big|_1^b$$

$$= \frac{1}{6}\arctan\left(\frac{2(b-1)}{3}\right) - \frac{1}{6}\arctan(0) = \frac{1}{6}\arctan\left(\frac{2(b-1)}{3}\right).$$

Therefore,

$$\int_1^\infty \frac{1}{4x^2 - 8x + 13}\,dx = \lim_{b \to +\infty} \int_1^b \frac{1}{4x^2 - 8x + 13}\,dx$$

$$= \lim_{b \to +\infty} \frac{1}{6}\arctan\left(\frac{2(b-1)}{3}\right) = \left(\frac{1}{6}\right)\frac{\pi}{2} = \frac{\pi}{12}.$$

Similarly,

$$\int_b^1 \frac{1}{4x^2 - 8x + 13}\,dx = \frac{1}{6}\arctan\left(\frac{2(x-1)}{3}\right)\Big|_b^1 = -\frac{1}{6}\arctan\left(\frac{2(b-1)}{3}\right),$$

so that

$$\int_{-\infty}^1 \frac{1}{4x^2 - 8x + 13}\,dx = \lim_{b \to -\infty} \int_b^1 \frac{1}{4x^2 - 8x + 13}\,dx$$

$$= \lim_{b \to -\infty}\left(-\frac{1}{6}\arctan\left(\frac{2(b-1)}{3}\right)\right) = -\frac{1}{6}\left(-\frac{\pi}{2}\right) = \frac{\pi}{12}.$$

## 6.6. IMPROPER INTEGRALS: PART 1

Therefore, the given improper integral converges, and

$$\int_{-\infty}^{\infty} \frac{1}{4x^2 - 8x + 13} dx = \int_{-\infty}^{1} \frac{1}{4x^2 - 8x + 13} dx + \int_{1}^{\infty} \frac{1}{4x^2 - 8x + 13} dx$$
$$= \frac{\pi}{12} + \frac{\pi}{12} = \frac{\pi}{6}.$$

Figure 5 shows the graph of

$$f(x) = \frac{1}{4x^2 - 8x + 13}.$$

The area of the region between the graph of $f$ and the number line is $\pi/6$. □

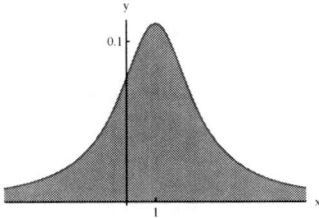

Figure 5

**Remark (Caution)** We have *not* defined the improper integral $\int_{-\infty}^{+\infty} f(x)\,dx$ as

$$\lim_{b \to +\infty} \int_{-b}^{b} f(x)\,dx.$$

Let's consider the improper integral $\int_{-\infty}^{+\infty} x\,dx$. Since

$$\lim_{b \to +\infty} \int_{0}^{b} x\,dx = \lim_{b \to +\infty} \frac{b^2}{2} = +\infty,$$

this improper integral diverges. On the other hand, we have

$$\lim_{b \to +\infty} \left( \int_{-b}^{b} x\,dx \right) = \lim_{b \to +\infty} (0) = 0.$$

Note that $f(x) = x$ defines an odd function, so that the graph of $f$ is symmetric with respect to the origin. Therefore, it is not surprising that $\int_{-b}^{b} f(x)\,dx = 0$.

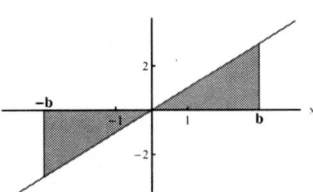

Figure 6: $\int_{-b}^{b} f(x)\,dx = 0$

Our definition of the convergence of the improper integral $\int_{-\infty}^{+\infty} f(x)\,dx$ amounts to the requirement that

$$\int_{b}^{c} f(x)\,dx$$

approaches a definite number as $b \to -\infty$ and $c \to +\infty$, with no prior restriction such as $b = -c$.
◇

**Example 7** Determine whether the improper integral
$$\int_{-\infty}^{\infty} \frac{x}{x^2+1} dx$$
converges or diverges, and its value if it converges.

**Solution**

We will select 0 as the intermediate point. Both improper integrals
$$\int_{-\infty}^{0} \frac{x}{x^2+1} dx \text{ and } \int_{0}^{\infty} \frac{x}{x^2+1} dx$$
must converge for the convergence of the given "two-sided" improper integral.
We set $u = x^2 + 1$ so that $du/dx = 2$. By the definite integral version of the substitution rule,
$$\int_0^b \frac{x}{x^2+1} dx = \frac{1}{2} \int_0^b \frac{x}{x^2+1} \frac{du}{dx} dx = \frac{1}{2} \int_1^{b^2+1} \frac{1}{u} du = \frac{1}{2} \ln\left(b^2 + 1\right).$$
Therefore,
$$\lim_{b \to +\infty} \int_0^b \frac{x}{x^2+1} dx = \lim_{b \to +\infty} \left(\frac{1}{2} \ln\left(b^2+1\right)\right) = +\infty,$$
Thus, the improper integral
$$\int_0^{\infty} \frac{x}{x^2+1} dx$$
diverges. This is sufficient to conclude that the improper integral
$$\int_{-\infty}^{\infty} \frac{x}{x^2+1} dx$$
diverges.

Note that $f(x) = x/\left(x^2 + 1\right)$ defines an odd function, so that the graph of $f$ is symmetric with respect to the origin, as shown in Figure 7. We have
$$\int_{-b}^{b} f(x) dx = \int_{-b}^{b} \frac{x}{x^2+1} dx = 0$$
for each $b \in R$, so that
$$\lim_{b \to +\infty} \int_{-b}^{b} \frac{x}{x^2+1} dx = 0.$$
Nevertheless, the improper integral $\int_{-\infty}^{+\infty} x/\left(x^2+1\right) dx$ does *not* converge. □

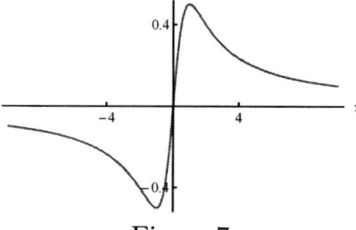

Figure 7

## 6.6. IMPROPER INTEGRALS: PART 1

### Improper Integrals that involve Discontinuous Functions

As we discussed in Section 6.2, $f$ is **piecewise continuous** on the interval $[a, b]$ if $f$ has at most finitely many **removable** or **jump discontinuities** in $[a, b]$. Thus, $f$ has (finite) one-sided limits at its discontinuities. In such a case we defined the integral of $f$ on $[a, b]$ as the sum of its integrals over the subintervals of $[a, b]$ that are separated from each other by the points of discontinuity of $f$. Now we will extend the definition of the integral to certain cases involving discontinuities other than jump discontinuities. Let's begin with a specific case:

**Example 8** Let
$$f(x) = \frac{1}{\sqrt{x}}, \ x > 0.$$

We have $\lim_{x \to 0+} f(x) = +\infty$. Since $f$ is continuous on any interval of the form $[\varepsilon, 1]$ where $0 < \varepsilon < 1$, the integral of $f$ on such an interval exists, and corresponds to the area between the graph of $f$ and the interval $[\varepsilon, 1]$, as indicated in Figure 8.

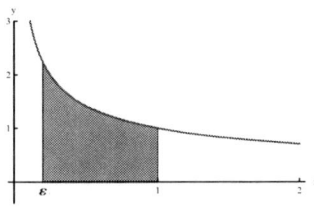

Figure 8

We have
$$\int_\varepsilon^1 f(x)\,dx = \int_\varepsilon^1 x^{-1/2}\,dx = \left.\frac{x^{1/2}}{1/2}\right|_\varepsilon^1 = 2 - 2\sqrt{\varepsilon}.$$

It is natural to consider the limit of the integral as $\varepsilon$ approaches 0 through positive values in an attempt to attach a meaning to the integral of $f$ on the interval $[0, 1]$. We have
$$\lim_{\varepsilon \to 0+} \int_\varepsilon^1 f(x)\,dx = \lim_{\varepsilon \to 0+} \left(2 - 2\sqrt{\varepsilon}\right) = 2.$$

We can interpret the result as the area between the graph of $f$ and the interval $[0, 1]$. $\square$

**Definition 4** Assume that $f$ is continuous on the interval $(a, b]$ and that $\lim_{x \to a+} f(x)$ does not exist. The **improper integral** $\int_a^b f(x)\,dx$ is said to converge if
$$\lim_{\varepsilon \to 0+} \int_{a+\varepsilon}^b f(x)\,dx$$
exists. In this case we define **the value of the improper integral** as the above limit and denote it by the same symbol:
$$\int_a^b f(x)\,dx = \lim_{\varepsilon \to 0+} \int_{a+\varepsilon}^b f(x)\,dx.$$

If
$$\lim_{\varepsilon \to 0+} \int_{a+\varepsilon}^b f(x)\,dx$$
does not exist, we say that the improper integral $\int_a^b f(x)\,dx$ **diverges**.

**Remark** (**An alternative expression**) With reference to Definition 4, we can set $c = a + \varepsilon$, so that $c$ approaches $a$ from the right as $\varepsilon$ approaches 0 through positive values. Therefore, we have

$$\int_a^b f(x)\,dx = \lim_{\varepsilon \to 0+} \int_{a+\varepsilon}^b f(x)\,dx = \lim_{c \to a+} \int_c^b f(x)\,dx.$$

The improper integral diverges if

$$\lim_{c \to a+} \int_c^b f(x)\,dx$$

does not exist.

**Example 9** Consider

$$\int_0^1 \ln(x)\,dx$$

a) Why is the integral an improper integral?
b) Determine whether the improper integral converges or diverges, and its value if it converges.

**Solution**

a) The given integral is improper since

$$\lim_{x \to 0+} \ln(x) = -\infty.$$

b) Let $0 < \varepsilon < 1$. We have

$$\int_\varepsilon^1 \ln(x)\,dx = x\ln(x) - x\big|_\varepsilon^1 = -1 - (\varepsilon \ln(\varepsilon) - \varepsilon) = -1 - \varepsilon \ln(\varepsilon) + \varepsilon.$$

Recall that $\lim_{\varepsilon \to 0+} \varepsilon \ln(\varepsilon) = 0$ (you can make use of L'Hôpital's rule). Therefore,

$$\lim_{\varepsilon \to 0+} \int_\varepsilon^1 \ln(x)\,dx = \lim_{\varepsilon \to 0+} (-1 - \varepsilon \ln(\varepsilon) + \varepsilon) = -1.$$

Thus, the given improper integral converges, and we have

$$\int_0^1 \ln(x)\,dx = \lim_{\varepsilon \to 0+} \int_\varepsilon^1 \ln(x)\,dx = -1.$$

We can interpret the value of the improper integral as the signed area of the region $G$ between the graph of the natural logarithm and the interval $[0, 1]$. $G$ is illustrated in Figure 9. □

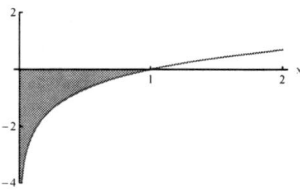

Figure 9

In Example 8 we showed that $\int_0^1 1/x^{1/2}\,dx$ converges. Let's state the following generalization for future reference:

## 6.6. IMPROPER INTEGRALS: PART 1

**Proposition 2** Assume that $p > 0$ and $a > 0$. The improper integral
$$\int_0^a \frac{1}{x^p} dx$$
converges if $p < 1$ and diverges if $p \geq 1$.

**Proof**

Let's begin with the case $p = 1$. Let $0 < \varepsilon < a$. We have
$$\lim_{\varepsilon \to 0+} \int_\varepsilon^a \frac{1}{x} dx = \lim_{\varepsilon \to 0+} (\ln(a) - \ln(\varepsilon)) = +\infty,$$
since $\lim_{\varepsilon \to 0+} \ln(\varepsilon) = -\infty$. Therefore, the improper integral $\int_0^a 1/x \, dx$ diverges.
Assume that $p \neq 1$ and $0 < \varepsilon < a$. We have
$$\int_\varepsilon^a \frac{1}{x^p} dx = \int_\varepsilon^a x^{-p} dx = \left. \frac{x^{-p+1}}{-p+1} \right|_\varepsilon^a = \frac{a^{-p+1}}{-p+1} - \frac{\varepsilon^{-p+1}}{-p+1}.$$
If $0 < p < 1$ then $1 - p > 0$, so that $\lim_{\varepsilon \to 0+} \varepsilon^{1-p} = 0$. Therefore,
$$\lim_{\varepsilon \to 0+} \int_\varepsilon^a \frac{1}{x^p} dx = \lim_{\varepsilon \to 0+} \left( \frac{a^{1-p}}{1-p} - \frac{\varepsilon^{1-p}}{1-p} \right) = \frac{a^{1-p}}{1-p}.$$
Thus, the improper integral $\int_0^a 1/x^p \, dx$ converges (and has the value $a^{1-p}/(1-p)$).
If $p > 1$ then $1 - p < 0$ so that $\lim_{\varepsilon \to 0+} \varepsilon^{1-p} = +\infty$. Therefore,
$$\lim_{\varepsilon \to 0+} \int_\varepsilon^a \frac{1}{x^p} dx = \lim_{\varepsilon \to 0+} \left( \frac{a^{1-p}}{1-p} - \frac{\varepsilon^{1-p}}{1-p} \right) = +\infty.$$
Thus, the improper integral $\int_0^a 1/x^p \, dx$ diverges. ∎

If $f$ is continuous on $[a, b)$ and $\lim_{x \to b-} f(x)$ does not exist, the integral $\int_a^b f(x) \, dx$ is improper:

**Definition 5** Assume that $f$ is continuous on the interval $[a, b)$ and that $\lim_{x \to b-} f(x)$ does not exist. The **improper integral** $\int_a^b f(x) \, dx$ is said to converge if
$$\lim_{\varepsilon \to 0+} \int_a^{b-\varepsilon} f(x) \, dx$$
exists. In this case **the value of the improper integral** is
$$\int_a^b f(x) \, dx = \lim_{\varepsilon \to 0+} \int_a^{b-\varepsilon} f(x) \, dx.$$
If
$$\lim_{\varepsilon \to 0+} \int_a^{b-\varepsilon} f(x) \, dx$$
does not exist, we say that the improper integral $\int_a^b f(x) \, dx$ diverges.

**Remark** (An alternative expression) With reference to Definition 5, we can set $c = b - \varepsilon$, so that $c$ approaches $b$ from the left as $\varepsilon$ approaches 0 through positive values. Therefore, we have
$$\int_a^b f(x) \, dx = \lim_{\varepsilon \to 0+} \int_a^{b-\varepsilon} f(x) \, dx = \lim_{c \to b-} \int_a^c f(x) \, dx.$$
The improper integral diverges if
$$\lim_{c \to b-} \int_a^c f(x) \, dx$$
does not exist.

**Example 10** Consider
$$\int_1^2 \frac{1}{(x-2)^{1/3}} dx$$

a) Why is the integral an improper integral?
b) Determine whether the improper integral converges or diverges, and its value if it converges.

**Solution**

a) The integral is improper since
$$\lim_{x \to 2^-} \frac{1}{(x-2)^{1/3}} = -\infty.$$

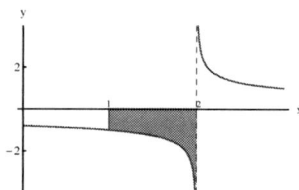

Figure 10: $y = (x-2)^{-1/3}$

b) If $0 < \varepsilon < 1$, we have
$$\int_1^{2-\varepsilon} \frac{1}{(x-2)^{1/3}} dx = \int_{-1}^{-\varepsilon} \frac{1}{u^{1/3}} du = \int_{-1}^{-\varepsilon} u^{-1/3} du = \left. \frac{3u^{2/3}}{2} \right|_{-1}^{-\varepsilon} = \frac{3\varepsilon^{2/3}}{2} - \frac{3}{2}.$$

Therefore
$$\lim_{\varepsilon \to 0+} \int_1^{2-\varepsilon} \frac{1}{(x-2)^{1/3}} dx = \lim_{\varepsilon \to 0+} \left( \frac{3\varepsilon^{2/3}}{2} - \frac{3}{2} \right) = -\frac{3}{2}.$$

Thus, the given improper integral converges, and we have
$$\int_1^2 \frac{1}{(x-2)^{1/3}} dx = -\frac{3}{2}.$$

We can interpret the result as the signed area of the region between the graph of $y = (x-2)^{-1/3}$ and the interval $[1, 2]$, as indicated in Figure 10. □

A function may have a discontinuity at an interior point of the given interval:

**Definition 6** Assume that $f$ is continuous at each point of the interval $[a, b]$, with the exception of the point $c$ between $a$ and $b$. Furthermore, assume that at least one of the limits, $\lim_{x \to c^-} f(x)$ or $\lim_{x \to c^+} f(x)$, fails to exists. In this case, the **improper integral** $\int_a^b f(x) \, dx$ converges if and only if
$$\int_a^c f(x) \, dx \text{ and } \int_c^b f(x) \, dx$$
exist as ordinary integrals or converge as improper integrals. If the improper integral converges, we define its value as
$$\int_a^b f(x) \, dx = \int_a^c f(x) \, dx + \int_c^b f(x) \, dx.$$

## 6.6. IMPROPER INTEGRALS: PART 1

The improper integral $\int_a^b f(x)\,dx$ **diverges** if $\int_a^c f(x)\,dx$ or $\int_c^b f(x)\,dx$ diverges.
The extension of Definition 6 to more general cases is obvious: We must take into account the discontinuities of the integrand and examine the improper integrals on the intervals that are separated from each other by the points of discontinuity of the function separately.

**Example 11** Consider the improper integral

$$\int_{-1}^{2} \frac{1}{x^{2/3}}\,dx$$

a) Why is the integral an improper integral?
b) Determine whether the improper integral converges or diverges, and its value if it converges.

**Solution**

a) The integral is improper since

$$\lim_{x \to 0} \frac{1}{x^{2/3}} = +\infty,$$

and $0 \in [-1, 2]$.
b) We must examine the improper integrals

$$\int_{-1}^{0} \frac{1}{x^{2/3}}\,dx \text{ and } \int_{0}^{2} \frac{1}{x^{2/3}}\,dx$$

separately. We have

$$\int_{-1}^{0} \frac{1}{x^{2/3}}\,dx = \lim_{\varepsilon \to 0+} \int_{-1}^{-\varepsilon} x^{-2/3}\,dx = \lim_{\varepsilon \to 0+} \left(3x^{1/3}\Big|_{-1}^{-\varepsilon}\right)$$
$$= \lim_{\varepsilon \to 0+} \left(-3\varepsilon^{1/3} + 3\right) = 3,$$

and

$$\int_{0}^{2} \frac{1}{x^{2/3}}\,dx = \lim_{\varepsilon \to 0+} \int_{\varepsilon}^{2} x^{-2/3} = \lim_{\varepsilon \to 0+} \left(3x^{1/3}\Big|_{\varepsilon}^{2}\right)$$
$$= \lim_{\varepsilon \to 0} \left(3\left(2^{1/3}\right) - 3\varepsilon^{1/3}\right) = 3\left(2^{1/3}\right).$$

Therefore, the given improper integral converges, and we have

$$\int_{-1}^{2} \frac{1}{x^{2/3}}\,dx = \int_{-1}^{0} \frac{1}{x^{2/3}}\,dx + \int_{0}^{2} \frac{1}{x^{2/3}}\,dx = 3 + 3\left(2^{1/3}\right).$$

We can interpret the result as the area of the region between the graph of $y = x^{-2/3}$ and the interval $[-1, 2]$, as illustrated in Figure 11. □

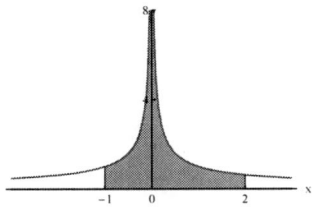

Fiure 11

**Remark** (**Caution**) The convergence of an improper integral on $[a, b]$ requires the convergence of the improper integrals on $[a, c]$ and $[c, b]$. We did *not* define $\int_a^b f(x)\, dx$ as

$$\lim_{\varepsilon \to 0+} \left( \int_a^{c-\varepsilon} f(x)\, dx + \int_{c+\varepsilon}^b f(x)\, dx \right).$$

Our definition of the convergence of the improper integral $\int_a^b f(x)\, dx$ requires that the limits

$$\lim_{\varepsilon \to 0+} \int_a^{c-\varepsilon} f(x)\, dx = \int_a^c f(x)\, dx$$

and

$$\lim_{\delta \to 0+} \int_{c+\delta}^b f(x)\, dx = \int_c^b f(x)\, dx$$

exist independently, without any relationship between $\varepsilon$ and $\delta$ (such as $\varepsilon = \delta$).

For example, the improper integral

$$\int_{-1}^1 \frac{1}{x}\, dx$$

does not converge since the improper integral

$$\int_0^1 \frac{1}{x}\, dx$$

diverges. On the other hand, $1/x$ defines an odd function, and we have

$$\int_{-1}^{-\varepsilon} \frac{1}{x}\, dx + \int_{\varepsilon}^1 \frac{1}{x}\, dx = 0$$

for each $\varepsilon > 0$, so that

$$\lim_{\varepsilon \to 0+} \left( \int_{-1}^{-\varepsilon} \frac{1}{x}\, dx + \int_{\varepsilon}^1 \frac{1}{x}\, dx \right) = 0.$$

◇

## Problems

In problems 1-14, determine whether the given improper integral converges or diverges, and the value of the improper integral in case of convergence.

1.
$$\int_0^\infty \frac{x}{(x^2+4)^2}\, dx$$

2.
$$\int_1^\infty \frac{1}{x+4}\, dx$$

3.
$$\int_6^\infty \frac{1}{(x-4)^2}\, dx$$

4.
$$\int_0^\infty \frac{1}{\sqrt{x^2+1}}\, dx$$

5.
$$\int_{\sqrt{3}/2}^\infty \frac{1}{4x^2+9}\, dx$$

6.
$$\int_4^\infty \frac{x}{\sqrt{x^2+1}}\, dx$$

7.
$$\int_0^\infty x^2 e^{-x}\, dx$$

8.
$$\int_{e^2}^{\infty} \frac{\ln(x)}{x^2} dx$$

9.
$$\int_{-\infty}^{0} xe^{-x} dx$$

10.
$$\int_{-\infty}^{0} e^x \sin(x) dx$$

Hint:
$$\int e^x \sin(x) dx = -\frac{1}{2} e^x \cos(x) + \frac{1}{2} e^x \sin(x)$$

11.
$$\int_{-\infty}^{0} \frac{x^2}{x^3 - 27} dx$$

12.
$$\int_{-\infty}^{-3} \frac{x^2 - 8x - 1}{(x^2 + 9)(x - 2)^2} dx$$

13.
$$\int_{-\infty}^{\infty} \frac{1}{x^2 - 8x + 17} dx$$

14.
$$\int_{-\infty}^{+\infty} \frac{x - 1}{x^2 - 2x + 5} dx$$

In problems 15-22,
a) Explain why the given integral is an improper integral,
b) Determine whether the given improper integral converges or diverges, and the value of the improper integral in case of convergence.

15.
$$\int_{0}^{2} \frac{1}{\sqrt{2 - x}} dx$$

16.
$$\int_{1}^{4} \frac{1}{(x - 1)^{2/3}} dx$$

17.
$$\int_{0}^{1} \frac{x}{1 - x^2} dx$$

18.
$$\int_{0}^{e} \frac{\ln(x)}{x^{1/3}} dx$$

19.
$$\int_{0}^{3} \frac{1}{(2 - x)^{1/3}} dx$$

20.
$$\int_{0}^{\pi} \tan(x) dx$$

21.
$$\int_{-1}^{2} \frac{1}{(x - 1)^{4/5}} dx$$

22.
$$\int_{1}^{3} \frac{1}{(x - 2)^{4/3}} dx$$

## 6.7 Improper Integrals: Part 2

In Section 6.6 we dealt with various types of improper integrals. In each case, we were able to decide whether a given improper is convergent or divergent by calculating the relevant integral exactly and then evaluating the relevant limit. It is possible to use **convergence tests** if it is impossible or impractical to perform such calculations.

### Improper integrals on unbounded intervals

Let's begin by considering the improper integrals of nonnegative functions.

**Proposition 1** Assume that $f$ is continuous on $[a, +\infty)$ and $f(x) \geq 0$ for each $x \geq a$. If there exists $M > 0$ such that
$$\int_{a}^{b} f(x) dx \leq M$$

for each $b \geq a$, the improper integral $\int_a^\infty f(x)dx$ converges, and

$$\int_a^b f(x)\,dx \leq \int_a^{+\infty} f(x)\,dx$$

for each $b \geq a$. **Otherwise,**

$$\lim_{b \to +\infty} \int_a^b f(x)\,dx = +\infty.$$

**so that $\int_a^\infty f(x)dx$ diverges.**

The proof of Proposition 1 it left to a course in advanced calculus. Proposition 1 leads to a useful "comparison test":

**Proposition 2 (A Comparison Test) Assume that $f$ and $g$ are continuous on $[a, +\infty)$.**

**a) (The convergence clause) If $0 \leq f(x) \leq g(x)$ for each $x \geq a$ and $\int_a^\infty g(x)\,dx$ converges, then the improper integral $\int_a^\infty f(x)\,dx$ converges as well, and we have**

$$\int_a^\infty f(x)\,dx \leq \int_a^\infty g(x)\,dx.$$

**b) (The divergence clause) If $f(x) \geq g(x) \geq 0$ for each $x \geq a$ and $\int_a^\infty g(x)$ diverges, then the improper integral $\int_a^\infty f(x)\,dx$ diverges as well.**

**Proof**

a) Let $b \geq a$. Since $0 \leq f(x) \leq g(x)$ for each $x \geq a$, we have

$$\int_a^b f(x)dx \leq \int_a^b g(x)dx \leq \int_a^\infty g(x)\,dx$$

for each $b \geq a$. We apply Proposition 1 with $M = \int_a^\infty g(x)\,dx$. Thus, the improper integral $\int_a^\infty f(x)\,dx$ converges and

$$\int_a^\infty f(x)\,dx \leq \int_a^\infty g(x)\,dx.$$

b) Since $g(x) \geq 0$ for each $x \geq a$ and the improper integral $\int_a^\infty g(x)$ diverges, we have

$$\lim_{b \to +\infty} \int_a^b g(x)dx = +\infty$$

by Proposition 1. Since $f(x) \geq g(x)$ for each $x \geq a$,

$$\int_a^b f(x)dx \geq \int_a^b g(x)dx$$

for each $b \geq a$. Therefore,

$$\lim_{b \to +\infty} \int_a^b f(x)dx = +\infty$$

as well. Thus, the improper integral $\int_a^b f(x)dx$ diverges. ∎

**Example 1** Show that the improper integral

$$\int_1^\infty e^{-x^2}\,dx$$

converges.

## 6.7. IMPROPER INTEGRALS: PART 2

**Solution**

If $x \geq 1$ then $x^2 \geq x$, so that $-x^2 \leq -x$. Since the natural exponential function is increasing on the entire number line, we have

$$e^{-x^2} \leq e^{-x} \text{ if } x \geq 1.$$

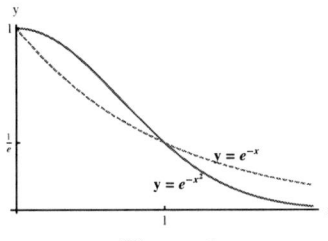

Figure 1

The improper integral $\int_1^\infty e^{-x} dx$ converges by direct calculation. Indeed,

$$\int_1^b e^{-x} dx = -e^{-x}\Big|_1^b = -e^{-b} + e^{-1} = \frac{1}{e} - \frac{1}{e^b},$$

so that

$$\int_1^\infty e^{-x} dx = \lim_{b \to +\infty} \left(\frac{1}{e} - \frac{1}{e^b}\right) = \frac{1}{e}.$$

By the convergence clause of the comparison test (Proposition 2, part a)), the improper integral $\int_1^\infty e^{-x^2} dx$ converges as well. □

Now let's consider the improper integrals of functions of arbitrary sign. In many cases the convergence of such integrals can be discussed within the framework of nonnegative functions.

**Definition 1** We say that the improper integral $\int_a^\infty f(x) dx$ **converges absolutely** if $\int_a^\infty |f(x)| dx$ converges.

Absolute convergence implies convergence:

**Proposition 3** Assume that $f$ is continuous on $[a, +\infty)$ and that $\int_a^\infty |f(x)| dx$ converges. Then $\int_a^\infty f(x) dx$ converges.

The proof of Proposition 3 is left to a course in advanced calculus.

**Example 2** Show that

$$\int_1^\infty e^{-x^2} \cos(3x) \, dx$$

converges.

**Solution**

We have

$$\left|e^{-x^2} \cos(3x)\right| = e^{-x^2} |\cos(3x)| \leq e^{-x^2}$$

since $|\cos(u)| \leq 1$ for each $u \in \mathbb{R}$. In Example 1 we showed that

$$\int_1^\infty e^{-x^2} dx$$

converges. Therefore, the improper integral $\int_1^\infty \left|e^{-x^2}\cos(3x)\right|dx$ converges by the comparison test (Proposition 2). By Proposition 3 the integral $\int_1^\infty e^{-x^2}\cos(3x)\,dx$ converges as well. $\square$

**Example 3** Show that the **Dirichlet integral**

$$\int_0^\infty \frac{\sin(x)}{x}dx$$

converges.

**Solution**

Since

$$\lim_{x\to 0}\frac{\sin(x)}{x} = 1,$$

the integrand has a continuous extension to the interval $[0,1]$. Therefore it is sufficient to show that the improper integral

$$\int_1^\infty \frac{\sin(x)}{x}dx$$

converges. The key to the solution is integration by parts. We set $u = 1/x$ and $dv = \sin(x)\,dx$, so that

$$du = -\frac{1}{x^2}dx \text{ and } v = \int \sin(x)\,dx = -\cos(x).$$

Therefore,

$$\int_1^b \sin(x)\frac{1}{x}dx = \int_1^b u\,dv = uv\Big|_1^b - \int_1^b v\,du$$

$$= -\frac{\cos(x)}{x}\Big|_1^b - \int_1^b \cos(x)\frac{1}{x^2}dx$$

$$= -\frac{\cos(b)}{b} + \cos(1) - \int_1^b \cos(x)\frac{1}{x^2}dx.$$

We have

$$\left|\frac{\cos(b)}{b}\right| = \frac{|\cos(b)|}{b} \leq \frac{1}{b},$$

and $\lim_{b\to\infty} 1/b = 0$. Therefore,

$$\lim_{b\to\infty}\frac{\cos(b)}{b} = 0.$$

Thus,

$$\lim_{b\to\infty}\left(-\frac{\cos(b)}{b} + \cos(1)\right) = \cos(1).$$

As for the integral,

$$\int_1^b \cos(x)\frac{1}{x^2}dx,$$

we claim that the limit as $b \to \infty$ exists, i.e., the improper integral

$$\int_1^\infty \cos(x)\frac{1}{x^2}dx$$

## 6.7. IMPROPER INTEGRALS: PART 2

converges. Indeed,
$$\left|\cos(x)\frac{1}{x^2}\right| = |\cos(x)|\frac{1}{x^2} \leq \frac{1}{x^2},$$
since $|\cos(x)| \leq 1$ for each $x \in R$. Since
$$\int_1^\infty \frac{1}{x^2}dx$$
converges, the improper integral
$$\int_1^\infty \left|\cos(x)\frac{1}{x^2}\right|dx$$
converges, by the comparison test. Therefore,
$$\int_1^b \cos(x)\frac{1}{x^2}dx$$
converges as well, by Proposition 3.
Thus,
$$\lim_{b\to\infty}\int_1^b \frac{1}{x}\sin(x)\,dx = \lim_{b\to\infty}\left(\frac{\cos(b)}{b^2} - \cos(1)\right) + \lim_{b\to\infty}\int_1^b \cos(x)\frac{1}{x^2}dx$$
$$= \cos(1) + \int_1^\infty \cos(x)\frac{1}{x^2}dx.$$

Therefore, the improper integral
$$\int_1^\infty \frac{\sin(x)}{x}dx$$
converges. □

It is known that
$$\int_0^\infty \frac{\sin(x)}{x}dx = \frac{\pi}{2}.$$
We can interpret this value as the signed area of the region between the graph of
$$y = \frac{\sin(x)}{x}$$
and the interval $[0, +\infty)$, as illustrated in Figure 2.

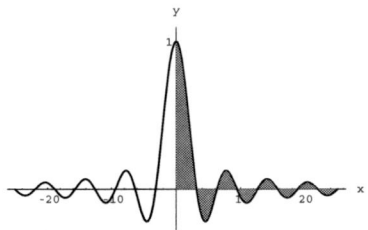

Figure 2

**Remark** The improper integral
$$\int_0^\infty \frac{\sin(x)}{x} dx$$
does not converge absolutely, i.e., the improper integral
$$\int_0^\infty \left|\frac{\sin(x)}{x}\right| dx$$
diverges. We leave the proof of these facts to a post-calculus course. ◊

**Definition 2** *The improper integral $\int_a^\infty f(x)\,dx$ is said to converge* **conditionally** *if it converges, but $\int_a^\infty |f(x)\,dx|$ diverges.*

Thus, the Dirichlet integral
$$\int_0^\infty \frac{\sin(x)}{x} dx$$
converges conditionally.

We have the obvious counterparts of the above definitions and propositions for improper integrals of the form
$$\int_{-\infty}^a f(x)\,dx \text{ and } \int_{-\infty}^\infty f(x)\,dx$$

## Improper Integrals that involve Discontinuous Functions

Now let's consider improper integrals which involve functions that have discontinuities on bounded intervals. The following **comparison test** is the counterpart of Proposition 2:

**Proposition 4 (The comparison test on bounded intervals)** *Assume that $f$ and $g$ are continuous on $(a, b)$.*

a) If $0 \leq f(x) \leq g(x)$ for each $x \in (a,b)$ and $g$ is integrable on $[a,b]$ or the improper integral $\int_a^b g(x)\,dx$ converges, then the improper integral $\int_a^b f(x)\,dx$ converges as well, and we have
$$\int_a^b f(x)\,dx \leq \int_a^b g(x)\,dx.$$

b) If $f(x) \geq g(x) \geq 0$ for each $x \in (a,b)$ and $\int_a^b g(x)$ diverges, then the improper integral $\int_a^b f(x)\,dx$ diverges as well.

We will leave the proof of Proposition 4 to a course in advanced calculus.

**Example 4** Consider the improper integral
$$\int_0^1 \frac{1}{\sqrt{(1-x^2)(1-\frac{1}{2}x^2)}} dx.$$

a) Why is the integral an improper integral?
b) Determine whether the improper integral converges or diverges.

**Solution**

## 6.7. IMPROPER INTEGRALS: PART 2

a) The integrand is continuous at each $x \in [0,1)$ (check). If $x < 1$ and $x \cong 1$, then

$$\frac{1}{\sqrt{(1-x^2)\left(1-\frac{1}{2}x^2\right)}} = \frac{1}{\sqrt{(1-x)(1+x)\left(1-\frac{1}{2}x^2\right)}} \cong \frac{1}{\sqrt{(1-x)(2)\left(\frac{1}{2}\right)}} = \frac{1}{\sqrt{1-x}}.$$

Therefore,

$$\lim_{x \to 1-} \frac{1}{\sqrt{(1-x^2)\left(1-\frac{1}{2}x^2\right)}} = \lim_{x \to 1-} \frac{1}{\sqrt{1-x}} = +\infty.$$

Thus, the integral

$$\int_0^1 \frac{1}{\sqrt{(1-x^2)\left(1-\frac{1}{2}x^2\right)}} dx$$

is an improper integral.

b) If $x \in (0,1)$, we have

$$(1-x^2)\left(1-\frac{1}{2}x^2\right) = (1-x)(1+x)\left(1-\frac{1}{2}x^2\right)$$

$$\geq (1-x)(1)\left(\frac{1}{2}\right)$$

$$= \frac{1}{2}(1-x),$$

so that

$$0 < \frac{1}{\sqrt{(1-x^2)\left(1-\frac{1}{2}x^2\right)}} \leq \frac{\sqrt{2}}{\sqrt{1-x}}.$$

The improper integral

$$\int_0^1 \frac{\sqrt{2}}{\sqrt{1-x}} dx$$

converges. Indeed,

$$\int \frac{\sqrt{2}}{\sqrt{1-x}} dx = -2\sqrt{2}\sqrt{1-x}$$

(check), so that

$$\int_0^{1-\varepsilon} \frac{\sqrt{2}}{\sqrt{1-x}} dx = -2\sqrt{2}\sqrt{1-x}\Big|_0^{1-\varepsilon} = -2\sqrt{2}\sqrt{\varepsilon} + 2\sqrt{2},$$

where $0 < \varepsilon < 1$. Therefore,

$$\int_0^1 \sqrt{\frac{2}{1-x}} dx = \lim_{\varepsilon \to 0+} \int_{-1}^{1-\varepsilon} \frac{1}{\sqrt{1-x}} dx = \lim_{\varepsilon \to 0+} \left(-2\sqrt{\varepsilon} + 2\sqrt{2}\right) = 2\sqrt{2}.$$

Since

$$0 < \frac{1}{\sqrt{(1-x^2)\left(1-\frac{1}{2}x^2\right)}} \leq \frac{\sqrt{2}}{\sqrt{1-x}},$$

the given improper integral also converges, by Proposition 4. □

**Definition 3** Assume that $f$ is continuous on $(a,b)$. We say that the improper integral $\int_a^b f(x)\,dx$ **converges absolutely** if the improper integral $\int_a^b |f(x)|\,dx$ converges.

Just as in the case of improper integrals on unbounded intervals, an improper integral that converges absolutely is convergent:

**Proposition 5** Assume that $f$ is continuous on $(a,b)$ and that $\int_a^b |f(x)|\,dx$ converges. Then $\int_a^b f(x)\,dx$ converges as well.

The proof of Proposition 5 is left to a course in advanced calculus.

**Example 5** Consider the improper integral

$$\int_0^1 \frac{\sin(1/x)}{\sqrt{x}}\,dx.$$

a) Why is the integral an improper integral?
b) Show that the improper integral converges absolutely.

**Solution**

a) Set

$$f(x) = \frac{\sin(1/x)}{\sqrt{x}}.$$

Then, $\lim_{x \to 0+} f(x)$ does not exist. Indeed, set

$$x_n = \frac{1}{\frac{\pi}{2} + 2n\pi},\ n = 0,1,2,3,\ldots.$$

Note that $\lim_{n \to \infty} x_n = 0$. We have

$$f(x_n) = \sqrt{\frac{\pi}{2} + 2n\pi}\,\sin\left(\frac{\pi}{2} + 2n\pi\right) = \sqrt{\frac{\pi}{2} + 2n\pi}.$$

Therefore,

$$\lim_{n \to \infty} f(x_n) = \lim_{n \to \infty} \sqrt{\frac{\pi}{2} + 2n\pi} = +\infty.$$

On the other hand, if

$$z_n = \frac{1}{\frac{3\pi}{2} + 2n\pi},\ n = 0,1,2,3,\ldots,$$

then $\lim_{n \to \infty} z_n = 0$, and

$$f(z_n) = \sqrt{\frac{3\pi}{2} + 2n\pi}\,\sin\left(\frac{3\pi}{2} + 2n\pi\right) = -\sqrt{\frac{3\pi}{2} + 2n\pi},$$

so that

$$\lim_{n \to \infty} f(z_n) = \lim_{n \to \infty} \left(-\sqrt{\frac{3\pi}{2} + 2n\pi}\right) = -\infty.$$

Therefore, $f$ does not have a right-limit at 0. Figure 3 is an attempt to give some idea about the graph of $f$. The integral is definitely improper.

## 6.7. IMPROPER INTEGRALS: PART 2

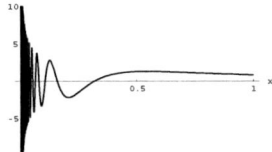

Figure 3

Since $\sin(1/x) \leq 1$,
$$\left|\frac{\sin(1/x)}{\sqrt{x}}\right| = \frac{|\sin(1/x)|}{\sqrt{x}} \leq \frac{1}{\sqrt{x}}$$
for any $x > 0$, and the improper integral
$$\int_0^1 \frac{1}{\sqrt{x}}dx$$
converges as we discussed before (Example 8 of Section 6.6). Therefore, the given improper integral converges absolutely by the comparison test. □

**Definition 4** Assume that $f$ is continuous on $(a,b)$. We say that the improper integral $\int_a^b f(x)\,dx$ **converges conditionally** if it converges but $\int_a^b f(x)\,dx$ diverges.

**Example 6** Show that the improper integral
$$\int_0^1 \frac{1}{x}\sin\left(\frac{1}{x}\right)dx$$
converges.

**Solution**

Set
$$f(x) = \frac{1}{x}\sin\left(\frac{1}{x}\right).$$
The integral is an improper integral, since
$$\lim_{x \to 0+} f(x) = \lim_{x \to 0+} \frac{1}{x}\sin\left(\frac{1}{x}\right)$$
does not exist (as in Example 7). Figure 4 attempts to display the graph of $f$.

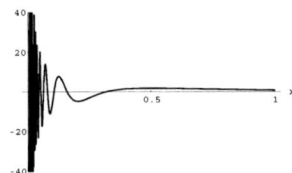

Figure 4

We will analyze the improper integral by transforming it to another integral which we have already discussed. If we make the substitution $u = 1/x$, we have

$$du = -\frac{1}{x^2}dx,$$

so that

$$\int_\varepsilon^1 \frac{1}{x}\sin\left(\frac{1}{x}\right)dx = -\int_\varepsilon^1 x\sin(1/x)\left(-\frac{1}{x^2}\right)dx = -\int_{1/\varepsilon}^1 \frac{1}{u}\sin(u)\,du$$

$$= -\int_{1/\varepsilon}^1 \frac{\sin(u)}{u}du = \int_1^{1/\varepsilon}\frac{\sin(u)}{u}du.$$

Thus,

$$\lim_{\varepsilon \to 0+}\int_\varepsilon^1 \frac{1}{x}\sin\left(\frac{1}{x}\right)dx \text{ exists} \Leftrightarrow \lim_{\varepsilon \to 0+}\int_1^{1/\varepsilon}\frac{\sin(u)}{u}du \text{ exists} \Leftrightarrow \int_1^\infty \frac{\sin(u)}{u}du \text{ converges}.$$

In Example 3 we showed that

$$\int_1^\infty \frac{\sin(u)}{u}du$$

converges. Therefore,

$$\int_0^1 \frac{1}{x}\sin\left(\frac{1}{x}\right)dx$$

converges.
It can be shown that the integral does not converge absolutely. Thus, the integral converges conditionally. □

## Problems

In problems 1-8, make use of **a comparison test** in order to determine whether the improper integral converges or diverges. Do not try to evaluate the integral (in the case of an integral on a bounded interval, you need to explain why the integral is an improper integral).

1.
$$\int_\pi^\infty \frac{\cos^2(4x)}{x^{3/2}}dx$$

2.
$$\int_1^\infty \frac{e^{-x}}{\sqrt{x}}dx$$

3.
$$\int_9^\infty \frac{\sqrt{x}}{x+4}dx$$

4.
$$\int_1^\infty \frac{1}{x^3+2x+9}dx$$

5.
$$\int_0^{+\infty} x^2 e^{-x^2}dx$$

6.
$$\int_1^\infty e^{-x}\sin\left(\frac{1}{x}\right)dx$$

7.
$$\int_1^4 \frac{x^{1/3}}{(4-x)^2}dx$$

8.
$$\int_0^1 \frac{1}{x^{1/3}}\cos\left(\frac{1}{x^2}\right)dx$$

# Chapter 7

# Applications of Integration

In this chapter we discuss the application of integration to the calculation of the **volume** of certain types of solids, the **length** of the graph of a function and the **surface area** of certain types of surfaces. We will also discuss some physical applications such as **mass** and **work**, and some applications to **probability**.

## 7.1 Volumes by Slices or Cylindrical Shells

In some cases we can express the volume of a solid as an integral that involves the areas of parallel cross sections. A special case involves a **solid of revolution** that can be obtained by revolving a region in the plane about an axis. The relevant cross sections are circular, and the technique of calculating the volume as the integral of the area of the cross section is referred to as **the method of disks**. The volume of a solid of a revolution can be determined by the **method of cylindrical shells** as well.

### Volume by Slices

Assume that $\{x_0, x_1, x_2, \ldots, x_{k-1}, x_k, \ldots, x_{n-1}, x_n\}$ is a partition of the interval $[a, b]$, so that

$$a = x_0 < x_1 < x_2 < \cdots < x_{k-1} < x_k < \cdots < x_{n-1} < x_n = b,$$

and let $x_k^* \in [x_{k-1}, x_k]$.

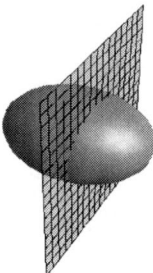

Figure 1: A vertical slice

Let $A(x_k^*)$ be the area of the vertical cross section of the solid that corresponds to $x = x_k^*$. If $\Delta x_k = x_k - x_{k-1}$ is small and $A(x)$ is continuous, the volume of the cylinder whose cross section has area $A\left(x_k^*\right)$ and thickness $\Delta x_k$ approximates the volume of the slice of the solid between the vertical planes $x = x_{k-1}$ and $x = x_k$. Since the volume of the cylinder is $A\left(x_k^*\right) \Delta x_k$, we obtain an approximation to the volume of the solid by summing the volumes of the cylinders. Thus,

$$\text{Volume} \cong \sum_{k=1}^{n} A\left(x_k^*\right) \Delta x_k$$

The above sum is a Riemann sum that approximates the integral

$$\int_a^b A(x)\, dx.$$

Therefore we will calculate the volume of the solid as

$$\int_a^b A(x)\, dx.$$

You can think of $A(x)\, dx$ as the volume of a cylinder whose cross section has area $A(x)$ and has "infinitesimal" thickness $dx$. We will refer to the technique that leads to the above formula as **the method of vertical slices.**

**Example 1** Determine the volume of a spherical ball of radius $R$ by the method of vertical slices.

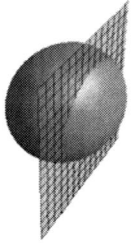

Figure 2

**Solution**

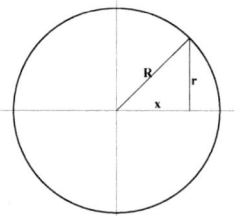

Figure 3

With reference to Figure 2 and Figure 3, the vertical cross section corresponding to $x$ is a disk of radius $r(x)$, so that its area $A(x)$ is $\pi \left(r(x)\right)^2$. By the Pythagorean Theorem, $R^2 = x^2 + \left(r(x)\right)^2$, so that $\left(r(x)\right)^2 = R^2 - x^2$. Therefore,

$$\text{Volume} = \int_{-R}^{R} A(x)\, dx = \int_{-R}^{R} \pi \left(r(x)\right)^2 dx = \int_{-R}^{R} \pi \left(R^2 - x^2\right) dx.$$

## 7.1. VOLUMES BY SLICES OR CYLINDRICAL SHELLS

We have
$$\int \left(R^2 - x^2\right) dx = \int R^2 dx - \int x^2 dx = R^2 x - \frac{1}{3}x^3$$
Therefore,
$$\int_{-R}^{R} \pi \left(R^2 - x^2\right) dx = \pi \int_{-R}^{R} \left(R^2 - x^2\right) dx$$
$$= \pi \left(R^2 x - \frac{1}{3}x^3 \Big|_{-R}^{R}\right)$$
$$= \pi \left(\left(R^3 - \frac{1}{3}R^3\right) - \left(-R^3 + \frac{1}{3}R^3\right)\right)$$
$$= \pi \left(2R^3 - \frac{2}{3}R^3\right) = \frac{4\pi}{3}R^3$$

Thus, the volume of a spherical ball of radius $R$ is
$$\frac{4\pi}{3}R^3,$$
as you have been told in precalculus courses. $\square$

In some cases, we can determine the volume of a solid by the **method of horizontal slices**.

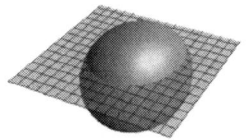

Figure 4

With reference to Figure 4, let $A(z)$ be the area of the horizontal cross section of the solid corresponding to $z$. Just as in the case of the vertical slices,
$$\text{Volume} = \int_a^b A(z) \, dz.$$

**Example 2** Consider the pyramid whose horizontal sections are squares. The base of the pyramid is a square with sides of length $a$ and the height of the pyramid is $h$. Determine the volume of the pyramid by the method of horizontal slices.

**Solution**

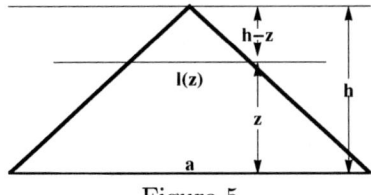

Figure 5

With reference to Figure 5, the horizontal cross section of the pyramid corresponding to $z$ is a square with sides of length $l(z)$. By similar triangles,
$$\frac{l(z)}{a} = \frac{h-z}{h} = 1 - \frac{1}{h}z,$$
so that
$$l(z) = a - \frac{a}{h}z.$$
Therefore, the area of the horizontal cross section corresponding to $z$ is
$$A(z) = (l(z))^2 = \left(a - \frac{a}{h}z\right)^2.$$
The volume of the pyramid is
$$\int_0^h A(z)\,dz = \int_0^h \left(a - \frac{a}{h}z\right)^2 dz.$$
Let's set
$$u = a - \frac{a}{h}z$$
so that
$$\frac{du}{dz} = -\frac{a}{h}.$$
Therefore,
$$\int_0^h \left(a - \frac{a}{h}z\right)^2 dz = -\frac{h}{a}\int_0^h u^2 \frac{du}{dz}dz = -\frac{h}{a}\int_{u(0)}^{u(h)} u^2\,du$$
$$= -\frac{h}{a}\int_a^0 u^2\,du = \frac{h}{a}\int_0^a u^2\,du = \frac{h}{a}\left(\frac{u^3}{3}\bigg|_0^a\right) = \frac{1}{3}a^2 h.$$
Thus, the volume of the pyramid is
$$\frac{1}{3} \times \text{(the area of the base)} \times \text{height},$$
as you have been told in precalculus courses. □

## The Method of Disks

Let $G$ be the region between the graph of $z = f(x)$ and the interval $[a, b]$ on the $x$-axis, as illustrated in Figure 6.

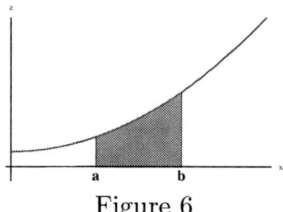

Figure 6

Consider the solid that is formed by revolving the region $G$ about the $x$-axis, as illustrated in Figure 7.

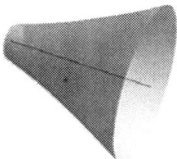

Figure 7

The vertical cross section of such a solid of revolution corresponding to $x$ is a disk of radius $f(x)$. Therefore, the area $A(x)$ of the cross section is $\pi(f(x))^2$. Thus, the volume of the solid is
$$\int_a^b A(x)\,dx = \int_a^b \pi(f(x))^2\,dx = \pi \int_a^b (f(x))^2\,dx.$$

You may think of $\pi(f(x))^2\,dx$ as the volume of a circular cylinder whose radius is $f(x)$ and whose "infinitesimal" thickness is $dx$. The technique that leads to the above formula for the calculation of the volume of a solid of revolution will be referred to as **the method of disks**.

**Example 3** Use the method of disks to determine the volume of the solid that is formed by revolving the region between the graph of $f(x) = x^2 + 1$ and the interval $[1, 2]$ about the $x$-axis.

**Solution**

The region $G$ between the graph of $f$ and the interval $[1, 2]$.is as in Figure 6 and the solid that is formed by revolving $G$ about the $x$-axis is as in Figure 7. The volume of the solid is
$$\pi \int_1^2 (f(x))^2\,dx = \pi \int_1^2 (x^2+1)^2\,dx = \pi\left(\frac{178}{15}\right) \cong 37.2802$$

□

Assume that the region $G$ is between the graphs of $f$ and $g$ and the lines $x = a$ and $x = b$, as illustrated in Figure 8.

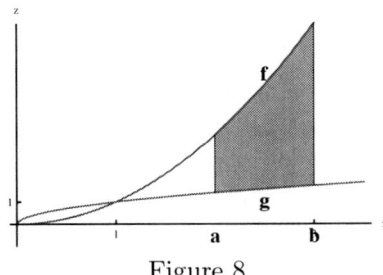

Figure 8

We would like to calculate the volume of the solid that is formed by revolving $G$ around the $x$-axis, as illustrated in Figure 9.

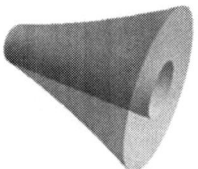

Figure 9

The volume of the solid can be calculated by subtracting the volume of the "inner solid" from the "outer solid". Thus, its volume is

$$\pi \int_a^b f^2(x)\,dx - \pi \int_a^b g^2(x)\,dx = \pi \int_a^b \left(f^2(x) - g^2(x)\right) dx.$$

We can also arrive at the above expression for the volume of the solid as follows: Let's subdivide the interval into subintervals $[x_{k-1}, x_k]$, $k = 1, 2, \ldots, n$. The slice of the solid that corresponds to the interval $[x_{k-1}, x_k]$ is approximately a **"washer"**, i.e., a solid which is bounded by two cylinders, the outer radius being $f(x_k^*)$ and the inner radius being $g(x_k^*)$. Here, $x_k^*$ can be any point between $x_{k-1}$ and $x_k$.

Figure 10: A "washer"

The volume of the washer is the difference between the volume of the outer cylinder and the volume of the inner cylinder, i.e.,

$$\pi f^2(x_k^*) \Delta x_k - \pi g^2(x_k^*) \Delta x_k = \pi \left(f^2(x_k^*) - g^2(x_k^*)\right) \Delta x_k.$$

We can approximate the volume of the slice of the solid between $x = x_{k-1}$ and $x = x_k$ by the volume of the corresponding washer, and the volume of the solid by the sum of the volumes of the washers:

$$\sum_{k=1}^n \pi \left(f^2(x_k^*) - g^2(x_k^*)\right) \Delta x_k = \pi \sum_{k=1}^n \left(f^2(x_k^*) - g^2(x_k^*)\right) \Delta x_k.$$

## 7.1. VOLUMES BY SLICES OR CYLINDRICAL SHELLS

Such a sum is a Riemann sum for the integral

$$\pi \int_a^b \left(f^2(x) - g^2(x)\right) dx.$$

You may think of $\pi \left(f^2(x) - g^2(x)\right) dx$ as the volume of a washer that has outer radius $f(x)$, inner radius $g(x)$, and "infinitesimal" thickness $dx$. The technique that leads to the above formula for the calculation of a solid of revolution will be referred to as **the method of washers.**

**Example 4** Use the method of washers to determine the volume of the solid that is formed by revolving the region between the graphs of $f(x) = x^2$, $g(x) = x$, the line $x = 1$ and the line $x = 3$ about the $x$-axis.

**Solution**

Figure 11 shows the region $G$ between the graphs of $f$ and $g$ and the lines $x = 1$ and $x = 3$.

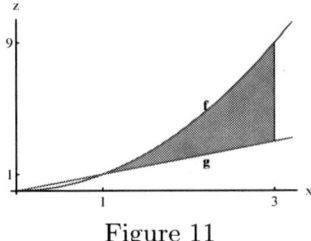

Figure 11

The solid that is formed by revolving $G$ about the $x$-axis is between the two surfaces that are shown in Figure 12.

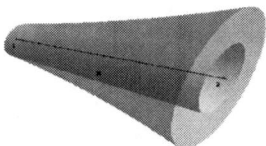

Figure 12

The volume of the solid is

$$\pi \int_1^2 \left(f^2(x) - g^2(x)\right) dx = \pi \int_1^2 \left((x^2)^2 - (x)^2\right) dx = \pi \int_1^2 \left(x^4 - x^2\right) dx = \frac{58}{15}\pi$$

(confirm). □

**Remark** The method of disks and the method of washers can be modified to calculate the volume of a solid that is obtained by revolving a region in the $xy$-plane about the $y$-axis. You merely interchange the roles of $x$ and $y$, as in some of the problems at the end of this section. In fact, the method can be modified easily if "the axis of revolution" is an arbitrary line that is parallel to one of the coordinate axes. ◊

## The Method of Cylindrical Shells

Now we will describe a method for the calculation of the volume of a solid that is obtained by revolving a region in the $xz$-plane around the $z$-axis. The method is not based on areas slices, unlike the method of disks. Assume that $G$ is the region between the graph of $z = f(x)$ and the interval $[a, b]$, as illustrated in Figure 13.

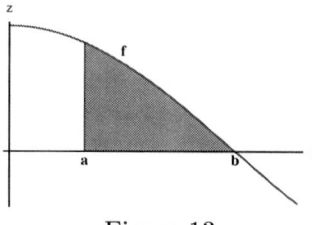

Figure 13

The solid that is formed by revolving the graph of $f$ around the $y$-axis is between the surfaces that are shown in Figure 14.

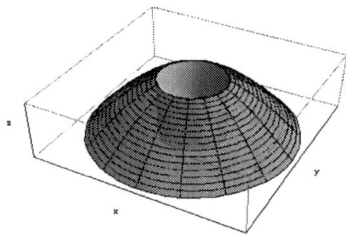

Figure 14

As usual, let us subdivide the interval $[a, b]$ into subintervals $[x_{k-1}, x_k]$, $k = 1, 2, \ldots, n$. The part of the solid which corresponds to the slice of the region between $x_{k-1}$ and $x_k$ is approximately a cylindrical shell. The inner radius of the shell is $x_{k-1}$ and the outer radius of the shell is $x_k$. We can take as the height of the shell $f(x_k^*)$, where $x_k^*$ is an arbitrary point in the interval $[x_{k-1}, x_k]$.

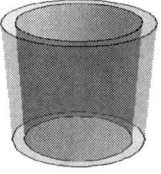

Figure 15: A cylindrical shell

The volume of the shell is

$$\pi x_k^2 f(x_k^*) - \pi x_{k-1}^2 f(x_k^*) = \pi \left(x_k^2 - x_{k-1}^2\right) f(x_k^*) = \pi (x_k + x_{k-1})(x_k - x_{k-1}) f(x_k^*)$$
$$= \pi (x_k + x_{k-1}) f(x_k^*) \Delta x_k$$
$$= 2\pi \left(\frac{x_k + x_{k-1}}{2}\right) f(x_k^*) \Delta x_k$$

## 7.1. VOLUMES BY SLICES OR CYLINDRICAL SHELLS

($2\pi \times$(**average radius**)$\times$**height**$\times$**thickness**). Thus, the sum of the volumes of the shells is

$$\sum_{k=1}^{n} 2\pi \left( \frac{x_k + x_{k-1}}{2} \right) f(x_k^*) \Delta x_k.$$

This is almost a Riemann sum corresponding to the integral

$$\int_a^b 2\pi x f(x)\, dx.$$

If we replace $x_k$ and $x_{k-1}$ by $x_k^*$, we do have a Riemann sum for the above integral. If $\Delta x_k$ is small, the points $x_{k-1}$, $x_k$ and $x_k^*$ are close to each other. Therefore, it is reasonable to expect that the above sum approximates the integral. This is indeed the case if $f$ is continuous on $[a, b]$. Thus, we will calculate the volume of the solid by the formula

$$\int_a^b 2\pi x f(x)\, dx = 2\pi \int_a^b x f(x)\, dx.$$

You may think of $2\pi x f(x)\, dx$ as the volume of a cylindrical shell of average radius $x$, height $f(x)$ and "infinitesimal" thickness $dx$. The method that has led to the above formula for the computation of the volume of the solid of revolution will be referred to as **the method of cylindrical shells**.

**Example 5** Use the method of cylindrical shells to determine the volume of the solid that is obtained by revolving the region between the graph of $z = \cos(x)$ and the interval $[\pi/6, \pi/2]$ about the $z$-axis.

**Solution**

Figure 16 shows the region $G$ between the graph of $f$ and the interval $[\pi/6, \pi/2]$.

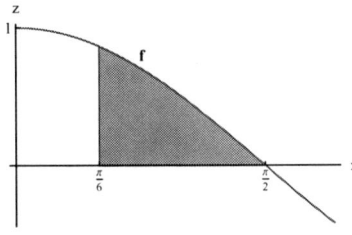

Figure 16

Figure 17 illustrates the solid that is formed by revolving $G$ about the $z$-axis.

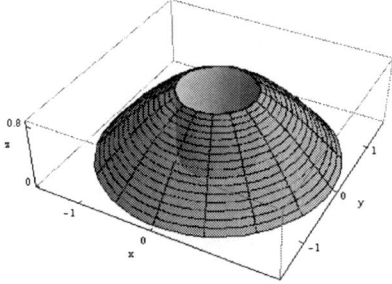

Figure 17

The volume of the solid is

$$2\pi \int_{\pi/6}^{\pi/2} x f(x)\, dx = 2\pi \int_{\pi/6}^{\pi/2} x \cos(x)\, dx$$

As in Section 7.1, we can apply integration by parts by setting $u = x$ and $dv = \cos(x)\, dx$, so that

$$du = dx \text{ and } v(x) = \int \cos(x)\, dx = \sin(x).$$

Thus,

$$\begin{aligned}\int x \cos(x)\, dx = \int u\, dv &= uv - \int v\, du \\ &= x \sin(x) - \int \sin(x)\, dx \\ &= x \sin(x) + \cos(x).\end{aligned}$$

Therefore,

$$\begin{aligned}2\pi \int_{\pi/6}^{\pi/2} x \cos(x)\, dx = 2\pi \left( x \sin(x) + \cos(x)|_{\pi/6}^{\pi/2} \right) &= 2\pi \left( \frac{\pi}{2} - \frac{\pi}{6} \sin\left(\frac{\pi}{6}\right) - \cos\left(\frac{\pi}{6}\right) \right) \\ &= 2\pi \left( \frac{\pi}{2} - \frac{\pi}{6} \left(\frac{1}{2}\right) - \frac{\sqrt{3}}{2} \right) \\ &= \frac{5}{6}\pi^2 - \sqrt{3}\pi.\end{aligned}$$

□

## Problems

**1.** Make use of the method of horizotal slices to confirm that the volume of a right circular cone with base radius $r$ and height $h$ is

$$\frac{1}{3}\pi r^2 h.$$

In problems 2-7, use **the method of disks** to determine the **volume** of the solid that is obtained by revolving the region between the graph of $z = f(x)$ and the given interval about **the $x$-axis**.

**2.**
$$f(x) = e^{-x};\ [-\ln(2), \ln(2)]$$

**3.**
$$f(x) = \sin\left(\frac{x}{4}\right);\ [2\pi, 3\pi]$$

**4.**
$$f(x) = \sqrt{\frac{x}{x^2+4}};\ \left[\sqrt{e^2-4}, \sqrt{e^3-4}\right]$$

**5.**
$$f(x) = \sqrt{\frac{1}{x^2+1}},\ [-1,1].$$

**6.**
$$f(x) = \frac{1}{\sqrt{x}};\ [1, e^2].$$

**7.**
$$f(x) = (16 - x^2)^{1/4};\ [2, 4].$$

In problems 8 and 9, use **the method of washers** to determine the **volume** of the solid that is obtained by revolving the region between the graphs of $z = f(x)$, $z = g(x)$, $x = a$ and $x = b$ about **the $x$-axis**.

## 7.2. LENGTH AND AREA

**8.**
$$f(x) = \frac{1}{x}, \ g(x) = \frac{1}{x^2}, \ x = 1, \ x = 2.$$

**9.**
$$f(x) = \sin(x), \ g(x) = \cos(x), \ x = \pi/4, \ x = \pi/2$$

In problems 10-15, use **the method of cylindrical shells** to determine the **volume** of the solid that is obtained by revolving the graph of $z = f(x)$ on the given interval about **the z-axis**.

**10.**
$$f(x) = \sin(x^2), \ \left[\sqrt{\frac{\pi}{6}}, \sqrt{\frac{\pi}{4}}\right]$$

**11.**
$$f(x) = e^{-x^2}, \ [1, 2]$$

**12.**
$$f(x) = \frac{1}{x^2+4}, \ [0, 2].$$

**13.**
$$f(x) = \frac{1}{\sqrt{16-x^2}}. \ [0, 3].$$

**14.**
$$f(x) = \cos(x), \ [\pi/4, \pi/2]$$

**15.**
$$f(z) = e^z, \ [\ln(2), \ln(4)]$$

In problems 16 and 17, use **the method of cylindrical shells** to determine the **volume** of the solid that is obtained by revolving the region between the graphs of $z = f(x)$, $z = g(x)$, $x = a$ and $x = b$ about **the z-axis**.

**16.**
$$f(x) = x^{1/3}, \ g(x) = x^2, \ x = 0, \ x = 1.$$

**17.**
$$f(x) = \sin(x), \ g(x) = \cos(x), \ x = 0, \ x = \frac{\pi}{4}$$

## 7.2 The Length of the Graph of a Function and the Area of a Surface of Revolution

We can express the length of the graph of a function and the area of a surface that is obtained by revolving the graph of a function about an axis as integrals.

### The Length of the Graph of a Function

Let us approximate the length of the graph of $y = f(x)$ on the interval $[a, b]$. As usual, we will subdivide the interval $[a, b]$ into subintervals $[x_{k-1}, x_k]$, $k = 1, 2, \ldots, n$, where $x_0 = a$ and $x_n = b$. If $\Delta x_k = x_k - x_{k-1}$ is small, we can approximate the length of the graph of $f$ corresponding to the interval $[x_{k-1}, x_k]$ by the length of the line segment that joins $(x_{k-1}, f(x_{k-1}))$ to $(x_k, f(x_k))$, as illustrated in Figure 1.

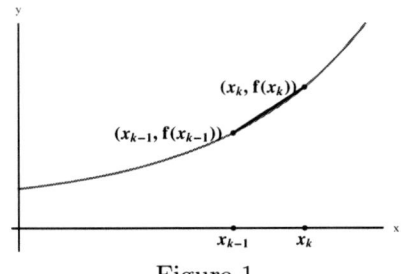

Figure 1

The length of that line segment is

$$\sqrt{(x_k - x_{k-1})^2 + (f(x_k) - f(x_{k-1}))^2} = \sqrt{(\Delta x_k)^2 + (f(x_k) - f(x_{k-1}))^2}.$$

We will assume that $f$ is continuous on $[a, b]$ and that $f'$ has a continuous extension to the interval $[a, b]$, as we discussed in Section 5.2. By the Mean Value Theorem,

$$f(x_k) - f(x_{k-1}) = f'(x_k^*)(x_k - x_{k-1}) = f'(x_k^*)\Delta x_k,$$

where $x_k^*$ is some point in the interval $(x_{k-1}, x_k)$. Therefore, the length of the line segment is

$$\sqrt{(\Delta x)^2 + (f(x_k) - f(x_{k-1}))^2} = \sqrt{(\Delta x_k)^2 + (f'(x_k^*))^2(\Delta x_k)^2} = \sqrt{1 + (f'(x_k^*))^2}\Delta x_k.$$

We will approximate the length of the graph of $f$ over the interval $[a, b]$ by the sum of the lengths of the line segments. Thus, the length of the graph of $f$ on $[a, b]$ is approximated by

$$\sum_{k=1}^{n} \sqrt{1 + (f'(x_k^*))^2}\Delta x_k.$$

This is a Riemann sum for the integral

$$\int_a^b \sqrt{1 + (f'(x))^2}\,dx,$$

and approximates the above integral if $f'$ is continuous on $[a, b]$. Therefore, we will compute the length of the graph of $f$ on the interval $[a, b]$ with the above formula:

**Definition 1** **The length of the graph of the function $f$ on the interval $[a, b]$ is defined as**

$$\int_a^b \sqrt{1 + (f'(x))^2}\,dx,$$

if $f'$ is continuous on $[a, b]$.

We will allow the possibility that the above integral is meaningful as an improper integral, provided that the improper integral converges.

**Example 1** Let $f(x) = x^2$. Compute the length of the graph of $f$ over the interval $[0, 1]$.

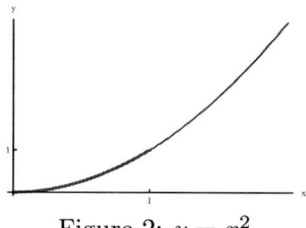

Figure 2: $y = x^2$

**Solution**

## 7.2. LENGTH AND AREA

The length of the graph of $f$ on $[0, 1]$ is

$$\int_0^1 \sqrt{1+(f'(x))^2}\,dx = \int_0^1 \sqrt{1+\left(\frac{d}{dx}(x^2)\right)^2}\,dx = \int_0^1 \sqrt{1+(2x)^2}\,dx = \int_0^1 \sqrt{1+4x^2}\,dx.$$

We can write

$$\int \sqrt{1+4x^2}\,dx = \int \sqrt{4\left(\frac{1}{4}+x^2\right)}\,dx = 2\int \sqrt{\left(\frac{1}{2}\right)^2+x^2}\,dx.$$

In Example 4 of Section 5.4 we derived the expression

$$\int \sqrt{x^2+a^2}\,dx = \frac{x}{2}\sqrt{x^2+a^2} + \frac{a^2}{2}\ln\left(x+\sqrt{x^2+a^2}\right)$$

with the help of the substitution $x = a\sinh(u)$. We apply the above formula with $a = 1/2$:

$$\int \sqrt{\left(\frac{1}{2}\right)^2+x^2}\,dx = \frac{x}{2}\sqrt{x^2+\frac{1}{4}} + \frac{1}{8}\ln\left(x+\sqrt{x^2+\frac{1}{4}}\right)$$

$$= \frac{x}{4}\sqrt{4x^2+1} + \frac{1}{8}\ln\left(x+\frac{1}{2}\sqrt{4x^2+1}\right).$$

Therefore,

$$\int \sqrt{1+4x^2}\,dx = 2\int \sqrt{\left(\frac{1}{2}\right)^2+x^2}\,dx = \frac{x}{2}\sqrt{4x^2+1} + \frac{1}{4}\ln\left(x+\frac{1}{2}\sqrt{4x^2+1}\right)$$

Thus, the length of the graph of $f$ over the interval $[0, 1]$ is

$$\int_0^1 \sqrt{1+4x^2}\,dx = \frac{1}{2}x\sqrt{4x^2+1} + \frac{1}{4}\ln\left(x+\frac{1}{2}\sqrt{4x^2+1}\right)\Big|_0^1$$

$$= \frac{1}{2}\sqrt{5} + \frac{1}{4}\ln\left(1+\frac{\sqrt{5}}{2}\right) - \frac{1}{4}\ln\left(\frac{1}{2}\right)$$

$$\cong 1.47894.$$

□

**Remark** If $y = f(x)$, we can express the length of the graph of $f$ on $[a, x]$ as

$$s(x) = \int_a^x \sqrt{1+\left(\frac{dy}{dt}\right)^2}\,dt.$$

This is the **arc length function**. By the Fundamental Theorem of Calculus,

$$\frac{ds}{dx} = \sqrt{1+\left(\frac{dy}{dx}\right)^2}.$$

You may find the following formalism appealing: If we set

$$ds = \frac{ds}{dx}dx = \sqrt{1+\left(\frac{dy}{dx}\right)^2}\,dx$$

and
$$dy = \frac{dy}{dx}dx, \ (dy)^2 = \left(\frac{dy}{dx}\right)^2 (dx)^2,$$
we can declare that
$$\sqrt{(dx)^2 + (dy)^2} = \sqrt{(dx)^2 + \left(\frac{dy}{dx}\right)^2 (dx)^2} = \sqrt{1 + \left(\frac{dy}{dx}\right)^2} \, dx = ds.$$

Thus, we can express the length of the graph of $f$ on the interval $[a, b]$ as
$$\int_a^b \sqrt{1 + \left(\frac{dy}{dx}\right)^2} \, dx = \int_a^b ds = \int_a^b \sqrt{(dx)^2 + (dy)^2} dx.$$

The above expression does not offer any theoretical or computational advantage, but it may serve as a reminder about the origin of the formula for the length of the graph of a function: You may imagine that the length of the graph of the function over the "infinitesimal interval" $[x, x+dx]$ is approximated by the length $ds = \sqrt{(dx)^2 + (dy)^2}$ of the line segment joining $(x, y)$ and $(x + dx, y + dy)$, as illustrated in Figure 3. The "summation" of these lengths is the integral that we use to compute the length of the graph of $f$. ◊

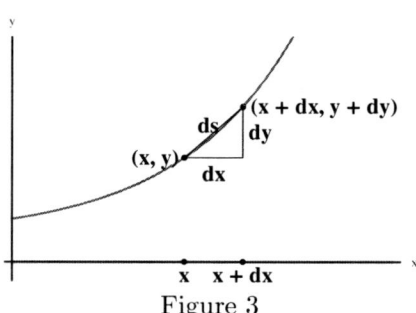

Figure 3

**Example 2** We have defined the radian measure geometrically. In particular, the length of the arc of the unit circle that is in the first quadrant should be $\pi/2$. Confirm this with the help of the formula for the length of the graph of a function.

**Solution**

In the $xy$-coordinate plane, the unit circle is the graph of the equation $x^2 + y^2 = 1$.

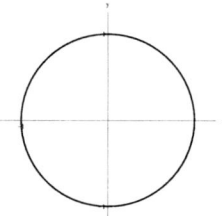

Figure 4: $x^2 + y^2 = 1$

## 7.2. LENGTH AND AREA

We have
$$y = \sqrt{1-x^2}, \quad -1 \le x \le 1,$$
for the points $(x,y)$ on the upper half of the unit circle. Thus, the length of the arc in question is

$$\int_0^1 \sqrt{1+\left(\frac{dy}{dx}\right)^2}\,dx = \int_0^1 \sqrt{1+\left(\frac{d}{dx}\sqrt{1-x^2}\right)^2}\,dx = \int_0^1 \sqrt{1+\left(-\frac{x}{\sqrt{(1-x^2)}}\right)^2}\,dx$$
$$= \int_0^1 \sqrt{1+\frac{x^2}{1-x^2}}\,dx$$
$$= \int_0^1 \frac{1}{\sqrt{1-x^2}}\,dx.$$

Since
$$\lim_{x \to 1-} \frac{1}{\sqrt{1-x^2}} = +\infty,$$
we have to interpret the above integral as an improper integral. If $0 < \varepsilon < 1$, we have
$$\int_0^{1-\varepsilon} \frac{1}{\sqrt{1-x^2}}\,dx = \arcsin(1-\varepsilon) - \arcsin(0) = \arcsin(1-\varepsilon).$$

Therefore,
$$\int_0^1 \frac{1}{\sqrt{1-x^2}}\,dx = \lim_{\varepsilon \to 0+} \int_0^{1-\varepsilon} \frac{1}{\sqrt{1-x^2}}\,dx = \lim_{\varepsilon \to 0+} \arcsin(1-\varepsilon) = \arcsin(1) = \frac{\pi}{2}.$$

□

In general, the calculation of the length of the graph of a function requires numerical integration, as in the following example:

**Example 3** Let $f(x) = 1/x$.

a) Express the length of the graph of $f$ on the interval $[1, 2]$ as an integral.

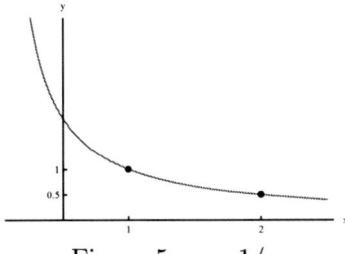

Figure 5: $y = 1/x$

b) Make use of the numerical integrator of your computational utility to approximate the length of the graph of $f$ on the interval $[1, 2]$.

**Solution**

a)
The length of the graph of $f$ on the interval $[1, 2]$ is

$$\int_1^2 \sqrt{1+(f'(x))^2}\,dx = \int_1^2 \sqrt{1+\left(\frac{d}{dx}\left(\frac{1}{x}\right)\right)^2}\,dx = \int_1^2 \sqrt{1+\left(-\frac{1}{x^2}\right)^2}\,dx = \int_1^2 \sqrt{1+\frac{1}{x^4}}\,dx.$$

b)
$$\int_1^2 \sqrt{1 + \frac{1}{x^4}}\,dx \cong 1.13209$$

Thus, the length of the part of the hyperbola $y = 1/x$ that corresponds to the interval $[1, 2]$ is approximately 1.13209. □

A computer algebra system may provide an expression for the integral of Example 3 in terms of special functions that you are not familiar with. Such an expression does not offer an advantage over the approximate result that is provided via numerical integration.

## The Area of a Surface of Revolution

Let's begin by computing the area of a the frustum of a right circular cone. This is a part of a right circular cone and resembles a lamp shade as illustrated in Figure 6.

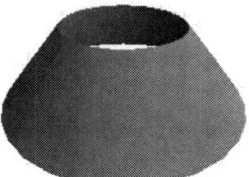

Figure 6

The slant height is measured as follows: Imagine a plane that is perpendicular to the base and passes through the center of the base. This plane cuts the surface along a line segment. The slant height of the surface is the length of that line segment.

**Lemma 1** Let $S$ be the frustum of a right circular cone such that the radius of the base is $r_1$, the radius of the top is $r_2$, and the slant height is $s$. The area of $S$ is

$$2\pi \left(\frac{r_1 + r_2}{2}\right) s.$$

Thus, if we declare

$$\frac{r_1 + r_2}{2}$$

to be the average radius, the area of $S$ is

$$2\pi \times \text{Average radius} \times \text{slant height}.$$

### The Proof of Lemma 1

Imagine that the surface is cut along a plane that passes through the center of the base and is perpendicular to the base, and then laid on a flat surface. The result is illustrated in Figure 7.

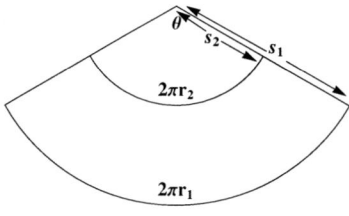

Figure 7

## 7.2. LENGTH AND AREA

With reference to Figure 7, the area of the circular sector of radius $s_1$ that subtends the angle $\theta$ (in radians) at the origin is
$$\frac{1}{2}s_1^2\theta,$$
and the area of the sector of radius $s_2$ that subtends the same angle $\theta$ at the origin is
$$\frac{1}{2}s_2^2\theta.$$
The angle $\theta$ can be expressed as the ratio of arc length to radius. Therefore,
$$\theta = \frac{2\pi r_1}{s_1} = \frac{2\pi r_2}{s_2}$$
Thus, the area of the surface is
$$\frac{1}{2}s_1^2\theta - \frac{1}{2}s_2^2\theta = \frac{1}{2}s_1^2\left(\frac{2\pi r_1}{s_1}\right) - \frac{1}{2}s_2^2\left(\frac{2\pi r_2}{s_2}\right) = \pi r_1 s_1 - \pi r_2 s_2.$$
Since
$$\theta = \frac{2\pi r_1}{s_1} = \frac{2\pi r_2}{s_2},$$
we have $r_1 s_2 = r_2 s_1$. Therefore, the area of the frustum of the cone can be expressed as
$$\pi r_1 s_1 - \pi r_2 s_2 = \pi(r_1 + r_2)(s_1 - s_2) = 2\pi\left(\frac{r_1 + r_2}{2}\right)(s_1 - s_2) = 2\pi\left(\frac{r_1 + r_2}{2}\right)s.$$

■

Now we will derive an expression for the area of a surface of revolution. Assume that $f$ is continuous on $[a,b]$ and that $f'$ has a continuous extension to $[a,b]$. Consider the surface $S$ that is formed by revolving the graph of $z = f(x)$ on $[a,b]$ about the $z$-axis. Let's subdivide the interval $[a,b]$ into subintervals $[x_{k-1}, x_k]$, $k = 1, 2, \ldots, n$. We will approximate the area of the part of the surface which lies between $x = x_{k-1}$ and $x = x_k$ by the area of the surface which is generated by revolving the line segment joining the points $(x_{k-1}, f(x_{k-1}))$ and $(x_k, f(x_k))$ about the $z$-axis. The approximating surface is a frustum of a cone such that the average radius is
$$\frac{f(x_{k-1}) + f(x_k)}{2}$$
and the slant height is the length of the line segment that joins the points $(x_{k-1}, f(x_{k-1}))$ and $(x_k, f(x_k))$, i.e.,
$$\sqrt{(\Delta x_k)^2 + (f(x_k) - f(x_{k-1}))^2}.$$

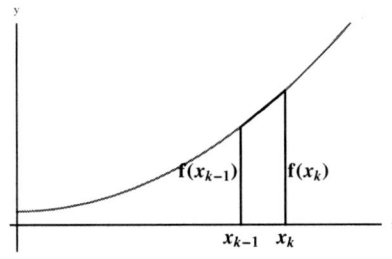

Figure 8

By Lemma 1 the area of this surface is

$$2\pi \left( \frac{f(x_{k-1}) + f(x_k)}{2} \right) \sqrt{(\Delta x_k)^2 + (f(x_k) - f(x_{k-1}))^2}.$$

By the Mean Value Theorem,

$$f(x_k) - f(x_{k-1}) = f'(x_k^*)(x_k - x_{k-1}) = f'(x_k^*)\Delta x_k,$$

for some $x_k^* \in (x_{k-1}, x_k)$. Therefore,

$$2\pi \left( \frac{f(x_{k-1}) + f(x_k)}{2} \right) \sqrt{(\Delta x_k)^2 + (f(x_k) - f(x_{k-1}))^2} = 2\pi \left( \frac{f(x_{k-1}) + f(x_k)}{2} \right) \sqrt{1 + \left(f'(x_k^*)^2\right)} \Delta x_k.$$

We approximate the total area of the surface by the sum of the areas of the approximating surfaces:

$$\sum_{k=1}^{n} 2\pi \left( \frac{f(x_{k-1}) + f(x_k)}{2} \right) \sqrt{1 + \left(f'(x_k^*)^2\right)} \Delta x_k.$$

If each $\Delta x_k$ is small,

$$\frac{f(x_{k-1}) + f(x_k)}{2} \cong f(x_k), \text{ and } f'(x_k^*) \cong f(x_k),$$

so that

$$\sum_{k=1}^{n} 2\pi \left( \frac{f(x_{k-1}) + f(x_k)}{2} \right) \sqrt{1 + \left(f'(x_k^*)^2\right)} \Delta x_k \cong \sum_{k=1}^{n} 2\pi \left( \frac{f(x_k) + f(x_k)}{2} \right) \sqrt{1 + \left(f'(x_k)^2\right)} \Delta x_k$$

$$= \sum_{k=1}^{n} 2\pi f(x_k) \sqrt{1 + \left(f'(x_k)^2\right)} \Delta x_k.$$

The last sum is a Riemann sum for

$$\int_a^b 2\pi f(x) \sqrt{1 + (f'(x))^2} dx.$$

We will use the above formula in order to calculate the area of the surface that is obtained by revolving the graph of $f$ on the interval $[a, b]$ about the $x$-axis.

**Definition 2** The area of the surface of revolution that is obtained by revolving the graph of $z = f(x)$ on the interval $[a.b]$ about the $x$-axis is defined as

$$\int_a^b 2\pi f(x) \sqrt{1 + (f'(x))^2} dx$$

if $f'$ is continuous on $[a, b]$.

**Example 4** Let $f(x) = x^2$.

Let $S$ be the surface that is obtained by revolving the graph of $z = f(x)$ corresponding to the interval $[1, 2]$ about the $x$-axis. Figure 9 illustrates the surface $S$.

## 7.2. LENGTH AND AREA

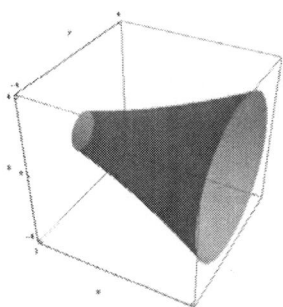

Figure 9

a) Express the area of $S$ as an integral.
b) Use the numerical integrator of your computational utility to approximate the integral of part a).

**Solution**

a)
We apply the formula
$$\int_a^b 2\pi f(x)\sqrt{1+(f'(x))^2}\,dx$$
for the surface area of $S$, with $f(x)=x^2$. Thus, the area of $S$ is
$$\int_1^2 2\pi x^2 \sqrt{1+\left(\frac{d}{dx}(x^2)\right)^2}\,dx = \int_1^2 2\pi x^2 \sqrt{1+4x^2}\,dx.$$

b)
$$\int_1^2 2\pi x^2 \sqrt{1+4x^2}\,dx \cong 49.4162$$

Note that you can compute the above integral exactly. Indeed, you can make use of the substitution $2x=\sinh(u)$ to derive the formula
$$\int 2\pi x^2 \sqrt{1+4x^2}\,dx = 2\pi\left(\frac{1}{16}x(1+4x^2)^{\frac{3}{2}} - \frac{1}{32}x\sqrt{(1+4x^2)} - \frac{1}{64}\operatorname{arcsinh}(2x)\right).$$

Thus,
$$\int 2\pi x^2 \sqrt{1+4x^2}\,dx = \frac{33}{8}\pi\sqrt{17} + \frac{1}{32}\pi\ln\left(-4+\sqrt{17}\right) - \frac{9}{16}\pi\sqrt{5} + \frac{1}{32}\pi\ln\left(2+\sqrt{5}\right)$$
$$\cong 49.416\,2.$$

□

**Remark** If we set $z=f(x)$, we can express the formula
$$\int_a^b 2\pi f(x)\sqrt{1+(f'(x))^2}\,dx$$
as
$$\int_a^b 2\pi z\sqrt{1+\left(\frac{dz}{dx}\right)^2}\,dx.$$

As in the case of the length of the graph of a function, we can use the formalism,

$$dz = \left(\frac{dz}{dx}\right) dx, \ (dz)^2 = \left(\frac{dz}{dx}\right)^2 (dx)^2 \ \text{and} \ ds = \sqrt{(dx)^2 + (dz)^2},$$

and express surface area as

$$\int_a^b 2\pi z \sqrt{(dx)^2 + (dz)^2} dx = \int_a^b 2\pi z \, ds.$$

The above expression does not offer any practical advantage over the previous expression for the surface area, except that you may find it to be convenient in order to remember the approximation procedure that led the computation of the surface area: The area of the "infinitesimal slice" of the surface between $x$ and $x + dx$ is approximately the area of the frustum of a cone with slant height $ds$. This area is approximately $2\pi z ds$ since $z$ can be considered to be the average radius if $dx$ is small. ◊

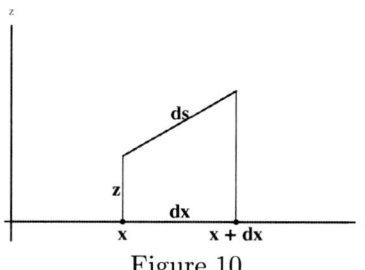

Figure 10

**Example 5** Let $f(x) = \sin(x)$ and let $S$ be the surface that is obtained by revolving the graph of $z = f(x)$ corresponding to the interval $[0, \pi/2]$ about the $x$-axis. Determine the area of $S$.

**Solution**

Figure 11 illustrates $S$.

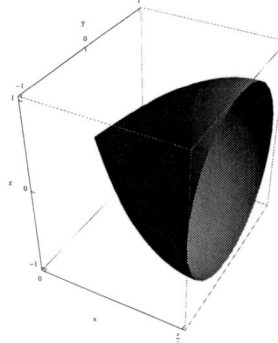

Figure 11

We set $z = \sin(x)$ so that

$$dz = \frac{dz}{dx} dx = \cos(x) \, dx \Rightarrow (dz)^2 = \cos^2(x)(dx)^2.$$

## 7.2. LENGTH AND AREA

Thus,
$$ds = \sqrt{(dx)^2 + (dz)^2} = \sqrt{(dx)^2 + \cos^2(x)(dx)^2} = \sqrt{1+\cos^2(x)}dx$$

Therefore, the area of $S$ is
$$\int_0^{\pi/2} 2\pi z \, ds = \int_0^{\pi/2} 2\pi \sin(x) \sqrt{1+\cos^2(x)} dx.$$

We set $u = \cos(x)$ so that $du = -\sin(x) dx$. Thus,
$$\int_0^{\pi/2} 2\pi \sin(x) \sqrt{1+\cos^2(x)} dx = -2\pi \int_{\cos(0)}^{\cos(\pi/2)} \sqrt{1+u^2} du = 2\pi \int_0^1 \sqrt{1+u^2} du.$$

We have evaluated such an integral in Example 4 of Section 7.4:
$$\int \sqrt{1+u^2} du = \frac{u}{2}\sqrt{u^2+1} + \frac{1}{2}\operatorname{arcsinh}(u).$$

Therefore, the area of $S$ is
$$2\pi \int_0^1 \sqrt{1+u^2} du = 2\pi \left( \frac{u}{2}\sqrt{u^2+1} + \frac{1}{2}\operatorname{arcsinh}(u) \Big|_0^1 \right) = 2\pi \left( \frac{\sqrt{2}}{2} + \frac{1}{2}\operatorname{arcsinh}(1) \right)$$
$$= \sqrt{2}\pi + \pi \ln\left(1+\sqrt{2}\right).$$

□

Similarly, if the surface $S$ is obtained by revolving the graph of $z = f(x)$ on the interval $[a,b]$ about the $z$-axis, we calculate its area by the formula
$$\int_a^b 2\pi x \, ds = \int_a^b 2\pi x \sqrt{1+\left(\frac{dz}{dx}\right)^2} dx.$$

You can think of the expression $2\pi x \, ds$ as the area of a frustum of a cone with average radius almost $x$ ($dx$ is small) and slant height $ds$.

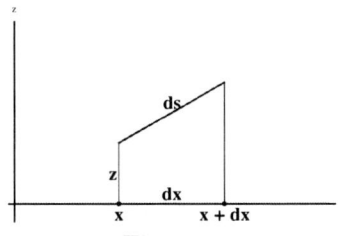

Figure 11

**Example 6** Assume that $S$ is the surface that is obtained by revolving the graph of $z = e^{-x^2}$ on the interval $[0,1]$ about the $z$-axis. Figure 12 shows $S$.

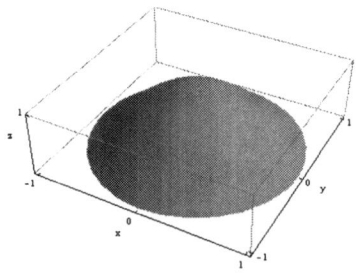

Figure 12

a) Express the area of $S$ as an integral.
b) Make use of the numerical integrator of your computational utility to approximate the area of the surface.

**Solution**

a) The area of the surface is

$$\int_0^1 2\pi x \sqrt{1 + \left(\frac{dz}{dx}\right)^2}\, dx = \int_0^1 2\pi x \sqrt{1 + \left(\frac{d}{dx}e^{-x^2}\right)^2}\, dx$$
$$= \int_0^1 2\pi x \sqrt{1 + \left(-2xe^{-x^2}\right)^2}\, dx$$
$$= \int_0^1 2\pi x \sqrt{1 + 4x^2 e^{-2x^2}}\, dx$$

b) The above integral cannot be expressed in terms of the values of familiar functions, but we can obtain an approximate value with the help of a computational utility. We have

$$\int_0^1 2\pi x \sqrt{1 + 4x^2 e^{-2x^2}}\, dx \cong 3.960\,25,$$

rounded to 6 significant digits. $\square$

## Problems

In problems 1-4,
a) Express the length of the graph of $f$ on the given interval as an integral. Do not attempt to evaluate the integral.
b) [C] Make use of your computational utility in order to determine an approximation to the integral of part a) (display 6 siginificant digits).
c) [CAS] Make use of your computer algebra system to determine an antiderivative for the integral of part a). Do you recogize the special functions that appear in the resulting expression? Evaluate the integral.

1.
$$f(x) = \sqrt{x^2 - 4},\ [3, 4]$$

2.
$$f(x) = \frac{2}{3}\sqrt{9 - x^2},\ [-3, 3]$$

3.
$$f(x) = \sin(x),\ [0, \pi/2]$$

4.
$$f(x) = \arctan(x),\ [1, 2].$$

In problems 5 and 6, determine the **length** of the graph of $f$ on the given interval:

**5.**
$$f(x) = x^{2/3},\ [8, 27]$$

**6.**
$$f(x) = \cosh(x),\ [0, 1]$$

In problems 7 and 8, express the **area** of the surface that is obtained by revolving the graph of $z = f(x)$ corresponding to given interval about the $x$-axis as an integral. **Do not attempt to evaluate the integral.**

**7.**
$$f(x) = x^2 + 4,\ [1, 2]$$

**8.**
$$f(x) = e^x,\ [0, 1].$$

In problems 9 and 10, determine the **area** of the surface that is obtained by revolving the graph of $z = f(x)$ corresponding to given interval about the $x$-axis:

**9.**
$$f(x) = \sqrt{1 + 4x},\ [1, 25]$$

**10.**
$$f(x) = \cosh(x),\ [0, 1].$$

In problems 11 and 12, express the **area** of the surface that is obtained by revolving the graph of $z = f(x)$ corresponding to given interval about the $z$-axis.as an integral. **Do not attempt to evaluate the integral.**

**11.**
$$f(x) = x^2 + 4,\ [1, 2]$$

**12.**
$$f(x) = e^{-x},\ [0, 1].$$

In problems 13 and 14, determine the **area** of the surface that is obtained by revolving the graph of $z = f(x)$ corresponding to given interval about the $z$-axis:

**13.**
$$f(x) = x^{1/3},\ [1, 2]$$

**14.**
$$f(x) = \sqrt{1 - x^2},\ [0, 1].$$

## 7.3 Some Physical Applications of the Integral

In this section we will discuss some applications of the integral to physics within contexts such as **mass**, **work** and **pressure**. The modeling aspects must be studied in detail in a physics course. Therefore our treatment is not deep, and our goal is modest: We will point out how the mathematical concept of the integral arises in some contexts other than area, volume or distance traveled. You may assume that metric units apply, even if that may not be stated explicitly in each case.

### Mass, Density and Center of Mass

Consider two objects $P_1$ and $P_2$ that are attached to a rod which is supported at the point $F$, as illustrated in Figure 1. We will refer to $F$ as the fulcrum. Let $d_1$ be the distance of $P_1$ from the fulcrum $F$ and let $d_2$ denote the distance of $P_2$ from $F$ (we will measure distance in meters).

Figure 1

Assume that the mass of $P_1$ is $m_1$ (kilograms) and the mass of $P_2$ is $m_2$. We will assume that the rod is light and neglect its mass. The **moment** of $P_1$ about the fulcrum $F$ is $m_1 d_1$ and the **moment** of $P_2$ about $F$ is $m_2 d_2$. The rod balances and stays horizontal if $m_1 d_1 = m_2 d_2$. In this case the support is at **the center of mass** of the two point masses. Let us place the rod along the $x$-axis. Assume that the object $P_1$ is at $x_1$, the object $P_2$ is at $x_2$ and the center of mass is at $\bar{x}$, as in Figure 2.

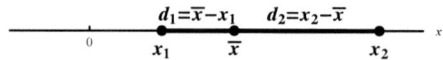

Figure 2

Thus, $d_1 = \bar{x} - x_1$ and $d_2 = x_2 - \bar{x}$. The equality $m_1 d_1 = m_2 d_2$ takes the form
$$m_1 (\bar{x} - x_1) = m_2 (x_2 - \bar{x}).$$
Therefore,
$$(m_1 + m_2) \bar{x} = m_1 x_1 + m_2 x_2 \Rightarrow \bar{x} = \frac{m_1 x_1 + m_2 x_2}{m_1 + m_2}.$$
The quantity $m_1 x_1$ is **the moment** of $P_1$ **about the origin**, and $m_2 x_2$ is the moment of $P_2$ about the origin. The above equality says that the center of mass is obtained by dividing the sum of the moments of the objects by the total mass of the objects. We can also say that the moment with respect to the origin of a single object whose mass is $m_1 + m_2$ is equal to the sum of the moments of $P_1$ and $P_2$. Similarly, if $n$ objects $P_1, P_2, \ldots, P_n$ have masses $m_1, m_2, \ldots, m_n$, and are placed at $x_1, x_2, \ldots, x_n$ respectively, the center of mass $\bar{x}$ is obtained by dividing the sum of the moments with respect to the origin by the sum of the masses of the objects. Thus,
$$\bar{x} = \frac{m_1 x_1 + m_2 x_2 + \cdots + m_n x_n}{m_1 + m_2 + \cdots + m_n} = \frac{\sum_{k=1}^{n} m_k x_k}{\sum_{k=1}^{n} m_k}.$$
As in the case of two objects, if we imagine that the objects are attached to a rod whose mass is negligible, the rod will stay horizontal if the fulcrum is placed at the center of mass $\bar{x}$.

Now assume that we have an object such as a metal rod that is modeled as the interval $[a, b]$ on the $x$-axis. We will not neglect the mass of the rod. In fact, we will assume that there is a continuous, positive-valued function $\rho(x)$ such that the mass of the rod can be obtained by integrating $\rho(x)$. Thus,
$$\text{Mass of the rod} = \int_a^b \rho(x)\, dx.$$
The function $\rho(x)$ is referred to as **mass density**. If we set
$$F(x) = \int_a^x \rho(u)\, du,$$
then $F(x)$ is the mass of the part of the rod corresponding to the interval $[a, x]$. By the Fundamental Theorem of Calculus,
$$\frac{dF}{dx} = \frac{d}{dx} \int_a^x \rho(u)\, du = \rho(x).$$

## 7.3. SOME PHYSICAL APPLICATIONS OF THE INTEGRAL

Therefore, **the mass density is the rate of change of mass with respect to $x$**. If $\Delta x$ is positive and small, then

$$F(x + \Delta x) - F(x) = \int_x^{x+\Delta x} \rho(u)\, du \cong \int_x^{x+\Delta x} \rho(x^*)\, du = \rho(x^*)\, \Delta x,$$

where $x^*$ is an arbitrary point in the interval $[x, x + \Delta x]$. Thus, the mass of a small piece of the rod can be assumed to have constant mass density, and the above approximation is consistent with the intuitive relationship between mass and mass density in the form

$$\text{mass} = (\text{mass density}) \times (\text{length}).$$

You can think of $\rho(x)\, dx$ as the mass of a piece of rod that has constant mass density $\rho(x)$ and "infinitesimal" length $dx$.

If mass is measured in kilograms and length is measured in meters, the unit of mass density is kilograms per meter. This is the case when we are considering a single dimension. In Calculus III we will consider the notions of mass and mass density in two and three dimensions.

We would like to arrive at the definition of **the center of mass of a rod** whose mass density is $\rho(x)$. Let us imagine that the rod is subdivided into small pieces corresponding to the partition $\{a = x_0, x_1, x_2, \ldots, x_{n-1}, x_n = b\}$ of the interval $[a, b]$. If $x_k^*$ is an arbitrary point in the $k$th subinterval $[x_{k-1}, x_k]$, we can assume that the part of the rod corresponding to $[x_{k-1}, x_k]$ is a point mass that is concentrated at $x_k^*$ has mass $\rho(x_k^*)\, \Delta x_k$, where $\Delta x_k = x_k - x_{k-1}$. The moment of this point mass with respect to the origin is $(\rho(x_k^*)\, \Delta x_k)\, x_k^*$. Therefore, the sum

$$\sum_{k=1}^{n} x_k^* \rho(x_k^*)\, \Delta x_k$$

can be considered to be an approximation to the moment of the rod with respect to the origin. The above sum is a Riemann sum that approximates the integral

$$\int_a^b x \rho(x)\, dx$$

with arbitrary accuracy, provided that the intervals are of sufficiently small length. Therefore, **we will define the moment of the rod with respect to the origin** to be

$$\int_a^b x \rho(x)\, dx.$$

You can think of $x\rho(x)\, dx$ as the moment with respect to the origin of a piece of rod whose center is at $x$, has constant mass density $\rho(x)$, and "infinitesimal" length $dx$.

**The center of mass $\bar{x}$** of the rod is defined to be the moment with respect to the origin divided by the mass, in analogy with the case of point masses. Thus,

$$\bar{x} = \frac{\int_a^b x \rho(x)\, dx}{\int_a^b \rho(x)\, dx}.$$

As in the case of point masses, if the rod is supported at the center of mass, it will be horizontal.

**Example 1** Assume that a metal rod of length 2 meters is placed along the $x$ axis and the left endpoint coincides with the origin. Calculate its center of mass if its mass density is $\sin(x) + 1$ kilograms per meter.

**Solution**

The mass of the rod is

$$\int_0^2 (\sin(x) + 1)\, dx = -\cos(x) + x \Big|_0^2 = -\cos(2) + 2 + \cos(0)$$
$$= 3 - \cos(2) \cong 3.416 \text{ (kilograms)}.$$

The moment of the rod with respect to the origin is

$$\int_0^2 x(\sin(x) + 1)\, dx = \int_0^2 (x\sin(x) + x)\, dx.$$

You can derive the expression

$$\int x\sin(x)\, dx = \sin(x) - x\cos(x)$$

as an exercise in integration by parts. Therefore,

$$\int_0^2 (x\sin(x) + x)\, dx = \sin(x) - x\cos(x) + x \Big|_0^2$$
$$= \sin(2) - 2\cos(2) + 2$$
$$\cong \sin(2) - 2\cos(2) + 2 \cong 3.74159 \text{ (kilogram} \times \text{meter)}$$

Thus, the center of mass is

$$\bar{x} = \frac{\int_0^2 x(\sin(x) + 1)\, dx}{\int_0^2 (\sin(x) + 1)\, dx} = \frac{\sin(2) - 2\cos(2) + 2}{3 - \cos(2)} \cong 1.09527 \text{ (meters)}.$$

□

## Work

Assume that an object is in one-dimensional motion. We model the object as a point on the number line that will be labeled as the $x$-axis. Assume that a constant force of $F$ newtons is acting on the object is along the $x$-axis. The sign of $F$ is nonnegative if the force is in the positive direction, and negative if it is in the negative direction. **The work done by the force as the object moves from the point $a$ to the point $b$ is $F(b-a)$, i.e., the product of force and displacement.** The work done against the force is $-F(b-a)$. In metric units, the unit of work is **newton×meter**, and is referred to as **joule**.

**Example 2** Assume that a suitcase of mass 20 kilograms is lifted vertically from a height of 0.5 meters to a height of 2 meters. Let's assume that gravitational acceleration is 9.8 meters/sec$^2$. The $x$-axis is placed vertically, and the positive direction is upward. The force acting on the suitcase is its weight, i.e.,

$$F = \text{mass} \times \text{gravitational acceleration} = 20 \times (-9.8) = -196 \text{ newtons}.$$

The sign is negative since the gravitational acceleration is downward. The work done by $F$ as the suitcase is lifted from 0.5 meters to 2 meters is

$$(-196) \times (2 - 0.5) = -196 \times 1.5 = -294 \text{ joules}.$$

The work done against $F$, i.e., against the weight of the suitcase is 294 joules. □

## 7.3. SOME PHYSICAL APPLICATIONS OF THE INTEGRAL

Now let's assume that the force acting on the object is not a constant but depends on the position of the object. Thus, assume that the force is the function $F$ that is not necessarily constant. How should we calculate the work done by the force as the object moves from $a$ to $b$? We will assume that the function $F$ is continuous and subdivide the interval $[a,b]$ to small subintervals by the partition $\{a = x_0, x_1, x_2, \ldots, x_{n-1}, x_n = b\}$. If $x_k^*$ is an arbitrary point in the interval $[x_{k-1}, x_k]$, we may treat the force as if it has the constant value $F(x_k^*)$ on $[x_{k-1}, x_k]$. Then, it is reasonable to set $F(x_k^*)(x_k - x_{k-1}) = F(x_k^*)\Delta x_k$ as an approximation to the work done by the force as the object moves from $x_{k-1}$ to $x_k$. Therefore, the sum,

$$\sum_{k=1}^{n} F(x_k^*) \Delta x_k$$

is an approximation to the work done by $F$ as the object moves from $a$ to $b$. Since the above sum is a Riemann sum for the integral

$$\int_a^b F(x)\,dx,$$

**we will calculate the work done by the force $F$ as the integral of $F$ from $a$ to $b$.** You can think of $F(x)dx$ as the work done by the constant force $F$ over an "infinitesimal" distance $dx$. The work done against the force is

$$-\int_a^b F(x)\,dx.$$

**Example 3 (Newton's Law of Gravitation)** Newton's law of gravitation provides an example where force varies with position. Let $M$ denote the mass of the earth (in kilograms), and let $m$ denote the mass of an object such as a rocket. We will assume that the object is modeled as a point that is in motion along a line that will be designated as the $x$-axis. We measure distances in meters. The origin coincides with the center of the earth and the positive direction is away from the center of the earth, as illustrated in Figure 3.

Figure 3

The force exerted on the object by the gravitational attraction of the earth when the object is at the point $x$ is

$$F(x) = -\frac{GMm}{x^2}.$$

We will use metric units, so that $F(x)$ is measured in newtons. Here, $G$ is the universal gravitational constant. The $(-)$ sign is due to the fact that $F(x)$ points towards the center of the earth. Note that $F(x)$ is the same as the force that treats the earth as a point mass at its origin. If $R$ is the radius of the earth, the force acting on the object when it is on the surface of the earth is

$$F(R) = -\frac{GMm}{R^2} = -\left(\frac{GM}{R^2}\right)m.$$

The weight of the object of mass $m$ (on the surface of the earth) is $-F(R)$ (newtons). If we set

$$k = \frac{GM}{R^2},$$

then $-F(R) = km$, and $k$ is approximately 9.8 meters/sec$^2$. Thus, $k$ is the gravitational acceleration on the surface of the earth.

Assume that the motion of the object starts on the surface of the earth, and that the object reaches a distance $r$ from the center of the earth. Thus, $x$ varies from $R$ to $r$. The work done by the earth's gravitational pull is

$$\int_R^r F(x)\,dx = \int_R^r -\frac{GMm}{x^2}\,dx = -GMm\int_R^r \frac{1}{x^2}\,dx = -GMm\left(-\frac{1}{x}\Big|_R^r\right) = -GMm\left(\frac{1}{R} - \frac{1}{r}\right).$$

Note that

$$r > R > 0 \Rightarrow \frac{1}{r} < \frac{1}{R} \Rightarrow \frac{1}{R} - \frac{1}{r} > 0,$$

so that work comes with a $(-)$ sign. The work done against earth's gravitational pull in taking the object to a distance $r > R$ from the center of the earth is

$$-\int_R^r F(x)\,dx = GMm\left(\frac{1}{R} - \frac{1}{r}\right) > 0.$$

**Example 4** We will follow up on Example 3 with some numerical data. Assume that the object in question is a rocket that has mass 1000 kilograms. Calculate the work done against the earth's gravitational pull in taking the rocket from the surface of the earth to a height of 100 kilometers above the earth. Assume that

$$\frac{GM}{R^2} = 9.8,$$

and that the radius of the earth is 6400 kilometers.

**Solution**

As in Example 3, the work done against the earth's gravitational pull is

$$GMm\left(\frac{1}{R} - \frac{1}{r}\right) = \frac{GMm}{R^2}\left(R - \frac{R^2}{r}\right),$$

where $m$ is the object in question, and $r$ is the point that the object reaches. We have

$$\frac{GMm}{R^2}\left(R - \frac{R^2}{r}\right) = \left(\frac{GM}{R^2}\right)m\left(R - \frac{R^2}{r}\right) = 9.8m\left(R - \frac{R^2}{r}\right).$$

The mass of the rocket is 1000 kilograms, $R = 6400 \times 1000 = 6.4 \times 10^6$ meters, and $r = (6400 + 100) \times 1000 = 6.5 \times 10^6$ meters. Therefore,

$$9.8m\left(R - \frac{R^2}{r}\right) = 9.8 \times 10^3\left(6.4 \times 10^6 - \frac{6.4^2 \times 10^{12}}{6.5 \times 10^6}\right) \cong 9.649\,23 \times 10^8 \text{ joules}.$$

This is the work done against earths gravitational pull in order to have the rocket reach a height of 100 kilometers above the surface of the earth. □

**Example 5 (The Elastic Spring)** Let us consider work within the framework of an elastic spring. Assume that a spring hangs vertically and that an object of mass $m$ is attached to the spring, as we discussed in Section 2.4. We place the $x$-axis vertically, the origin coincides with the object at equilibrium, and the positive direction is downward, as illustrated in Figure 4.

## 7.3. SOME PHYSICAL APPLICATIONS OF THE INTEGRAL

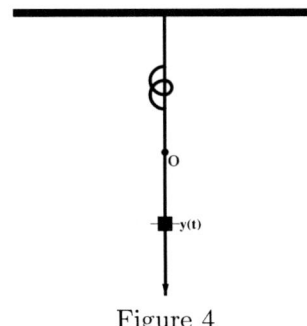

Figure 4

We assume that **Hooke's law** is valid: If the position of the object is $x$, the spring exerts the force $F(x) = -kx$, where $k > 0$ is **the spring constant**. Thus, $F(x)$ is negative if the spring is pulled downward the distance $x$ ( the spring exerts a force upward), and $F(x)$ is positive if the spring is pushed upward a distance $-x$ (the spring exerts a force downward). The work done by the force that is exerted by the spring as the object is moved from position $a$ to position $b$ is

$$\int_a^b -kx\,dx = -\frac{k}{2}x^2 \Big|_a^b = \frac{k}{2}\left(a^2 - b^2\right).$$

In particular, if the object attached to the spring is pulled from the position of equilibrium downward a distance $b$, the work done by the spring force is

$$\int_0^b -kx\,dx = -\frac{k}{2}b^2.$$

The work done against the spring is $kb^2/2$.

If the spring is compressed by an amount $|-b| = b$, where $b > 0$, the work done by the spring force is the same as above:

$$\int_0^{-b} -kx\,dx = -\frac{k}{2}b^2.$$

In either case, the work done against the spring is $kb^2/2$. □

**Example 6** With the notation of Example 5, assume that the spring is stretched a distance of 4 centimeters by a mass of 2 kilogram. Assume that gravitational acceleration is 9.8 meters/sec$^2$. Determine the work done against the spring in order to stretch it a distance of 8 centimeters from equilibrium.

**Solution**

The weight of the object of mass 2 kilograms is $9.8 \times 2 = 19.6$ newtons. Since a force of 19.6 newtons stretches the spring by 4 centimeters, i.e., 0.04 meter, by Hooke's law, the force exerted by the spring is

$$-19.6 = -k \times 0.04.$$

Therefore,

$$k = \frac{19.6}{0.04} = 490.$$

Therefore,

$$F(x) = -490x$$

is the force exerted by the spring when the object is at position $x$. The work done by the force of the spring as its stretched a distance of 8 centimeters, i.e., 0.08 meters from equilibrium is

$$\int_0^{0.08} F(x)\,dx = \int_0^{0.08} -490x\,dx = -490 \int_0^{0.08} x\,dx = -490 \left( \frac{x^2}{2} \Big|_0^{0.08} \right)$$

$$= -490 \left( \frac{64 \times 10^{-4}}{2} \right) = -\frac{196}{125} = -1.568$$

joules. The work done against the spring is 1.568 joules. $\square$

### Kinetic Energy and Potential Energy

As in our discussion of work, let's consider an object in one-dimensional motion. Let $F(x)$ be the net force that acts on the object if the position of the object corresponds to the point $x$. Let $x(t)$, $v(t)$ and $a(t)$ denote the position, velocity and acceleration of the object, respectively, at time $t$. By Newton's Second Law of Motion, $F(x(t)) = ma(t)$. Since acceleration is the rate of change of velocity, we have

$$F(x(t)) = m\frac{dv(t)}{dt}.$$

Therefore,

$$F(x(t))\frac{dx(t)}{dt} = m\frac{dv(t)}{dt}\frac{dx(t)}{dt} = m\frac{dv(t)}{dt}v(t) = \frac{d}{dt}\left(\frac{1}{2}mv^2(t)\right).$$

We integrate from some instant $t_0$ to another instant $t_1$:

$$\int_{t_0}^{t_1} F(x(t))\frac{dx(t)}{dt}\,dt = \int_{t_0}^{t_1} \frac{d}{dt}\left(\frac{1}{2}mv^2(t)\right)\,dt.$$

By the substitution rule,

$$\int_{t_0}^{t_1} F(x(t))\frac{dx(t)}{dt}\,dt = \int_{x(t_0)}^{x(t_1)} F(x)\,dx.$$

By the Fundamental Theorem of Calculus,

$$\int_{t_0}^{t_1} \frac{d}{dt}\left(\frac{1}{2}mv^2(t)\right)\,dt = \frac{1}{2}mv^2(t_1) - \frac{1}{2}mv^2(t_0).$$

Therefore,

$$\int_{x(t_0)}^{x(t_1)} F(x)\,dx = \frac{1}{2}mv^2(t_1) - \frac{1}{2}mv^2(t_0)$$

Note that the expression on the left-hand side of the above equality is the work done by $F$ as the object moves from $x(t_0)$ to $x(t_1)$. The **kinetic energy** of the object at the instant $t$ is defined as

$$\frac{1}{2}mv^2(t).$$

Thus, the expression on the right-hand side of the above equality is the difference between the kinetic energy of the object at time $t_1$ and and at time $t_0$. Therefore, **the work done by the resultant of all the forces on the object as the object moves from $x(t_0)$ to $x(t_1)$ is equal to the change in the kinetic energy of the object.**

## 7.3. SOME PHYSICAL APPLICATIONS OF THE INTEGRAL

Assume that $-V$ **is an antiderivative of** $F$, so that

$$F(x) = -\frac{dV(x)}{dx}$$

(for each $x$ in some interval). In this case, $V(x)$ may referred to as **the potential energy** of the object at the position $x$ and the function $V$ may be referred to as a potential function for the force function $F$ (some people may refer to $-V$ as the potential for $F$). Since an antiderivative of $F$ is not unique, the potential function corresponding to the same force function may be different in different accounts. Nevertheless, any two potentials corresponding to the same force function $F$ can differ at most be an additive constant, as antiderivatives of the same function. By the Fundamental Theorem of Calculus,

$$\int_{x(t_0)}^{x(t_1)} F(x)\,dx = \int_{x(t_0)}^{x(t_1)} -\frac{dV(x)}{dx}\,dx = -\int_{x(t_0)}^{x(t_1)} \frac{dV(x)}{dx}\,dx$$

$$= \int_{x(t_1)}^{x(t_0)} \frac{dV(x)}{dx}\,dx = V(x(t_0)) - V(x(t_1)).$$

We have derived the equality

$$\int_{x(t_0)}^{x(t_1)} F(x)\,dx = \frac{1}{2}mv^2(t_1) - \frac{1}{2}mv^2(t_0).$$

Therefore,

$$\frac{1}{2}mv^2(t_1) - \frac{1}{2}mv^2(t_0) = V(x(t_0)) - V(x(t_1))$$

Thus,

$$\frac{1}{2}mv^2(t_0) + V(x(t_0)) = \frac{1}{2}mv^2(t_1) + V(x(t_1)).$$

**The total energy is the sum of the kinetic energy and the potential energy:**

$$E(t) = \frac{1}{2}mv^2(t) + V(x(t)).$$

The above equality says that **the total energy is conserved if there is a potential function for the force function:**

$$E(t_1) = \frac{1}{2}mv^2(t_1) + V(x(t_1)) = \frac{1}{2}mv^2(t_0) + V(x(t_0)) = E(t_0).$$

**Example 7** With reference to Newton's law of gravitation, as in Example 3, we will consider motion that is close to the surface and assume that the gravitational acceleration is the constant $g$ (meters/second$^2$). Let the $x$-axis be vertical and let the positive direction be upwards. Therefore, the force $F$ due to gravity that acts on a body of mass $m$ (kilograms), i.e., the weight of the object, is $-mg$ (newtons = kilogram meter / second$^2$). Since a potential $V$ corresponding to $F$ is an antiderivative of $-F$, we have

$$V(x) = \int -F(x)\,dx = \int mg\,dx = mgx + C,$$

where $C$ is an arbitrary constant. Let us set $C = 0$, so that the potential energy of an object at height 0 is 0. The total energy at time $t$ is

$$E(t) = \text{Kinetic Energy} + \text{Potential Energy} = \frac{1}{2}mv^2(t) + V(x(t)) = \frac{1}{2}mv^2(t) + mgx(t).$$

Confirm that total energy is conserved.

**Solution**

$$\frac{d}{dt}E(t) = \frac{d}{dt}\left(\frac{1}{2}mv(t)^2 + mgx\right) = mv(t)\frac{dv(t)}{dt} + mg\frac{d}{dt}x(t) = mv(t)a(t) + mgv(t).$$

We have
$$ma(t) = F = -mg.$$

Therefore,
$$\frac{d}{dt}E(t) = mv(t)a(t) + mgv(t) = mv(t)(-mg) + mgv(t) = 0.$$

Since a function whose derivative is 0 on an interval is constant on that interval, $E(t)$ is a constant for each $t$. $\square$

**Example 8** The setting is **Newton's law of gravitation**, as in Example 3. Thus,
$$F(x) = -\frac{GMm}{x^2}.$$

Then
$$V(x) = \int -F(x)\,dx = GMm\int \frac{1}{x^2}dx = -\frac{GMm}{x} + C.$$

We may set $C = 0$, so that $\lim_{x\to+\infty} V(x) = 0$. Confirm that total energy is conserved.

**Solution**

The total energy at time $t$ is
$$E(t) = \frac{1}{2}mv^2(t) + V(x(t)) = \frac{1}{2}mv^2(t) - \frac{GMm}{x(t)}.$$

With the help of the chain rule,
$$\frac{dE(t)}{dt} = \frac{d}{dt}\left(\frac{1}{2}mv^2(t) - \frac{GMm}{x(t)}\right) = mv(t)\left(\frac{d}{dt}v(t)\right) + \frac{GMm}{x^2(t)}\left(\frac{dx(t)}{dt}\right)$$
$$= mv(t)a(t) + \frac{GMm}{x^2(t)}v(t) = v(t)\left(ma(t) + \frac{GMm}{x^2(t)}\right).$$

By Newton's second law of motion,
$$ma(t) = F(x(t)) = -\frac{GMm}{x^2(t)}.$$

Thus,
$$\frac{dE(t)}{dt} = v(t)\left(ma(t) + \frac{GMm}{x^2(t)}\right) = v(t)\left(-\frac{GMm}{x^2(t)} + \frac{GMm}{x^2(t)}\right) = 0.$$

Therefore, the energy is a conserved quantity. $\square$

**Example 9** The setting is the case of an **elastic spring**, as in Example 5. The force exerted by the spring is $F(x) = -kx$ if the position of the attached object corresponds to $x$. Since a potential $V$ is an antiderivative of $F$,
$$V(x) = \int -F(x)\,dx = \int kx\,dx = \frac{k}{2}x^2 + C.$$

Let us set $C = 0$. The total energy at time $t$ is
$$E(t) = \frac{1}{2}mv^2(t) + \frac{k}{2}x^2(t).$$

Confirm that total energy is conserved.

## Solution

With the help of the chain rule,

$$\frac{dE(t)}{dt} = mv(t)\frac{dv(t)}{dt} + kx(t)\frac{dx(t)}{dt} = mv(t)a(t) + kx(t)v(t).$$

By Newton's second law of motion,

$$ma(t) = F(x(t)) = -kx(t).$$

Therefore,

$$\frac{dE(t)}{dt} = v(t)ma(t) + kx(t)v(t) = v(t)(-kx(t)) + v(t)kx(t) = 0.$$

Thus, $E$ is a constant. $\square$

## Problems

In problems 1 and 2, assume that a metal rod of length $L$ meters is placed along the $x$ axis and the left endpoint coincides with the point $a$. Calculate its mass if its mass density is $f(x)$ kilograms per meter.

**1.**
$$L = \pi/2, \ a = 3, \ f(x) = \cos^2(x-3).$$

**2.**
$$L = 0.5, \ a = 0, \ f(x) = \sqrt{x^2 + 1}.$$

In problems 3 and 4, an object that is subjected to the force $F(x)$ moves from the point $a$ to the point $b$ along a line that coincides with the $x$-axis. Calculate the work done by $F$.

**3.**
$$F(x) = -\frac{1}{4}x, \ a = 3, \ b = 5.$$

**4.**
$$F(x) = \sin(4x), \ a = 0, \ b = \pi/8.$$

**5.** Assume that the mass of a suitcase is 25 kilograms is lifted vertically, and gravitational acceleration is 9.8 meters/sec$^2$. Calculate the work done against gravity when the suitcase is lifted from a height of 50 centimeters to 2 meters.

**6.** Assume that Hooke's law is valid for a spring that is stretched a distance of 3 centimeters by a mass of 4 kilogram and that gravitational acceleration is 9.8 meters/sec$^2$. Determine the work done against the spring in order to stretch it a distance of 12 centimeters from equilibrium.

## 7.4 The Integral and Probability

The mathematical theory of probability is a vast subject. You will learn about certain aspects of probability in a course other than calculus, depending on your field of study. You cannot learn about probability in a single section of a calculus book. This section has the modest goal of indicating how integrals arise in probability. In particular, you will see that improper integrals on unbounded intervals are indispensable in probability.

Let's begin with two specific examples. These examples will serve as case studies for the illustration of the general concepts that will be discussed in this section.

**Example 1** Assume that a certain company manufactures hard disks for personal computers. Each hard disk has a certain life span $X$ that cannot be predicted exactly. On the other hand, we may have an idea about the probability that $X$ is between 2 and 2.5 years, for example. □

**Example 2** Assume that all students in the 10th grade in California take a standard Math test with a total score of 800. We may not predict what an individual student will score exactly. On the other hand, we may have some idea based on past data as to whether the probability that the student's score $X$ will be between 600 and 700, for example. □

The quantity $X$ in each of the above cases is an example of a **random variable**: Even though we cannot know the value of $X$ that corresponds to a particular outcome of an experiment exactly, we have knowledge about the probability that the value of $X$ is in some interval. In Example 1, the experiment consists of the selection of a hard disk manufactured by the company at random. The corresponding value of the random variable $X$ is the life span of the selected hard disk. In Example 2, the experiment consists of the selection of a 10th-grade student in California at random. The corresponding value of the random variable $X$ is the test score of the selected student.

We will denote the probability that the random variable $X$ has values between $a$ and $b$ as $P(a < X < b)$. The probability that the value of $X$ is less than $b$ will be denoted by $P(X < b)$, and the probability that the value of $X$ is greater than $b$ is denoted by $P(X > b)$.

**Definition 1** The continuous function $f$ is the **probability density function** of the random variable $X$ if
$$P(a < X < b) = \int_a^b f(x)\,dx.$$

Note that $f(x) \geq 0$ for each $x$,
$$P(-\infty < X < +\infty) = \int_{-\infty}^{+\infty} f(x)\,dx = 1$$

and
$$P(X = a) = \int_a^a f(x)\,dx = 0.$$

Thus, if a random variable $X$ has a continuous probability density function the probability that $X$ attains a specific value $a$ is zero. In this section we will discuss only those random variables that have continuous probability density functions.

**Example 3** With reference to Example 1, let's assume that the life span of a hard disk can be any number greater than 0, even though there is an upper bound such as 20 years on the life span, and that
$$P(a < X < b) = \int_a^b \frac{1}{3} e^{-x/3}\,dx$$

if $0 \leq a < b$. We must set $P(a < X < b) = 0$ if $b < 0$. Thus,
$$P(a < X < b) = \int_a^b f(x)\,dx,$$

where
$$f(x) = \begin{cases} 0 & \text{if } x < 0, \\ \frac{1}{3} e^{-x/3} & \text{if } x \geq 0. \end{cases}$$

Therefore, $f$ is the probability density function of the random variable of Example 1. Figure 1 shows the graph of $f$.

## 7.4. THE INTEGRAL AND PROBABILITY

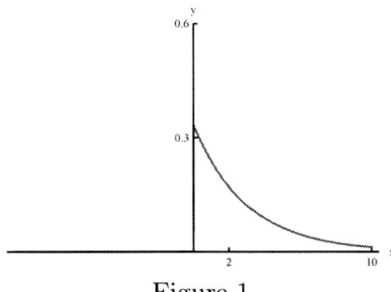

Figure 1

For example,
$$P(0 < X < 2) = \int_0^2 \frac{1}{3} e^{-x/3} dx = -e^{-\frac{2}{3}} + 1 \cong 0.486\,583.$$

This is the area between the graph of $f$ and the interval $[0, 2]$.

We can consider the probability that $X$ has any value greater than 4, for example. In this case the probability is calculated as an improper integral:

$$P(X > 4) = \lim_{b \to +\infty} P(4 < X < b) = \lim_{b \to +\infty} \int_4^b f(x)\, dx = \int_4^{+\infty} f(x)\, dx.$$

We have

$$\lim_{b \to +\infty} \int_4^b f(x)\, dx = \lim_{b \to +\infty} \int_4^b \frac{1}{3} e^{-x/3} dx = \lim_{b \to +\infty} \left( -e^{-x/3} \Big|_4^b \right)$$
$$= \lim_{b \to +\infty} \left( -e^{-b/3} + e^{-4/3} \right) = e^{-4/3} \cong 0.263\,597.$$

The probability that $X$ takes on an arbitrary value should be 1. Indeed,

$$P(-\infty < X < +\infty) = \int_{-\infty}^{+\infty} f(x)\, dx = \int_0^{\infty} \frac{1}{3} e^{-x/3} dx$$
$$= \lim_{b \to +\infty} \left( -e^{-x/3} \Big|_0^b \right) = \lim_{b \to +\infty} \left( -e^{-b/3} + 1 \right) = 1.$$

The above example illustrates an **exponential probability density function**: A function of the form
$$f(x) = \begin{cases} 0 & \text{if } x < 0, \\ \frac{1}{a} e^{-x/a} & \text{if } x \geq 0, \end{cases}$$

where $a$ is a positive constant, is an exponential probability density function. If $X$ is a random variable with the probability density function $f$, we must have $P(-\infty < X < +\infty) = 1$. Indeed,

$$P(-\infty < X < +\infty) = \int_{-\infty}^{+\infty} f(x)\, dx = \int_0^{+\infty} \frac{1}{a} e^{-x/a} dx$$
$$= \lim_{b \to +\infty} \left( -e^{-x/a} \Big|_0^b \right) = \lim_{b \to +\infty} \left( -e^{-b/a} + 1 \right) = 1.$$

□

**Example 4** With reference to Example 2, we will make the simplifying assumptions that an exam score can have any value whatsoever, even though the actual values are integers between 0 and 800, and that

$$P(a < X < b) = \frac{1}{100\sqrt{2\pi}} \int_a^b e^{-(x-600)^2/2(100)^2} dx.$$

Thus, the probability density function of the random variable $X$ is

$$f(x) = \frac{1}{100\sqrt{2\pi}} e^{-(x-600)^2/2(100)^2}.$$

Figure 2 shows the graph of $f$.

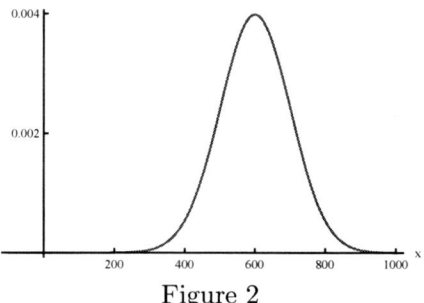

Figure 2

For example, the probability that the exam score of a student is between 500 and 700 is
$f(x) = \frac{1}{\sigma\sqrt{2\pi}} e^{-(x-\mu)^2/2\sigma^2}$

$$P(500 < X < 700) = \int_{500}^{700} f(x)\, dx = \frac{1}{100\sqrt{2\pi}} \int_{500}^{700} e^{-(x-600)^2/2(100)^2} dx.$$

The integral can be expressed in terms of the error function

$$\text{erf}(x) = \int_0^x \frac{2}{\sqrt{\pi}} e^{-t^2} dt$$

that was discussed in Section 5.4. We have

$$\frac{1}{100\sqrt{2\pi}} \int_{500}^{700} e^{-(x-600)^2/2(100)^2} dx = \text{erf}\left(\frac{1}{2}\sqrt{2}\right)$$

(confirm as an exercise). Even if you don't have access to a CAS, you can obtain an approximation with the help of the numerical integrator of our calculator:

$$\frac{1}{100\sqrt{2\pi}} \int_{500}^{700} e^{-(x-600)^2/2(100)^2} dx \cong 0.682\,689$$

The probability that $X$ has an arbitrary value is 1. Indeed,

$$\int_{-\infty}^{+\infty} f(x)\, dx = \frac{1}{100\sqrt{2\pi}} \int_{-\infty}^{+\infty} e^{-(x-600)^2/2(100)^2} dx = 1,$$

with the help of a CAS. You can obtain an approximation with the numerical integrator of your calculator that should indicate that the integral is 1. □

## 7.4. THE INTEGRAL AND PROBABILITY

**Definition 2 The distribution function** $F$ corresponding to a random variable $X$ is

$$F(x) = P(X \leq x) \text{ for each } x \in \mathbb{R}.$$

Thus, if $X$ has the probability density $f$, we have

$$F(x) = \int_{-\infty}^{x} f(u)\, du.$$

Thus, $F(x)$ corresponds to the area between the graph of the probability density function $f$ and the interval $(-\infty, a]$, as illustrated in Figure 3.

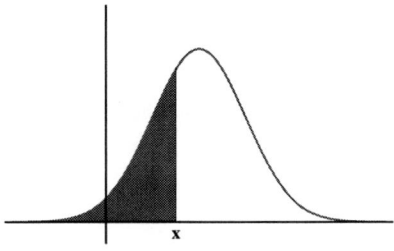

Figure 3

Since we are considering continuous probability density functions, we have

$$F'(x) = f(x).$$

Indeed, if we fix $a < x$, we have

$$F(x) - F(a) = \int_{-\infty}^{x} f(u)\, du - \int_{-\infty}^{a} f(u)\, du = \int_{a}^{x} f(u)\, du.$$

Therefore,

$$F'(x) = \frac{d}{dx}(F(x) - F(a)) = \frac{d}{dx}\int_{a}^{x} f(u)\, du = f(x),$$

by the Fundamental Theorem of Calculus.

A distribution function $F$ is nondecreasing. Indeed, if $b > a$,

$$F(b) - F(a) = \int_{a}^{b} f(x)\, dx \geq 0,$$

since $f(x) \geq 0$ for each $x \in \mathbb{R}$. Furthermore,

$$\lim_{x \to -\infty} F(x) = \lim_{x \to -\infty} \int_{-\infty}^{x} f(u)\, du = 0,$$

and

$$\lim_{x \to +\infty} F(x) = \lim_{x \to +\infty} \int_{-\infty}^{x} f(u)\, du = \int_{-\infty}^{+\infty} f(x)\, dx = 1.$$

Figure 4 shows the graph of a typical distribution function.

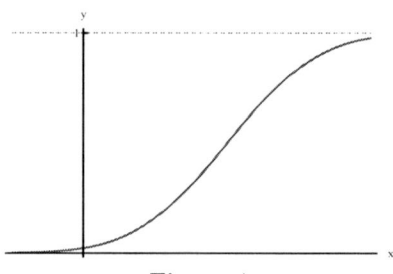

Figure 4

**Example 5** Let
$$f(x) = \begin{cases} 0 & \text{if } x < 0, \\ \frac{1}{3}e^{-x/3} & \text{if } x \geq 0, \end{cases}$$
be the probability density function of the random variable of Example 3.
a) Determine the distribution $F$ of the random variable $X$. Sketch the graph of $F$.
b) Confirm that $\lim_{x \to -\infty} F(x) = 0$ and $\lim_{x \to +\infty} F(x) = 1$.

**Solution**

a) If $x \leq 0$, then
$$F(x) = \int_{-\infty}^{x} f(u)\, du = \int_{0}^{x} 0\, dx = 0,$$
since $f(u) = 0$ if $u \leq 0$. If $x \geq 0$, then
$$F(x) = \int_{-\infty}^{x} f(u)\, du = \int_{0}^{x} \frac{1}{3} e^{-u/3}\, du = \left(-e^{-u/3}\Big|_{0}^{x}\right) = -e^{-x/3} + 1.$$

Thus,
$$F(x) = \begin{cases} 0 & \text{if } x < 0, \\ 1 - e^{-x/3} & \text{if } x \geq 0. \end{cases}$$

Figure 5 shows the graph of $F$.

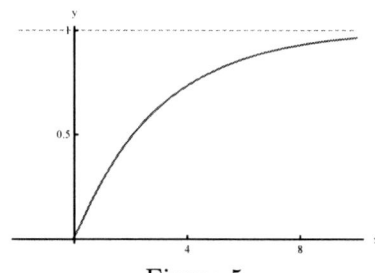

Figure 5

b) Since $F(x) = 0$ if $x \leq 0$, we have
$$\lim_{x \to -\infty} F(x) = \lim_{x \to -\infty} (0) = 0.$$

On the other hand,
$$\lim_{x \to +} F(x) = \lim_{x \to +\infty} \left(-e^{-x/3} + 1\right) = 1.$$

□

## 7.4. THE INTEGRAL AND PROBABILITY

**Definition 3** The **expectation** (or **mean**) of a random variable $X$ with probability density function $f$ is

$$\mu(X) = \int_{-\infty}^{+\infty} x f(x)\, dx,$$

assuming that the improper integral converges.

**Example 6** Let

$$f(x) = \begin{cases} 0 & \text{if } x < 0, \\ \frac{1}{3} e^{-x/3} & \text{if } x \geq 0, \end{cases}$$

as in Example 3. Calculate the expectation of $X$.

**Solution**

$$\mu(X) = \int_0^\infty x \left(\frac{1}{3} e^{-x/3}\right) dx = \frac{1}{3} \int_0^\infty x e^{-x/3}\, dx.$$

We have

$$\int x e^{-x/3}\, dx = -3x e^{-x/3} - 9 e^{-x/3}$$

(an exercise in integration by parts). Therefore,

$$\frac{1}{3} \int_0^\infty x e^{-x/3}\, dx = \frac{1}{3} \lim_{b \to +\infty} \int_0^b x e^{-x/3}\, dx = \frac{1}{3} \lim_{b \to +\infty} \left( -3x e^{-x/3} - 9 e^{-x/3} \Big|_0^b \right)$$

$$= \frac{1}{3} \lim_{b \to +\infty} \left( -3b e^{-b/3} - 9 e^{-b/3} + 9 \right) = 3,$$

since $\lim_{b \to +\infty} b e^{-b/3} = 0$ and $\lim_{b \to +\infty} e^{-b/3} = 0$. Thus, the expectation of the random variable that corresponded to the life span of the hard disk turned out to be 3. Roughly speaking, the "average" life span of a hard disk that is manufactured by the company in question is 3. $\square$

**Definition 4** The **variance** of the random variable $X$ with the probability density function $f$ is

$$\sigma^2(X) = \int_{-\infty}^{+\infty} (x - \mu(X))^2 f(x)\, dx,$$

where $\sigma(X) > 0$ is the **standard deviation** of $X$.

Roughly speaking, the variance and the standard deviation of a random variable $X$ are measures of how much the values of $X$ spread about its mean.

**Example 7** Assume that $X$ has the exponential probability density

$$f(x) = \begin{cases} 0 & \text{if } x < 0, \\ \frac{1}{3} e^{-x/3} & \text{if } x \geq 0, \end{cases}$$

as in Example 6. We showed that the mean of $X$ is 3. Calculate the variance and the standard deviation of $X$.

**Solution**

The variance of $X$ is

$$\sigma^2(X) = \int_{-\infty}^{+\infty} (x-3)^2 f(x)\, dx = \int_0^{+\infty} (x-3)^2 \frac{1}{3} e^{-x/3}\, dx = 9$$

(Check with the help of integration by parts). Thus, the standard deviation of $X$ is 3. $\square$

## The Normal Distribution

Let $\mu$ and $\sigma$ be real numbers with $\sigma > 0$. A random variable $X$ is said to have a **normal distribution** if its density function is of the form

$$f(x) = \frac{1}{\sigma\sqrt{2\pi}} e^{-(x-\mu)^2/2\sigma^2}.$$

Thus, the distribution function of $X$ is

$$F(x) = \frac{1}{\sigma\sqrt{2\pi}} \int_0^x e^{-(t-\mu)^2/2\sigma^2} dt.$$

Figure 6 shows the probability density function $f$ for a typical normal distribution, and Figure 7 shows the corresponding distribution function.

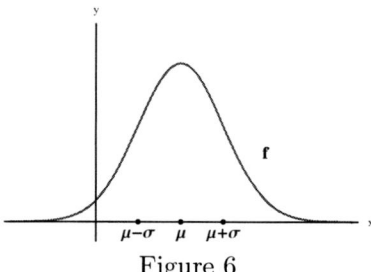

Figure 6

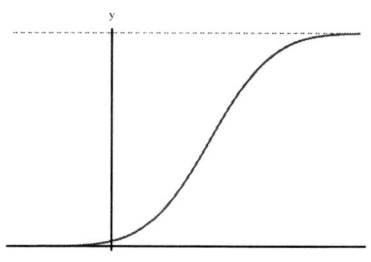

Figure 7

**Proposition 1** The mean of the normal distribution with probability density function

$$f(x) = \frac{1}{\sigma\sqrt{2\pi}} e^{-(x-\mu)^2/2\sigma^2}$$

**is $\mu$ and its standard deviation is $\sigma$.**

We will not be able to give a complete proof for Proposition 1 since certain integrals will have to be dealt with much later. Let's indicate the main steps, though.

The mean of $X$ is

$$\frac{1}{\sigma\sqrt{2\pi}} \int_{-\infty}^{+\infty} x e^{-(x-\mu)^2/2\sigma^2} dx$$

Let's set

$$u = \frac{x-\mu}{\sigma\sqrt{2}} \Rightarrow du = \frac{1}{\sigma\sqrt{2}} dx.$$

## 7.4. THE INTEGRAL AND PROBABILITY

Therefore,

$$\frac{1}{\sigma\sqrt{2\pi}}\int_{-\infty}^{+\infty} xe^{-(x-\mu)^2/2\sigma^2}\,dx = \frac{1}{\sqrt{\pi}}\int_{-\infty}^{+\infty}\left(\sigma\sqrt{2}u + \mu\right)e^{-u^2}\,du$$

$$= \frac{\sigma\sqrt{2}}{\sqrt{\pi}}\int_{-\infty}^{+\infty} ue^{-u^2}\,du + \frac{\mu}{\sqrt{\pi}}\int_{-\infty}^{+\infty} e^{-u^2}\,du.$$

Since $ue^{-u^2}$ defines an odd function, the first integral on the right-hand side is 0. It can be shown that

$$\int_{-\infty}^{+\infty} e^{-u^2}\,du = \sqrt{\pi}$$

(as you can confirm with the help of a CAS). Therefore,

$$\frac{1}{\sigma\sqrt{2\pi}}\int_{-\infty}^{+\infty} xe^{-(x-\mu)^2/2\sigma^2}\,dx = \frac{\mu}{\sqrt{\pi}}\int_{-\infty}^{+\infty} e^{-u^2}\,du = \mu,$$

so that the mean is $\mu$.

The variance is

$$\frac{1}{\sigma\sqrt{2\pi}}\int_{-\infty}^{+\infty} (x-\mu)^2\, e^{-(x-\mu)^2/2\sigma^2}\,dx.$$

Let's make use of the substitution

$$u = \frac{x-\mu}{\sigma\sqrt{2}}$$

again. Then,

$$\frac{1}{\sigma\sqrt{2\pi}}\int_{-\infty}^{+\infty} (x-\mu)^2\, e^{-(x-\mu)^2/2\sigma^2}\,dx = \frac{1}{\sqrt{\pi}}\int_{-\infty}^{+\infty} 2\sigma^2 u^2 e^{-u^2}\,du = \frac{2\sigma^2}{\sqrt{\pi}}\int_{-\infty}^{+\infty} u^2 e^{-u^2}\,du.$$

It can be shown that

$$\int_{-\infty}^{+\infty} u^2 e^{-u^2}\,du = \frac{\sqrt{\pi}}{2}$$

(as you can confirm with the help of a CAS). Therefore, the variance is

$$\frac{2\sigma^2}{\sqrt{\pi}}\int_{-\infty}^{+\infty} u^2 e^{-u^2}\,du = \sigma^2,$$

so that the standard deviation is $\sigma$, as claimed. ∎

**Example 8** Let

$$f(x) = \frac{1}{100\sqrt{2\pi}} e^{-(x-600)^2/2(100)^2}$$

be the probability density function of the random variable $X$ of Example 4. Thus $X$ has a normal distribution with mean $\mu = 600$ and standard deviation $\sigma = 10$. The variance is $\sigma^2 = 100$. □

The special case where the mean $\mu = 0$ and the standard deviation $\sigma = 1$ is referred to as **the standard normal distribution**. In this case, we will denote the distribution function by $\Phi$. Thus,

$$\Phi(x) = \frac{1}{\sqrt{2\pi}}\int_{-\infty}^{x} e^{-u^2/2}\,du.$$

Note that we can express the distribution function corresponding to an arbitrary normal distribution in terms of $\Phi$. We have
$$F(x) = \frac{1}{\sigma\sqrt{2\pi}} \int_{-\infty}^{x} e^{-(u-\mu)^2/2\sigma^2} du = \Phi\left(\frac{x-\mu}{\sigma}\right).$$

Indeed, let's set
$$v = \frac{u-\mu}{\sigma},$$
and apply the substitution rule for (definite) integrals: We have
$$dv = \frac{1}{\sigma} du,$$
so that
$$F(x) = \frac{1}{\sigma\sqrt{2\pi}} \int_{0}^{x} e^{-(u-\mu)^2/2\sigma^2} du = \frac{1}{\sigma\sqrt{2\pi}} \int_{-\infty}^{(x-\mu)/\sigma} e^{-v^2/2} \sigma dv$$
$$= \frac{1}{\sqrt{2\pi}} \int_{-\infty}^{(x-\mu)/\sigma} e^{-v^2/2} dv = \Phi\left(\frac{x-\mu}{\sigma}\right),$$
as claimed.

Integrals that involve the probability density function of a Gaussian normal distribution can be expressed in terms of
$$\operatorname{erf}(x) = \int_{0}^{x} \frac{2}{\sqrt{\pi}} e^{-t^2} dt$$

For example, let's express the distribution function $\Phi$ for the standard normal distribution in terms of erf:
$$\Phi(x) = \int_{-\infty}^{x} \frac{1}{\sqrt{2\pi}} e^{-t^2/2} dt = \lim_{b \to -\infty} \int_{b}^{x} \frac{1}{\sqrt{2\pi}} e^{-t^2/2} dt.$$

We will set $u = t/\sqrt{2}$, so that $du = dt/\sqrt{2}$. Therefore,
$$\int_{b}^{x} \frac{1}{\sqrt{2\pi}} e^{-t^2/2} dt = \int_{b/\sqrt{2}}^{x/\sqrt{2}} \frac{1}{\sqrt{2\pi}} e^{-u^2} \sqrt{2} du = \int_{b/\sqrt{2}}^{x/\sqrt{2}} \frac{1}{\sqrt{\pi}} e^{-u^2} du$$
$$= \frac{1}{2} \int_{b/\sqrt{2}}^{x/\sqrt{2}} \frac{2}{\sqrt{\pi}} e^{-u^2} du$$
$$= \frac{1}{2} \left( \int_{0}^{x/\sqrt{2}} \frac{2}{\sqrt{\pi}} e^{-u^2} du - \int_{0}^{b/\sqrt{2}} \frac{2}{\sqrt{\pi}} e^{-u^2} du \right)$$
$$= \frac{1}{2} \operatorname{erf}\left(\frac{x}{\sqrt{2}}\right) - \frac{1}{2} \operatorname{erf}\left(\frac{b}{\sqrt{2}}\right).$$

Thus,
$$\Phi(x) = \lim_{b \to -\infty} \int_{b}^{x} \frac{1}{\sqrt{2\pi}} e^{-t^2/2} dt = \lim_{b \to -\infty} \left( \frac{1}{2} \operatorname{erf}\left(\frac{x}{\sqrt{2}}\right) - \frac{1}{2} \operatorname{erf}\left(\frac{b}{\sqrt{2}}\right) \right)$$
$$= \frac{1}{2} \operatorname{erf}\left(\frac{x}{\sqrt{2}}\right) - \frac{1}{2} \lim_{b \to -\infty} \operatorname{erf}\left(\frac{b}{\sqrt{2}}\right).$$

Now,
$$\lim_{b \to -\infty} \operatorname{erf}\left(\frac{b}{\sqrt{2}}\right) = \lim_{b \to -\infty} \int_{0}^{b} \frac{2}{\sqrt{\pi}} e^{-t^2} dt = -\lim_{b \to -\infty} \int_{b}^{0} \frac{2}{\sqrt{\pi}} e^{-t^2} dt$$
$$= -\int_{-\infty}^{0} \frac{2}{\sqrt{\pi}} e^{-t^2} dt = -\frac{1}{\sqrt{\pi}} \int_{-\infty}^{+\infty} e^{-t^2} dt.$$

## 7.4. THE INTEGRAL AND PROBABILITY

We have
$$\int_{-\infty}^{+\infty} e^{-t^2}\,dt = \sqrt{\pi}$$
as stated before. Therefore,
$$\Phi(x) = \frac{1}{2}\operatorname{erf}\left(\frac{x}{\sqrt{2}}\right) - \frac{1}{2}\lim_{b\to-\infty}\operatorname{erf}\left(\frac{b}{\sqrt{2}}\right) = \frac{1}{2}\operatorname{erf}\left(\frac{x}{\sqrt{2}}\right) - \frac{1}{2}(-1)$$
$$= \frac{1}{2}\operatorname{erf}\left(\frac{x}{\sqrt{2}}\right) + \frac{1}{2}.$$

□

## Problems

**1.** Assume that the life span (in years) of the motherboard of a laptop is determined by a probability density function
$$f(x) = \begin{cases} 0 & \text{if } x < 0, \\ \frac{1}{a}e^{-x/a} & \text{if } x \geq 0, \end{cases}$$
where $a$ is a positive constant.
a) Determine $f$ if the probability that the life span of the motherboard is more than 2 years is 90%.
b) [C] Having determined $f$, calculate the probability that the life span of the motherboard is more than 3 years.

**2.** Assume that the life span (in months) of light bulb is determined by a probability density function
$$f(x) = \begin{cases} 0 & \text{if } x < 0, \\ \frac{1}{a}e^{-x/a} & \text{if } x \geq 0, \end{cases}$$
where $a$ is a positive constant.
a) Determine $f$ if the probability that the life span of the light bulb is more than 10 months is 80%.
b) [C] Having determined $f$, calculate the life span of the light bulb is more than 20 months.

**3.** Assume that the IQ score of an individual is a random variable with normal distribution that has mean 105 and standard deviation 16.
a) Determine the associated probability density function.
b) [C] Make use of the numerical integrator of your computational utility to calculate the probability that the IQ score of an individual is higher than 120.

**4.** Assume that the speed of a car that travels on a certain California freeway is a random variable with normal distribution that has mean 60 (miles/hour) and standard deviation 10 (miles/hour).
a) Determine the associated probability density function.
b) [C] Make use of the numerical integrator of your computational utility to calculate the probability that the speed of a car on that freeway is between 65 and 75?

**5.** Let
$$f(x) = \frac{1}{\sigma\sqrt{2\pi}} e^{-(x-\mu)^2/2\sigma^2},$$
so that $f$ is the density function of a normal distribution with mean $\mu$ and standard deviation $\sigma$.
a) Show that $f$ attains its absolute maximum at $x = \mu$. Determine the absolute maximum value of $f$.
b) Show that the graph of $f$ has inflection points that correspond to $\mu \pm \sigma$.

# Chapter 8

# Differential Equations

In this chapter we discuss the solutions of **linear first-order differential equations**, **separable differential equations** and their applications.

## 8.1 First-Order Linear Differential Equations

In this section we will consider differential equations of the type

$$\frac{dy}{dt} = p(t) y + q(t),$$

where $p(t)$ and $q(t)$ are given functions, and $y = y(t)$ is the function that is to be determined. Such a differential equation is referred to as a **first-order linear differential equation**. If we set $f(t, y) = p(t) y + q(t)$, the differential equation can be expressed as

$$\frac{dy}{dt} = f(t, y).$$

The term "linear" corresponds to the fact that $f(t, y)$ defines a linear function of $y$ for fixed $t$. The term "first-order" corresponds to the fact that only the first derivative of the unknown function $y$ appears in the differential equation.

We have covered the special case

$$\frac{dy}{dt} = ky,$$

where $k$ is a constant, in Section 4.6 ($p(t) = k$, $q(t) = 0$). We saw that the general solution is a constant multiple of $e^{kt}$, and that the solution corresponding to an initial condition of the form $y(t_0) = y_0$ is uniquely determined.

We have covered the special case

$$\frac{dy}{dt} = q(t)$$

in Section 5.7 ($p(t) = 0$). The general solution is simply the indefinite integral of $q(t)$:

$$y(t) = \int q(t) \, dt.$$

The expression conceals an arbitrary additive constant. The solution corresponding to an initial condition of the form $y(t_0) = y_0$ is unique.

In this section you will learn how to find **the general solution of any first-order differential equation** and the solution of **an initial-value problem** in the form

$$\frac{dy}{dt} = p(t) y + q(t), \ y(t_0) = y_0,$$

133

where $t_0$ and $y_0$ are given numbers.

## Cases where the Coefficient of $y$ is Constant

We will begin with first-order linear differential equations where the coefficient of $y$ is a constant. Thus,
$$\frac{dy}{dt} = ky + q(t),$$
where $k$ is a constant. Let's write the equation as
$$\frac{dy}{dt} - ky = q(t).$$
If $q(t)$ were identically 0, the solution would have been $e^{kt}$. Let us divide both sides of the equation by $e^{kt}$, i.e., multiply by $e^{-kt}$. Thus,
$$e^{-kt}\frac{dy}{dt} - e^{-kt}ky = e^{-kt}q(t).$$
Notice that the left-hand side is the derivative of $e^{-kt}y$. Indeed, with the help of the product rule and the chain rule,
$$\frac{d}{dt}\left(e^{-kt}y\right) = e^{-kt}\left(\frac{dy}{dt}\right) + \left(\frac{d}{dt}e^{-kt}\right)y = e^{-kt}\frac{dy}{dt} + (-ke^{-kt})y$$
$$= e^{-kt}\frac{dy}{dt} - ke^{-kt}y.$$
Therefore,
$$\frac{d}{dt}\left(e^{-kt}y\right) = e^{-kt}q(t).$$
This means that $e^{-kt}y$ is an antiderivative of $e^{-kt}q(t)$:
$$e^{-kt}y = \int e^{-kt}q(t)\,dt.$$
Therefore,
$$y(t) = e^{kt}\left(\int e^{-kt}q(t)\,dt\right).$$

**The indefinite integral involves an arbitrary constant.** The above expression is **the general solution** of the differential equation $y'(t) = ky(t) + q(t)$. The constant is uniquely determined by **an initial condition** in the form $y(t_0) = y_0$, so that the solution of an initial-value problem for the differential equation $y' = ky + q(t)$ is unique. The technique that led to the above expression for the solution is referred to as **the technique of an integrating factor**, and $e^{-kt}$ is referred to as **an integrating factor**.

**Example 1**

a) Determine the general solution of the differential equation
$$\frac{dy}{dt} = \frac{1}{4}y + 3$$
by using the technique of an integrating factor.
b) Determine the solution of the initial-value problem
$$\frac{dy}{dt} = \frac{1}{4}y + 3,\ y(2) = 24.$$

## Solution

a) We express the equation as
$$\frac{dy}{dt} - \frac{1}{4}y = 3.$$

We multiply both sides of the equation by the integrating factor $e^{-t/4}$ ($k = 1/4$):
$$e^{-t/4}\frac{dy}{dt} - \frac{1}{4}e^{-t/4}y = 3e^{-t/4}.$$

Therefore,
$$\frac{d}{dt}\left(e^{-t/4}y\right) = 3e^{-t/4}.$$

Thus,
$$e^{-t/4}y = \int 3e^{-t/4}dt.$$

In order to evaluate the indefinite integral, we set $w = -t/4$ so that $dw/dt = -1/4$. Therefore,
$$\int 3e^{-t/4}dt = -12\int e^{-t/4}\frac{dw}{dt}dt = -12\int e^w dw = -12e^w + C = -12e^{-t/4} + C,$$

where $C$ is an arbitrary constant. Therefore,
$$e^{-t/4}y = -12e^{-t/4} + C$$

so that
$$y(t) = -12 + Ce^{t/4}.$$

Since $C$ can be a real number, the differential equation
$$\frac{dy}{dt} = \frac{1}{4}y + 3$$

has infinitely many solutions. Figure 1 shows the solutions corresponding to $C = 0$, $\pm 36e^{-1/2}$, and $\pm 48$. Note that the solution corresponding to $C = 0$ is a constant function with value $-12$.

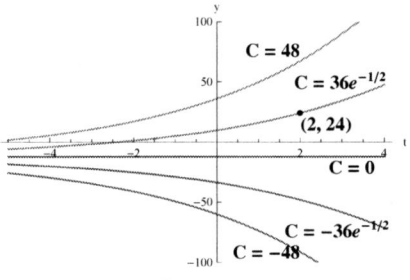

Figure 1

b) Since the general solution of the given differential equation is $y(t) = -12 + Ce^{t/4}$, we must determine $C$ so that $y(2) = 24$. Thus,
$$24 = -12 + Ce^{2/4} \Leftrightarrow 36 = Ce^{1/2} \Leftrightarrow C = 36e^{-1/2}.$$

Therefore, the solution of the given initial-value problem is
$$y(t) = -12 + 36e^{-1/2}e^{t/4}.$$
This is the only solution of the differential equation
$$\frac{dy(t)}{dt} = \frac{1}{4}y(t) + 3$$
whose graph passes through the point $(2, 24)$ (corresponding to the initial condition $y(2) = 24$), as indicated in Figure 1. $\square$

**Example 2**

a) Determine the general solution of the differential equation
$$\frac{d}{dt}y(t) = -\frac{1}{10}y(t) + 2.$$

b) Determine the solutions of the initial-value problems
$$\frac{d}{dt}y(t) = -\frac{1}{10}y(t) + 2, \ y(0) = y_0,$$
where $y_0 = 0, 20, \pm 40$.

**Solution**

a) Let's rewrite the given differential equation as
$$\frac{d}{dt}y(t) + \frac{1}{10}y(t) = 2,$$
and multiply both sides of the equation by the integrating factor $e^{t/10}$:
$$e^{t/10}\frac{d}{dt}y(t) + e^{t/10}\frac{1}{10}y(t) = 2e^{t/10}.$$
Thus
$$\frac{d}{dt}\left(e^{t/10}y(t)\right) = 2e^{t/10}.$$
Therefore,
$$e^{t/10}y(t) = \int 2e^{t/10}dt = 20e^{t/10} + C.$$
Thus,
$$y(t) = e^{-1/10}\left(20e^{t/10} + C\right) = 20 + Ce^{-t/10}$$
describes the general solution of the given differential equation.

b) If the initial condition is specified as $y(0) = y_0$, we must have
$$y_0 = 20 + C \Leftrightarrow C = y_0 - 20,$$
so that
$$y(t) = 20 + (y_0 - 20)e^{-t/10}.$$
Therefore, the solutions that correspond to $y_0 = 0, 20, 40$ and $-40$ are
$$20 - 20e^{-t/10}, \ 20, \ 20 + 20e^{-t/10} \text{ and } 20 - 60e^{-t/10},$$

## 8.1. FIRST-ORDER LINEAR DIFFERENTIAL EQUATIONS

respectively. Figure 2 shows these solutions. Note that the solution that corresponds to $y_0 = 20$ is constant. □

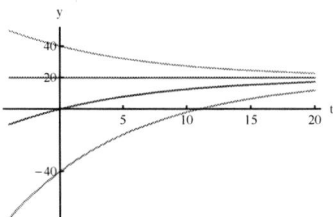

Figure 2

**Definition 1** A constant solution of a differential equation is a **steady-state solution** (or **equilibrium solution**) of the differential equation.

**Example 3** In Example 1 we observed that the constant $-12$ is a solution of the differential equation
$$\frac{dy(t)}{dt} = \frac{1}{4}y(t) + 3.$$
Thus, $-12$ is a steady-state solution of the above differential equation. Indeed, $-12$ is the only steady-state solution: If the constant $c$ is a solution of the given differential equation, then
$$0 = \frac{dc}{dt} = \frac{1}{4}c + 3 \Rightarrow c = -12.$$
In Example 2 we noted that the constant $20$ is a solution of the differential equation
$$\frac{d}{dt}y(t) = -\frac{1}{10}y(t) + 2.$$
Indeed, the constant $c$ is a solution of the above differential equation if and only if
$$0 = \frac{dc}{dt} = -\frac{1}{10}c + 2 \Leftrightarrow c = 20.$$

□

**Definition 2** Assume that $c$ is a steady-state solution of a differential equation of the form $y'(t) = f(y(t))$. The steady-state $c$ is **stable** if there exists an open interval $J$ containing $c$ such that $\lim_{t \to +\infty} y(t) = c$ for any solution $y(t)$ where $y(t_0) = y_0$, $t_0$ is arbitrary and $y_0 \in J$. The steady-state $c$ is **unstable** if, given any open interval $J$ containing $c$, there exists $y_0 \in J$ such that the solution $y(t)$ with $y(t_0) = y_0$ does not approach $c$ as $t$ tends to infinity.

**Example 4** Consider the differential equation
$$y'(t) = -\frac{1}{10}y(t) + 2$$
as in Example 2. Show that the steady-state solution $20$ is stable.

**Solution**

Note that the differential equation can be expressed as $y'(t) = f(y(t))$, where
$$f(y) = -\frac{1}{10}y + 2.$$

As we determined in Example 2, the general solution of the differential equation is
$$y(t) = 20 + Ce^{-t/10}.$$
Therefore,
$$y(t_0) = y_0 \Leftrightarrow 20 + Ce^{-t_0/10} = y_0 \Leftrightarrow C = (y_0 - 20)e^{t_0/10}.$$
Thus, the solution corresponding to the initial condition $y(t_0) = y_0$ is
$$y(t) = 20 + (y_0 - 20)e^{t_0/10}e^{-t/10} = 20 + (y_0 - 20)e^{-(t-t_0)/10}.$$
Therefore,
$$\lim_{t \to +\infty} y(t) = \lim_{t \to +\infty} \left(20 + (y_0 - 20)e^{-(t-t_0)/10}\right) = 20 + (y_0 - 20)\lim_{t \to +\infty} e^{-(t-t_0)/10} = 20.$$
$$\lim_{t \to +\infty} y(t) = 20 + C \lim_{t \to +\infty} e^{-t/10} = 20.$$
Thus, the steady-state solution 20 is stable. Note that
$$|y(t) - 20| = \left|(y_0 - 20)\lim_{t \to +\infty} e^{-(t-t_0)/10}\right| = |y - y_0|\left|e^{-(t-t_0)/10}\right|,$$
so that $|y(t) - 20|$ decreases exponentially as $t \to +\infty$, i.e., any solution approaches the steady-state solution as $t \to +\infty$.any solution of the differential equation approaches the steady-state solution of the differential equation exponentially. □

**Example 5** Consider the differential equation
$$y'(t) = \frac{1}{4}y(t) + 3$$
as in Example 1. Show that the steady-state solution $-12$ is unstable.

**Solution**

Note that the differential equation can be expressed as $y'(t) = f(y(t))$, where
$$f(y) = \frac{1}{4}y + 3.$$
As we showed in Example 1, the general solution of the differential equation
$$y'(t) = \frac{1}{4}y(t) + 3$$
is
$$y(t) = -12 + Ce^{t/4},$$
Therefore, $y(t_0) = y_0$ if and only if
$$-12 + Ce^{t_0/4} = y_0 \Leftrightarrow Ce^{t_0/4} = y_0 + 12 \Leftrightarrow C = (y_0 + 12)e^{-t_0/4}.$$
Thus, the solution of the corresponding initial-value problem is
$$y(t) = -12 + (y_0 + 12)e^{-t_0/4}e^{t/4} = -12 + (y_0 + 12)e^{(t-t_0)/4}.$$
If $y_0 = -12$, the solution is the steady-state solution $-12$. Otherwise,
$$\lim_{t \to +\infty} |y(t) - (-12)| = \lim_{t \to +\infty} \left|(y_0 + 12)e^{(t-t_0)/4}\right| = |y_0 + 12|e^{(t-t_0)/4} = +\infty.$$
This is the case no matter how close $y_0$ is to $-12$, as long as $y_0 \neq -12$. Therefore, the steady-state solution $-12$ is unstable. Note that $|y(t) - (-12)|$ grows exponentially as $t \to +\infty$ if $y_0 \neq -12$. □

## 8.1. FIRST-ORDER LINEAR DIFFERENTIAL EQUATIONS

**Example 6**

a) Determine the general solution of the differential equation
$$\frac{dy}{dt} = -\frac{1}{2}y(t) + \cos(t).$$

b) Determine the solution of the initial-value problem
$$\frac{dy}{dt} = -\frac{1}{2}y(t) + \cos(t),\ y(\pi) = 3.$$

**Solution**

a) We express the equation as
$$\frac{dy}{dt} + \frac{1}{2}y(t) = \cos(t),$$
and multiply both sides the equation by the integrating factor $e^{t/2}$. Thus,
$$e^{t/2}\frac{dy}{dt} + \frac{1}{2}e^{t/2}y(t) = e^{t/2}\cos(t),$$
so that
$$\frac{d}{dt}\left(e^{t/2}y(t)\right) = e^{t/2}\cos(t).$$
Therefore,
$$e^{t/2}y(t) = \int e^{t/2}\cos(t)\,dt.$$

As in Section 6.1, the indefinite integral on the right-hand side can be evaluated with the help of integration by parts:
$$\int e^{t/2}\cos(t)\,dt = \frac{2}{5}e^{t/2}\cos(t) + \frac{4}{5}e^{t/2}\sin(t) + C$$

(check). Therefore,
$$y(t) = e^{-t/2}\left(\frac{2}{5}e^{t/2}\cos(t) + \frac{4}{5}e^{t/2}\sin(t) + C\right) = \frac{2}{5}\cos(t) + \frac{4}{5}\sin(t) + Ce^{-t/2}.$$

b) We have
$$y(\pi) = 3 \Leftrightarrow -\frac{2}{5} + Ce^{-\pi/2} = 3 \Leftrightarrow C = \frac{17}{5}e^{\pi/2}.$$
Therefore, the solution of the given initial-value problem is
$$y(t) = \frac{2}{5}\cos(t) + \frac{4}{5}\sin(t) + \frac{17}{5}e^{\pi/2}e^{-t/2}$$

□

Note that the differential equation of Example 6 does not have steady-states: If $y(t)$ were a constant $c$ and were a solution, we would have to have
$$0 = \frac{dy}{dt} = -\frac{1}{2}y + \cos(t) \Leftrightarrow 0 = -\frac{1}{2}c + \cos(t) \Leftrightarrow \cos(t) = \frac{1}{2}c,$$

but $\cos(t)$ is not a constant. On the other hand, the general solution is
$$y(t) = \frac{2}{5}\cos(t) + \frac{4}{5}\sin(t) + Ce^{-t/2}.$$
Therefore,
$$\lim_{t\to\infty}\left(y(t) - \left(\frac{2}{5}\cos(t) + \frac{4}{5}\sin(t)\right)\right) = \lim_{t\to+\infty} Ce^{-t/2} = 0.$$
Thus, the difference between an arbitrary solution of the differential equation and the solution
$$\frac{2}{5}\cos(t) + \frac{4}{5}\sin(t)$$
approaches 0 as $t$ tends to infinity. The expression $Ce^{-t/2}$ is **"the transient part"** of the solution and dies down very rapidly as $t \to +\infty$. Figure 3 displays the graphs of some solutions of differential equations. All the graphs approach the graph of
$$y = \frac{2}{5}\cos(t) + \frac{4}{5}\sin(t)$$
as $t$ increases. $\square$

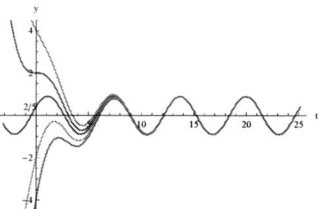

Figure 3

In the above examples, we were able to express the relevant indefinite integral, hence the general solution of the differential equation, in terms of familiar functions. The technique of an integrating factor enables us to express the solution in terms of an integral, even if we are unable to express the required indefinite integral in terms of familiar functions. Here is the strategy: We will express the general solution of the differential equation
$$\frac{dy}{dt} = ky(t) + q(t)$$
in terms of an integral from $t_0$ to $t$, where $t_0$ is arbitrary. Therefore, we will need to introduce a dummy integration variable. The letter $w$ will do:
$$\frac{dy}{dw} - ky(w) = q(w).$$
As before, we multiply both sides of the equation by the integrating factor $e^{-kw}$:
$$e^{-kw}\frac{dy}{dw} - ke^{-kw}y(w) = e^{-kw}q(w),$$
so that
$$\frac{d}{dw}\left(e^{-kw}y(w)\right) = e^{-kw}q(w).$$
We integrate from $t_0$ to $t$:
$$\int_{t_0}^{t}\frac{d}{dw}\left(e^{-kw}y(w)\right)dw = \int_{t_0}^{t}e^{-kw}q(w)\,dw.$$

## 8.1. FIRST-ORDER LINEAR DIFFERENTIAL EQUATIONS

By the Fundamental Theorem of Calculus,

$$e^{-kw}y(w)\Big|_{t_0}^{t} = \int_{t_0}^{t} e^{-kw}q(w)\,dw.$$

Therefore,

$$e^{-kt}y(t) - e^{-kt_0}y(t_0) = \int_{t_0}^{t} e^{-kw}q(w)\,dw,$$

so that

$$y(t) = e^{kt}e^{-kt_0}y(t_0) + e^{kt}\int_{t_0}^{t} e^{-kw}q(w)\,dw.$$

If we set $C = e^{-kt_0}y(t_0)$, the general solution of the differential equation $y'(t) = ky(t) + q(t)$ can be expressed as

$$y(t) = Ce^{kt} + e^{kt}\int_{t_0}^{t} e^{-kw}q(w)\,dw.$$

If the initial condition is in the form $y(t_0) = y_0$, the solution is

$$y(t) = e^{kt}e^{-kt_0}y_0 + e^{kt}\int_{t_0}^{t} e^{-kw}q(w)\,dw = e^{k(t-t_0)}y_0 + e^{kt}\int_{t_0}^{t} e^{-kw}q(w)\,dw.$$

There is no need to memorize the above expressions. The important point is the strategy that leads to the expressions. You should implement the strategy in each specific case, as in the following example:

**Example 7**

a) Express the general solution of the differential equation

$$\frac{dy(t)}{dt} = -\frac{1}{4}y(t) + \frac{1}{1+t^2}$$

in terms of an integral.

b) Express the general solution of the initial-value problem

$$\frac{dy(t)}{dt} = -\frac{1}{4}y(t) + \frac{1}{1+t^2},\ y(1) = 2$$

in terms of an integral.

**Solution**

a) Since the initial condition in part b) is specified as $y(1) = 2$, it will be convenient to express the general solution in terms of an integral from 1 to $t$. We introduce the integration variable $w$:

$$\frac{dy}{dw} + \frac{1}{4}y(w) = \frac{1}{1+w^2}.$$

We multiply both sides by the integrating factor $e^{w/4}$. Thus,

$$e^{w/4}\frac{dy}{dw} + \frac{1}{4}e^{w/4}y(w) = \frac{e^{w/4}}{1+w^2},$$

so that

$$\frac{d}{dw}\left(e^{w/4}y(w)\right) = \frac{e^{w/4}}{1+w^2}.$$

Therefore,
$$\int_1^t \frac{d}{dw}\left(e^{w/4}y(w)\right)dw = \int_1^t \frac{e^{w/4}}{1+w^2}dw.$$

By the Fundamental Theorem of Calculus,
$$e^{t/4}y(t) - e^{1/4}y(1) = \int_1^t \frac{e^{u/4}}{1+w^2}dw.$$

Therefore,
$$y(t) = e^{-t/4}e^{1/4}y(1) + e^{-t/4}\int_1^t \frac{e^{w/4}}{1+w^2}du$$

We can set $C = e^{1/4}y(1)$ and express the general solution as
$$y(t) = Ce^{-t/4} + e^{-t/4}\int_1^t \frac{e^{w/4}}{1+w^2}du.$$

b) Since $y(1) = 2$, the solution of the given initial-value problem is
$$y(t) = e^{-t/4}e^{1/4}y(1) + e^{-t/4}\int_1^t \frac{e^{w/4}}{1+w^2}dw = e^{-t/4}e^{1/4}(2) + e^{-t/4}\int_1^t \frac{e^{w/4}}{1+w^2}dw$$
$$= 2e^{-(t-1)/4} + e^{-t/4}\int_1^t \frac{e^{w/4}}{1+w^2}dw.$$

The integral which appears on the right-hand side cannot be expressed in terms of familiar special functions. In any case, it is possible to compute approximations to the values of the solution via numerical integration. For example,
$$y(2) = 2e^{-(2-1)/4} + e^{-2/4}\int_1^2 \frac{e^{w/4}}{1+w^2}du \cong 1.8369$$

Figure 4 shows the graph of $y(t)$ on the interval $[0, 4]$. $\square$

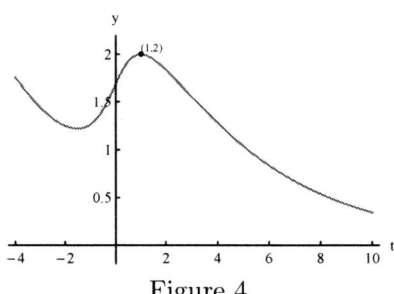

Figure 4

## The General Case

Now we will consider first-order linear differential equations of the form
$$\frac{dy}{dt} = p(t)y(t) + q(t),$$

## 8.1. FIRST-ORDER LINEAR DIFFERENTIAL EQUATIONS

where $p$ and $q$ are continuous on an interval $J$ and $p$ is not necessarily a constant. We will also solve initial-value problems for such equations that specify initial conditions in the form $y(t_0) = y_0$.

As in the case where $p(t)$ is a constant, we write the equation as

$$\frac{dy}{dt} - p(t)y(t) = q(t).$$

Let $P(t)$ be an antiderivative of $p(t)$:

$$P(t) = \int p(t)\,dt$$

(any antiderivative of $p$ will do, so that there is no need to involve an arbitrary constant). We will use the expression $e^{-P(t)}$ as **an integrating factor**. Note that $P(t)$ can be taken to be $-kt$ if $p(t) = k$, where $k$ is a constant, since

$$P(t) = \int k\,dt = kt$$

(we have set the constant of integration to be 0).

We multiply both sides of the equation by $e^{-P(t)}$:

$$e^{-P(t)} - e^{-P(t)}p(t)y(t) = e^{-P(t)}q(t).$$

The left-hand side can be expressed as

$$\frac{d}{dt}\left(e^{-P(t)}y(t)\right).$$

Indeed, by the product rule and the chain rule,

$$\frac{d}{dt}\left(e^{-P(t)}y(t)\right) = e^{-P(t)}\frac{dy}{dt} + \left(-\frac{dP}{dt}e^{-P(t)}\right)y(t).$$

We have

$$P(t) = \int p(t)\,dt \Leftrightarrow \frac{dP}{dt} = p(t).$$

Therefore,

$$\frac{d}{dt}\left(e^{-P(t)}y(t)\right) = e^{-P(t)}\frac{dy}{dt} + \left(-\frac{dP}{dt}e^{-P(t)}\right)y(t)$$
$$= e^{-P(t)}\frac{dy}{dt} - p(t)e^{-P(t)}y(t),$$

as claimed. Thus,

$$\frac{d}{dt}\left(e^{-P(t)}y(t)\right) = e^{-P(t)}q(t),$$

so that

$$e^{-P(t)}y(t) = \int e^{-P(t)}q(t)\,dt.$$

As in the case of a constant coefficient $k$ of $y(t)$, there is no need to memorize the above expression. You should adapt the technique that led to the expression to a specific case, as in the following example.

**Example 8**

a) Determine the general solution of the differential equation
$$y'(t) = -ty(t) + t$$
by implementing the technique of an integrating factor.
b) Determine the solution of the initial-value problems
$$y'(t) = -ty(t) + t,\ y(0) = y_0,$$
with $y_0 = -10$ and $y_0 = 10$.

**Solution**

a) Let's rewrite the differential equation as
$$\frac{dy(t)}{dt} + ty(t) = t.$$

The integrating factor is in the form
$$e^{\int t\,dt}.$$

We have
$$\int t\,dt = \frac{1}{2}t^2 + C,$$

where $C$ is an arbitrary constant. The choice $C = 0$ will do. Thus, we will use
$$e^{\int t\,dt} = e^{t^2/2}$$

as an integrating factor. We multiply both sides of the equation by $e^{t^2/2}$:
$$e^{t^2/2}\frac{dy(t)}{dt} + te^{t^2/2}y(t) = te^{t^2/2}.$$

The left-hand side of the equation can be expressed as
$$\frac{d}{dt}\left(e^{t^2/2}y(t)\right)$$

(check). Therefore,
$$\frac{d}{dt}\left(e^{t^2/2}y(t)\right) = te^{t^2/2},$$

so that
$$e^{t^2/2}y(t) = \int te^{t^2/2}dt = e^{t^2/2} + C,$$

where $C$ is an arbitrary constant (check). Thus,
$$y(t) = 1 + Ce^{-t^2/2}$$

is the general solution of the given differential equation.

b) We have $y(0) = 10$ if and only if
$$1 + C = 10 \Leftrightarrow C = 9.$$

Therefore,
$$f(t) = 1 + 9e^{-t^2/2}$$

## 8.1. FIRST-ORDER LINEAR DIFFERENTIAL EQUATIONS

is the solution of the initial-value problem
$$y'(t) = -ty(t) + t, \; y(0) = 10.$$

We have $y(0) = -10$ if and only if
$$1 + C = -10 \Leftrightarrow C = -11.$$

Therefore,
$$g(t) = 1 - 11e^{-t^2/2}$$
is the solution of the initial-value problem
$$y'(t) = -ty(t) + t, \; y(0) = -10.$$

Figure 5 shows the graphs of $f$ and $g$. □

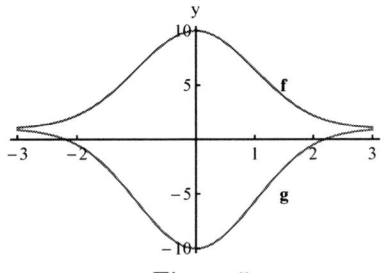

Figure 5

In Example 8 we were able to express the integrals in terms of familiar functions. Just as in the case of a constant coefficient for $y$, we can express the solution of
$$\frac{dy}{dt} = p(t)y(t) + q(t)$$
in terms of an integral, provided that we can determine an integrating factor.

**Example 9**

a) Express the general solution of the differential equation
$$\frac{dy}{dt} = -2ty(t) + \frac{1}{1+t^2}$$
in terms of an integral.

b) Express the solutions of the initial value problem
$$\frac{dy}{dt} = -2ty(t) + \frac{1}{1+t^2}, \; y(0) = y_0,$$
in terms of an integral.

**Solution**

a) We rewrite the equation as
$$\frac{dy}{dw} + 2wy(w) + \frac{1}{1+w^2}.$$

An integrating factor is
$$e^{\int 2w\,dw} = e^{w^2}.$$
We multiply both sides of the equation by $e^{w^2}$:
$$e^{w^2}\frac{dy}{dw} + 2we^{w^2}y(w) = \frac{e^{w^2}}{1+w^2}.$$
Thus,
$$\frac{d}{dw}\left(e^{w^2}y(w)\right) = \frac{e^{w^2}}{1+w^2}.$$
We integrate from 0 to $t$:
$$\int_0^t \frac{d}{dw}\left(e^{w^2}y(w)\right)dw = \int_0^t \frac{e^{w^2}}{1+w^2}dw.$$
By the Fundamental Theorem of Calculus,
$$e^{t^2}y(t) - y(0) = \int_0^t \frac{e^{w^2}}{1+w^2}dw.$$
Therefore,
$$y(t) = e^{-t^2}y(0) + e^{-t^2}\int_0^t \frac{e^{w^2}}{1+w^2}dw.$$
If we set $C = y(0)$, then $C$ is an arbitrary constant and we can express the general solution of the given differentia equation as
$$y(t) = Ce^{-t^2} + e^{-t^2}\int_0^t \frac{e^{w^2}}{1+w^2}dw.$$

b) By part a), the solution of the initial-value problem
$$y'(t) = -2ty(t) + \frac{1}{1+t^2},\ y(0) = y_0$$
is
$$y(t) = e^{-t^2}y_0 + e^{-t^2}\int_0^t \frac{e^{w^2}}{1+w^2}dw.$$
The integral cannot be expressed in terms of the familiar special functions. Nevertheless, the above solution is quite respectable, and we can approximates its values via numerical integration. Figure 6 shows the graphs of the solutions corresponding to $y_0 = 1$ and $y_0 = 2$. □

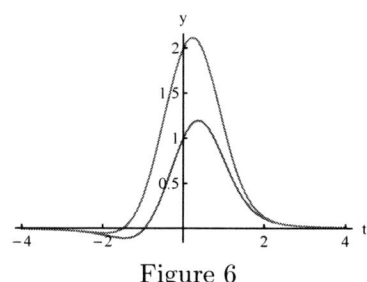

Figure 6

In the next section we will discuss some mathematical models that lead to first-order linear differential equations.

## 8.1. FIRST-ORDER LINEAR DIFFERENTIAL EQUATIONS

## Problems

In problems 1 and 2,
a) Determine the general solution of the given differential equation, the solution of the initial value problem such that $y(t_0) = y_0$, and $\lim_{t \to \infty} y(t)$.
b) Sketch the graph of the solution.

**1.**
$$\frac{dy}{dt} = -\frac{1}{4}y(t), \ t_0 = 0, \ y_0 = 6.$$

**2.**
$$\frac{dy}{dt} = 3y(t), \ t_0 = 1, \ y_0 = 2.$$

In problems 3 and 4,
a) Determine the steady-state solution of the differential equation.
b) Determine the general solution of the given differential equation by making use of the technique of an integrating factor. What is the limit of any solution at infinity?
c) Determine the solutions of the initial value problems such that $y(0)$ has the given values,
d) [C] Plot the graph of the solutions in part c) with the help of your graphing utility.

**3.**
$$\frac{dy}{dt} = -\frac{1}{10}y(t) + 5, \ y(0) = 20, \ 40, \ 60, \ 80.$$

**4.**
$$\frac{dy}{dt} = \frac{1}{5}y(t) + 10, \ y(0) = -80, \ -60, \ 10, \ 80.$$

In problems 5-10, determine the general solution of the given differential equation by making use of the technique of an integrating factor and the solution of the initial value problem such that $y(t_0) = y_0$.

**5.**
$$\frac{dy}{dt} = \frac{1}{5}y(t) + t, \ t_0 = 0, \ y_0 = 3.$$

**6.**
$$\frac{dy}{dt} = -\frac{1}{2}y(t) + t^2, \ t_0 = 1, \ y_0 = 4$$

**7.**
$$\frac{dy}{dt} = ty(t) - t, \ t_0 = 0, \ y_0 = 4$$

**8.**
$$\frac{dy}{dt} = \sin(t) - y(t)\sin(t), \ t_0 = \frac{\pi}{3}, \ y_0 = 2$$

**9.**
$$\frac{dy}{dt} = \frac{y(t)}{t} + t^2, \ t_0 = 1, \ y_0 = 4$$

**10.**
$$\frac{dy}{dt} + \frac{y(t)}{t} = \sin(t), \ t_0 = \pi, \ y_0 = 3$$

(Assume that $t > 0$).

In problems 11 and 12,
a) Determine the solution of the initial value problem such that $y(0) = y_0$,
b) Express the solution as $y = y_p + y_{tran}$, where $\lim_{t \to \infty} y_{tran}(t) = 0$.
c) [C] Plot the graph of the solution and $y_p$ with your graphing utility. Does the picture support your determination of $y_{tran}$?

**11.**
$$\frac{dy}{dt} = -\frac{1}{4}y(t) + \sin(4t), \quad y_0 = 2.$$

Hint:
$$\int e^{at} \sin(bt)\, dt = \frac{a\sin(bt)e^{at} - b\cos(bt)e^{at}}{a^2 + b^2}$$

**12.**
$$\frac{dy}{dt} = -\frac{1}{2}y(t) + \cos(3t), \quad y_0 = 5$$

Hint:
$$\int e^{at} \cos(bt)\, dt = \frac{a\cos(bt)e^{at} + b\sin(bt)e^{at}}{a^2 + b^2}$$

In problems 13 and 14, determine an expression for the solution of the given initial value problem where $y(t_0) = y_0$. The expression may contain an integral that you may not be able to evaluate. Do not attempt to evaluate that integral.

**13.**
$$\frac{dy}{dt} = -\frac{1}{3}y(t) + \frac{1}{\sqrt{t^2 + 1}}, \quad t_0 = 1, \ y_0 = 10$$

**14**
$$\frac{dy}{dt} = -\frac{1}{2}y(t) + \sin(t^2), \quad t_0 = 2, \ y_0 = 4$$

**15.**
a) Determine an expression for the solution of the initial value problem
$$\frac{dy}{dt} = -ty + \sin(t), \quad y(0) = 3.$$

The expression may contain an integral that you may not be able to evaluate. Do not attempt to evaluate that integral.
b) [C] Make use of your computational utility in order to approximate $y(1)$.
c) [C] Make use of your computational/graphing utility in order to plot the solution on the interval $[-2, 2]$.

## 8.2 Applications of First-Order Linear Differential Equations

The mathematical models of diverse phenomena involve differential equations. In this section we will discuss some examples that lead to first-order linear differential equations.

## 8.2. APPLICATIONS OF FIRST-ORDER LINEAR DIFFERENTIAL EQUATIONS

### Falling Body Subject to Viscous Damping

Assume that an object of mass $m$ kilograms is dropped from a great height, and that its initial velocity is 0. There is the downward force $mg$ newtons on the body, where $g$ is the gravitational acceleration (in meters/second$^2$) and is assumed to be constant. We will assume that the resistance of the air exerts a force on the object that is directed upwards and has magnitude proportional to velocity of the object. Thus, there is a constant $\gamma > 0$ such that the magnitude of the resisting force is $\gamma v(t)$, where $v(t)$ is the velocity at the instant $t$. This kind of resistive force is called **viscous damping** and $\gamma$ is the **viscous damping constant**. The model of viscous damping is relevant to falling bodies that have low density, such as feathers and snowflakes, and not to a parachutist (a model that is relevant to the latter case will be discussed in Section 8.4). The net downward force acting on the body is $mg - \gamma v(t)$ at time $t$. By Newton's second law of motion, the net force acting on a body is equal to $ma(t)$, where $a(t)$ is the acceleration at time $t$ (in meters/second$^2$). Since acceleration is the rate of change of velocity, $a(t) = v'(t)$. Therefore, we have the equality,

$$m\frac{dv}{dt} = mg - \gamma v(t).$$

We can rewrite the above relationship as

$$\frac{dv}{dt} = -\frac{\gamma}{m}v(t) + g.$$

Thus, $v(t)$ satisfies a first-order linear differential equation, as we discussed in Section 8.1. We will assume that at the body is initially at rest and $t = 0$ corresponds to the instant the body is dropped. Thus, the initial condition is that $v(0) = 0$.

We will implement the technique of an integrating factor. We write the equation as

$$\frac{dv}{dt} + \frac{\gamma}{m}v(t) = g$$

and multiply both sides the equation by the integrating factor $e^{\gamma t/m}$:

$$e^{\gamma t/m}v'(t) + e^{\gamma t/m}\frac{\gamma}{m}v(t) = e^{\gamma t/m}g,$$

Thus,

$$\frac{d}{dt}\left(e^{\gamma t/m}v(t)\right) = ge^{\gamma t/m}.$$

Therefore,

$$e^{\gamma t/m}v(t) = \int ge^{\gamma t/m}dt = \frac{mg}{\gamma}e^{\gamma t/m} + C,$$

where $C$ is a constant. Since $v(0) = 0$,

$$\frac{mg}{\gamma} + C = 0 \Rightarrow C = -\frac{mg}{\gamma}.$$

Therefore,

$$v(t) = e^{-\gamma t/m}\left(\frac{mg}{\gamma}e^{\gamma t/m} - \frac{mg}{\gamma}\right) = \frac{mg}{\gamma}\left(1 - e^{-\gamma t/m}\right).$$

Since $\gamma > 0$, we have $\lim_{t \to \infty} e^{-\gamma t/m} = 0$. Thus,

$$\lim_{t \to \infty} v(t) = \frac{mg}{\gamma}.$$

The limit of $v(t)$ as $t \to \infty$ is called **the terminal velocity** of the object. Since

$$v(t) - \frac{mg}{\gamma} = \frac{mg}{\gamma}e^{-\gamma t/m},$$

$v(t)$ approaches the terminal velocity exponentially. The body is virtually at terminal velocity not too long after the start of the fall. The graph of the velocity function is as in Figure 1. Note that the terminal velocity $mg/\gamma$ is a steady-state of the differential equation

$$\frac{dv}{dt} = -\frac{\gamma}{m}v(t) + g.$$

This is a stable steady-state, since

$$\lim_{t \to \infty} v(t) = \frac{mg}{\gamma}$$

for any solution $v(t)$ of the differential equation. □

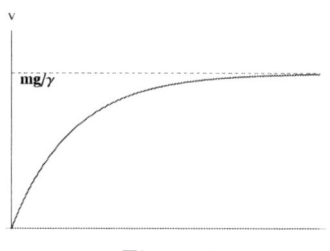

Figure 1

**Example 1** Assume that the viscous damping model is applicable to the fall of a whiffle ball. With the notation of the preceding discussion, assume that $g = 9.8$ meters/sec$^2$, $\gamma = 0.1$ and $m = 200$ grams. Determine the velocity of the whiffle ball as a function of time and its terminal velocity.

**Solution**

We have

$$\frac{dv}{dt} = -\frac{0.1}{0.2}v(t) + 9.8 = -\frac{1}{2}v(t) + 9.8$$

Therefore,

$$v(t) = \frac{mg}{\gamma}\left(1 - e^{-\gamma t/m}\right) = \frac{0.2 \times 9.8}{0.1}\left(1 - e^{-0.1t/0.2}\right) = 19.6\left(1 - e^{-t/2}\right).$$

In particular,

$$\lim_{t \to +\infty} v(t) = \lim_{t \to +\infty} 19.6\left(1 - e^{-t/2}\right) = 19.6 \text{ (meters per second)},$$

so that the terminal velocity of the whiffle ball is 19.6 meters per second. Figure 2 shows the graph of the velocity function. Note that the velocity of the whiffle ball is very close to the terminal velocity after 8 seconds. □

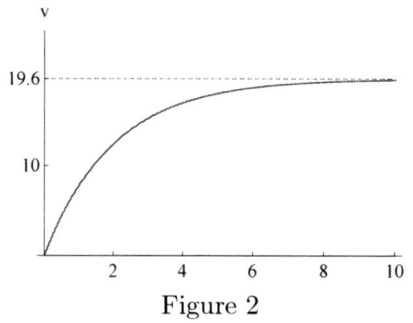

Figure 2

## Newton's Law of Cooling

Assume that a hot body is immersed in a medium such water in a tank. We will assume that the surrounding medium is at a lower temperature and is big enough so that we can assume its temperature to be constant, say $T_m$ ($^\circ C$). Let $T(t)$ denote the temperature of the hot body at time $t$ (in minutes). We will assume that **Newton's law of cooling** is valid: **The rate at which the temperature of the body changes is proportional to the difference between the temperature of the body and the temperature of the surrounding medium.** Thus, there exists a constant $k > 0$ such

$$\frac{d}{dt}T(t) = -k(T(t) - T_m)$$

at any time $t$ (the $(-)$ sign corresponds to the fact that the body is cooling, i.e., $T(t)$ is decreasing). The initial temperature of the body is $T_0 > T_m$. Let us determine $T(t)$.
We can write the equation as

$$\frac{d}{dt}T(t) + kT(t) = kT_m.$$

This is a first-order linear differential equation, and we will implement the technique of an integrating factor. We begin by multiplying both sides of the equation by the integrating factor is $e^{kt}$:

$$e^{kt}\frac{d}{dt}T(t) + ke^{kt}T(t) = ke^{kt}T_m,$$

so that

$$\frac{d}{dt}\left(e^{kt}T(t)\right) = ke^{kt}T_m.$$

Therefore,

$$e^{kt}T(t) = \int ke^{kt}T_m dt = kT_m \int e^{kt} dt = kT_m \left(\frac{1}{k}e^{kt} + C\right) = T_m e^{kt} + CkT_m,$$

where $C$ is a constant. Since $C$ denotes a constant that will be determined by the given initial condition, it does not hurt to relabel $CkT_m$ simply as $C$. Thus,

$$e^{kt}T(t) = T_m e^{kt} + C \Rightarrow T(t) = T_m + Ce^{-kt}.$$

In order to have $T(0) = T_0$, we must have

$$T_m + C = T_0 \Leftrightarrow C = T_0 - T_m.$$

Therefore,

$$T(t) = T_m + (T_0 - T_m)e^{-kt}.$$

We see that $T(t)$ decays towards the temperature of the surrounding medium exponentially :

$$\lim_{t \to \infty} T(t) = \lim_{t \to \infty} \left(T_m + (T_0 - T_m)e^{-kt}\right) = T_m + (T - T_0)\lim_{t \to \infty} e^{-kt} = T_m$$

(The constant $k > 0$). The graph of $T$ is as in Figure 3.

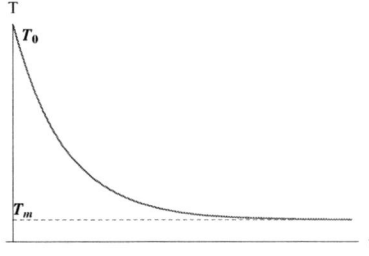

Figure 3

**Example 2** We will assume Newton's law of cooling and the use the notation of the preceding discussion. Assume that a hot body is immersed in a big tank that contains water at temperature $20°C$ and cools from $200°C$ to $40°C$ in 2 hours. Determine the temperature $T$ at any time $t$.

**Solution**

We have not been given the constant $k$, but we are given the information that $T(2) = 40$, in addition to the initial temperature, $T_0 = 200$. We know that $T(t)$ is in the form,

$$T(t) = T_m + (T(0) - T_m)e^{-kt} = 20 + (200 - 20)\,e^{-kt} = 20 + 180e^{-kt}.$$

We can determine $k$ by making use of the information that $T(2) = 40$:

$$40 = T(2) = 20 + 180e^{-2k}.$$

Therefore,

$$e^{-2k} = \frac{20}{180} \Leftrightarrow -2k = \ln(\frac{1}{9}) \Leftrightarrow -2k = -\ln(9) \Leftrightarrow k = \frac{1}{2}\ln(9).$$

Thus,

$$T(t) = 20 + 180e^{-\ln(9)t/2} = 20 + 180\left(e^{\ln(1/9)}\right)^{t/2} = 20 + 180\left(\frac{1}{9}\right)^{t/2}.$$

Figure 4 shows the graph of $T$.

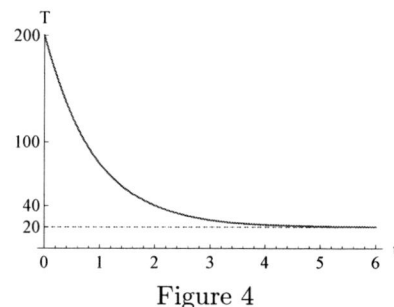

Figure 4

Note that

$$\lim_{t \to \infty} T(t) = \lim_{t \to \infty} \left(20 + 180\left(\frac{1}{9}\right)^{t/2}\right) = 20,$$

i.e., the temperature of the hot body decays to the temperature of the surrounding medium as $t \to +\infty$. The temperature of the hot body is virtually the same as the temperature of the water after 5 hours. □

## A Mixing Problem

Certain mixing problems lead to first-order linear differential equations as in the following example:

**Example 3** Assume that a reservoir contains clean water initially, and that the volume of the water in the reservoir is $10^5$ cubic meters. A stream carries water into the reservoir at the rate of 200 cubic meters per hour, and contains a pollutant with concentration 0.2 grams per cubic meter. The outflow of water is at the same rate as inflow. Determine the amount of the pollutant in the reservoir as a function of time.

## Solution

Let $y(t)$ denote the amount of pollutant in the reservoir at time $t$. The concentration of the pollutant in the reservoir at time $t$ is

$$\frac{\text{amount of pollutant}}{\text{total volume of water}} = \frac{y(t)}{10^5} \text{ (grams/meter}^3\text{)}.$$

The rate at which the pollutant flows out is

$$\frac{y(t)}{10^5} \text{ (grams/meter}^3\text{)} \times 200 \text{ (meter}^3\text{/hour)} = 2 \times 10^{-3} y(t) \text{ (grams/hour)}.$$

The rate at which the pollutant enters the reservoir is

$$0.2 \text{ (grams/meter}^3\text{)} \times 200 \text{ (meter}^3\text{/hour)} = 40 \text{ (grams per hour)}.$$

Therefore, the rate at which $y$ changes at time $t$ is

$$\frac{dy}{dt} = \text{rate in } - \text{ rate out} = 40 - 2 \times 10^{-3} y(t) \text{ (grams/hour)}$$

Thus,

$$\frac{dy}{dt} + 2 \times 10^{-3} y(t) = 40.$$

We multiply both sides of the equation by the integrating factor is $e^{2 \times 10^{-3} t}$, and obtain

$$\frac{d}{dt}\left(e^{2 \times 10^{-3} t} y(t)\right) = 40 e^{2 \times 10^{-3} t}.$$

Therefore,

$$e^{2 \times 10^{-3} t} y(t) = \int 40 e^{2 \times 10^{-3} t} dt = \frac{40}{2 \times 10^{-3}} e^{2 \times 10^{-3} t} + C = 2 \times 10^4 e^{2 \times 10^{-3} t} + C.$$

Thus,

$$y(t) = 2 \times 10^4 + C e^{-2 \times 10^{-3} t}.$$

We have $y(0) = 0$. Therefore,

$$0 = 2 \times 10^4 + C \Rightarrow C = -2 \times 10^4.$$

Thus,

$$y(t) = 2 \times 10^4 \left(1 - e^{-2 \times 10^{-3} t}\right)$$

Figure 5 shows the graph of the solution.

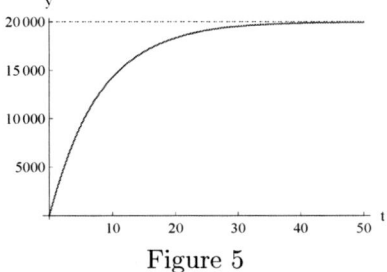

Figure 5

Note that
$$\lim_{t \to +\infty} y(t) = 2 \times 10^4.$$
The concentration of the pollutant in the reservoir corresponding to total amount $2 \times 10^4$ is
$$\frac{2 \times 10^4}{10^5} = 0.2 \text{ grams}/m^3.$$
This is the concentration of the pollutant in the inflow. This result might have been anticipated: The concentration of the pollutant in the reservoir approaches the concentration of the pollutant in the inflow as time goes by. Figure 5 indicates that the concentration of the pollutant is virtually the same as the concentration of the pollutant in the stream within two days.

## Simple Electrical Circuits

Consider a simple **electrical circuit** which involves the resistance $R$, and the inductance $L$. Assume that the power source supplies the constant voltage $E$. The current $I$ is a function of time $t$. If you take a course which covers electrical circuits, it will be shown that $I$ satisfies the following differential equation under appropriate assumptions:
$$L\frac{dI}{dt} + RI(t) = E,$$
i.e.,
$$\frac{dI}{dt} + \left(\frac{R}{L}\right)I(t) = \frac{E}{L}.$$
If we set
$$k = \frac{R}{L} \text{ and } q = \frac{E}{L},$$
The differential equation can be expressed as
$$\frac{dI}{dt} + kI(t) = q.$$
We multiply by the integrating factor $e^{kt}$ so that
$$\frac{d}{dt}\left(e^{kt}I(t)\right) = e^{kt}q.$$
Therefore,
$$e^{kt}I(t) = \int qe^{kt}dt = \frac{q}{k}e^{kt} + C,$$
where $C$ is a constant. Thus,
$$I(t) = e^{-kt}\left(\frac{q}{k}e^{kt} + C\right) = \frac{q}{k} + Ce^{-kt}.$$
If $I(0) = I_0$, we have
$$I_0 = \frac{q}{k} + C \Rightarrow C = I_0 - \frac{q}{k}.$$
Therefore,
$$I(t) = \frac{q}{k} + \left(I_0 - \frac{q}{k}\right)e^{-kt}.$$
Since
$$q = \frac{E}{L} \text{ and } k = \frac{R}{L},$$

## 8.2. APPLICATIONS OF FIRST-ORDER LINEAR DIFFERENTIAL EQUATIONS

we have
$$\frac{q}{k} = \frac{E}{R}.$$

Thus,
$$I(t) = \frac{q}{k} + \left(I_0 - \frac{q}{k}\right)e^{-kt} = \frac{E}{R} + \left(I_0 - \frac{E}{R}\right)e^{-Rt/L}.$$

Since $\lim_{t\to+\infty} e^{-Rt/L} = 0$,
$$\lim_{t\to\infty} I(t) = \frac{E}{R}.$$

The quantity $E/R$ is the steady-state solution of the equation
$$L\frac{dI}{dt} + RI(t) = E.$$

Indeed, if the current $I$ has the constant value $E/R$, we have $I'(t) = 0$, and $I = E/R$ is the solution of the equation $RI = E$. The term
$$\left(I_0 - \frac{E}{R}\right)e^{-Rt/L}$$

is called the transient part of the solution. This term decays very rapidly due the factor $\exp((-R/L)t)$, and the current is hardly distinguishable from the steady-state in a short time.

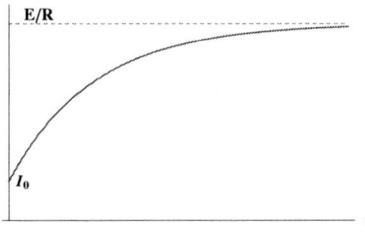

Foigure 6

Now let us assume that the voltage is sinusoidal. Thus, we have a simple electrical circuit which involves the resistance $R$, the inductance $L$ and the voltage $E$ is of the form $E_0 \sin(\omega t)$. The quantity $E_0$ is the amplitude of the voltage and $2\pi/\omega$ is its frequency. The current $I$ satisfies the differential equation,
$$L\frac{dI}{dt} + RI(t) = E_0 \sin(\omega t) \Rightarrow \frac{dI}{dt} + \frac{R}{L}I(t) = \frac{E_0}{L}\sin(\omega t).$$

If we set $k = R/L$, as in the case of a constant voltage, and $q = E_0/L$,
$$\frac{dI}{dt} + kI(t) = q\sin(\omega t),$$

Therefore,
$$\frac{d}{dt}\left(e^{kt}I(t)\right) = qe^{kt}\sin(\omega t).$$

Thus,
$$e^{kt}I(t) = \int qe^{kt}\sin(\omega t)\,dt = q\left(-\frac{\omega}{k^2+\omega^2}e^{kt}\cos(\omega t) + \frac{k}{k^2+\omega^2}e^{kt}\sin(\omega t)\right) + C,$$

where $C$ is a constant. Thus,

$$I(t) = q\left(-\frac{\omega}{k^2+\omega^2}\cos(\omega t) + \frac{k}{k^2+\omega^2}\sin(\omega t)\right) + Ce^{-kt}.$$

If $I(0) = I_0$, we must have

$$-\frac{q\omega}{k^2+\omega^2} + C = I_0 \Leftrightarrow C = I_0 + \frac{q\omega}{k^2+\omega^2}.$$

Therefore,

$$I(t) = q\left(-\frac{\omega}{k^2+\omega^2}\cos(\omega t) + \frac{k}{k^2+\omega^2}\sin(\omega t)\right) + \left(I_0 + \frac{q\omega}{k^2+\omega^2}\right)e^{-kt}$$

$$= \frac{E_0}{L\left(\left(\frac{R}{L}\right)^2+\omega^2\right)}\left(-\omega\cos(\omega t) + \frac{R}{L}\sin(\omega t)\right) + e^{-\frac{R}{L}t}\left(I_0 + \frac{E_0}{L}\frac{\omega}{\left(\frac{R}{L}\right)^2+\omega^2}\right).$$

The second term represents **the transient part of the solution**:

$$\lim_{t\to\infty} e^{-\frac{R}{L}t}\left(I_0 + \frac{E_0}{L}\frac{\omega}{\left(\frac{R}{L}\right)^2+\omega^2}\right) = 0.$$

The first term is **oscillatory** and has the same frequency as the driving voltage. Since the transient part of the solution decays exponentially, the current is essentially sinusoidal after a short time.

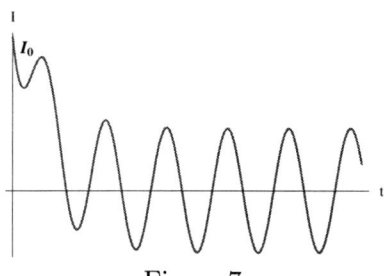

Figure 7

## Problems

**1.** Assume that the viscous damping model is applicable to a falling feather so that

$$\frac{dv}{dt} = -\frac{\gamma}{m}v(t) + g.$$

where $g = 9.8$ meters/sec$^2$, $\gamma = 0.01$ and $m = 1$ gram.
a) Determine the velocity of the feather as a function of time and its terminal velocity.
b) [C] Plot the graph of the velocity function. Indicate the asymptotes.

**2.** Assume that Newton's law of cooling is valid. A hot piece of iron is immersed in a big tank that contains water at temperature $15°C$ and cools from $300°C$ to $80°C$ in 1 hour.
a) Determine the temperature $T$ of the piece of iron at any time $t$. What is $\lim_{t\to\infty} T(t)$?
b) [C] Plot the graph of the velocity function. Indicate the asymptotes.

**3.** Assume that a reservoir contains clean water initially, and that the volume of the water in the reservoir is $10^4$ cubic meters. A stream carries water into the reservoir at the rate of 180 cubic meters per hour, and contains a pollutant with concentration 0.15 grams per cubic meter. The outflow of water is at the same rate as inflow.
a) Determine the amount of the pollutant in the reservoir as a function of time and the limit of the amount of the pollutant in the reservoir as time tends to infinity.
c) [C] Plot the graph of the pollutant. Indicate the asymptotes.

**4.** Consider a simple **electrical circuit** which involves the resistance $R$, and the inductance $L$. Assume that the power source supplies the constant voltage $E$. The current $I$ is a function of time $t$. We have
$$L\frac{dI}{dt} + RI(t) = E.$$
a) Determine $I(t)$ (amperes) if $E = 4$ volts, $L = 10$ henry, $R = 4$ ohm and $I(0) = 0$.
b) Determine the steady-state solution and the transient part of $I(t)$. Show that $I(t)$ tends to the steady-state as $t$ tends to infinity.
c) [C] Plot the graph of $I$. Indicate the asymptotes.

**5.** Consider a simple **electrical circuit** which involves the resistance $R$, and the inductance $L$. Assume that the power source supplies $E(t) = E_0 \sin(\omega t)$. The current $I$ is a function of time $t$. We have
$$L\frac{dI}{dt} + RI(t) = E_0 \sin(\omega t).$$
a) Determine $I(t)$ (amperes) if $E_0 = 240$ volts, $\omega = 60$, $L = 10$ henry, $R = 4$ ohm and $I(0) = 0$.
b) Determine the sinusoidal and the transient parts of $I(t)$. Show that the difference between $I(t)$ and its sinusoidal part tends to 0 as $t$ tends to infinity.
c) [C] Plot the graph of $I$. Indicate the sinusoidal part of $I$.

## 8.3 Separable Differential Equations

In sections 8.1 and 8.2 we discussed first-order linear differential equations and obtained the solutions by the technique of an integrating factor. In this section we will discuss first-order nonlinear differential equations that can be solved by the method of separation of variables.

A differential equation such as
$$\frac{dy}{dt} = \frac{t}{t^2 + 1} y^2$$
is a first-order differential equation since it involves only the first derivative of the unknown function $y(t)$. If we set
$$f(t, y) = \frac{t}{t^2 + 1} y^2,$$
we can express the differential equation as
$$\frac{dy}{dt} = f(t, y).$$

For fixed $t$, $f(t, y)$ is a nonlinear function of $y$. Therefore, the differential equation is not a linear differential equation, and we cannot implement the technique of an integrating factor in order to find its solutions. On the other hand, we can express $f(t, y)$ as the product of $t/(t^2+1)$ and $y^2$. The first expression depends only on $t$, and the second expression depends only on $y$. Such a differential equation is referred to as a separable differential equation:

**Definition 1** A differential equation of the form

$$\frac{dy}{dt} = g(t)h(y),$$

where $y = y(t)$ is **a separable differential equation.**

**Remark** A first-order linear differential equation of the type

$$\frac{dy}{dt} = g(t)y$$

is separable, since we can express the equation as $y' = g(t)h(y)$, where $h(y) = y$. On the other hand, a linear differential equation in the form

$$\frac{dy}{dt} = p(t)y + q(t),$$

where $q(t)$ is not identically 0 is not separable. $\Diamond$

We will implement **the technique of the separation of variables** in order to solve nonlinear separable differential equations:

Assume that $y(t)$ is a solution of the separable differential equation $y' = g(y)h(y)$. We divide both sides by $h(y)$ (without worrying about $t$ such that $h(y(t)) = 0$). Thus,

$$\frac{1}{h(y)}\frac{dy}{dt} = g(t).$$

We antidifferentiate:

$$\int \frac{1}{h(y)}\frac{dy}{dt}dt = \int g(t)\,dt.$$

By the substitution rule,

$$\int \frac{1}{h(y)}\frac{dy}{dt}dt = \int \frac{1}{h(y)}dy.$$

Therefore,

$$\int \frac{1}{h(y)}dy = \int g(t)\,dt.$$

Conversely, a differentiable function that satisfied the above relationship satisfies the differential equation $y' = g(t)h(y)$:
Let's set

$$H(y) = \int \frac{1}{h(y)}dy$$

so that

$$\frac{dH}{dy} = \frac{1}{h(y)},$$

and

$$G(t) = \int g(t)\,dt$$

so that

$$\frac{dG}{dt} = g(t).$$

Thus,

$$\int \frac{1}{h(y)}dy = \int g(t)\,dt \Leftrightarrow H(y) = G(t).$$

## 8.3. SEPARABLE DIFFERENTIAL EQUATIONS

Since $y(t)$ is defined implicitly by the relationship $H(y) = G(t)$, we have $H(y(t)) = G(t)$ for each $t$ in some interval $J$. By the chain rule,

$$\left(\frac{dH}{dy}\bigg|_{y=y(t)}\right)\frac{dy}{dt} = \frac{dG}{dt} \Rightarrow \frac{1}{h(y(t))}\frac{dy}{dt} = g(t) \Rightarrow \frac{dy}{dt} = g(t)h(y(t))$$

for each $t \in J$. Thus, $y(t)$ is a solution of the differential equation $y'(t) = g(t)h(y)$.

In practice, given the separable differential equation

$$\frac{dy}{dt} = g(t)h(y),$$

we can treat the symbolic fraction $dy/dt$ as a genuine fraction and write

$$\frac{1}{h(y)}dy = g(t)\,dt \Rightarrow \int \frac{1}{h(y)}dy = \int g(t)\,dt.$$

There is no magic! The symbolic manipulation is justified by the previous discussion that is based on the substitution rule.

Since the indefinite integrals are determined only up to an additive constant, we can add an arbitrary constant to either side of the equality

$$\int \frac{1}{h(y)}dy = \int g(t)\,dt.$$

It is sufficient to add a constant to the right-hand side. Therefore, the solution of the above expression for $y$ involves an arbitrary constant, so that we determine **the general solution** of the differential equation $y' = g(t)h(y)$. If an **initial condition** of the form $y(t_0) = y_0$ is specified, the solution of **the initial-value problem**

$$\frac{dy}{dt} = g(t)h(y),\ y(t_0) = y_0$$

is uniquely determined.

**Example 1**

a) Determine the general solution of the differential equation

$$\frac{dy}{dt} = \frac{t}{t^2+1}y^2$$

b) Determine the solution of the initial-value problem

$$\frac{dy}{dt} = \frac{t}{t^2+1}y^2,\ y(0) = \frac{1}{2}.$$

**Solution**

a) We divide both sides of the equation by $y^2$:

$$\frac{1}{y^2}\frac{dy}{dt} = \frac{t}{t^2+1}.$$

Therefore,

$$\int \frac{1}{y^2}\frac{dy}{dt}dt = \int \frac{t}{t^2+1}dt.$$

By the substitution rule,
$$\int \frac{1}{y^2} \frac{dy}{dt} dt = \int \frac{1}{y^2} dy = -\frac{1}{y}.$$
As for the right-hand side, if we set $u = t^2 + 1$, then $du/dt = 2t$. Therefore,
$$\int \frac{t}{t^2+1} dt = \frac{1}{2} \int \frac{1}{t^2+1} \frac{du}{dt} dt = \frac{1}{2} \int \frac{1}{u} du = \frac{1}{2} \ln(|u|) + C = \frac{1}{2} \ln(t^2+1) + C,$$
where $C$ is an arbitrary constant. Thus,
$$-\frac{1}{y} = \frac{1}{2} \ln(t^2+1) + C,$$
We solve the above relationship for $y$:
$$y(t) = -\frac{1}{\frac{1}{2}\ln(t^2+1) + C}.$$
We have shown that a solution of the given differential equation must be expressible as above. Conversely, if $y(t)$ is given by the above expression, $y(t)$ solves the given equation (confirm). Thus, the general solution of the given differential equation is
$$y(t) = -\frac{1}{\frac{1}{2}\ln(t^2+1) + C}.$$

b) We will determine $C$ so that $y(0) = 1/2$:
$$\frac{1}{2} = -\frac{1}{\frac{1}{2}\ln(1) + C} = -\frac{1}{C} \Leftrightarrow C = -2.$$
Therefore, the unique solution of the initial-value problem
$$\frac{dy}{dt} = \frac{t}{t^2+1} y^2(t), \; y(0) = \frac{1}{2}$$
is
$$y(t) = -\frac{1}{\frac{1}{2}\ln(t^2+1) - 2} = \frac{2}{4 - \ln(t^2+1)}.$$
Note that
$$4 - \ln(t^2+1) = 0 \Leftrightarrow \ln(t^2+1) = 4 \Leftrightarrow t^2 + 1 = e^4 \Leftrightarrow t = \pm\sqrt{e^4 - 1}.$$
Thus, the solution is not defined at $\pm\sqrt{e^4 - 1}$. Figure 1 shows the graph of $y(t)$.

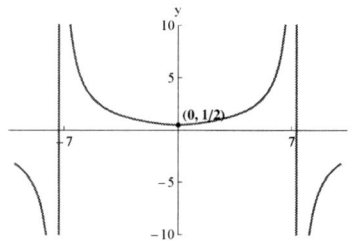

Figure 1

## 8.3. SEPARABLE DIFFERENTIAL EQUATIONS

As the picture indicates,

$$\lim_{t \to \sqrt{e^4-1}-} y(t) = -\infty \text{ and } \lim_{t \to -\sqrt{e^4-1}+} y(t) = +\infty$$

(check). The solution is continuous on the open interval

$$(-\sqrt{e^4-1}, \sqrt{e^4-1}) \cong (-7.321\,08, 7.321\,08).$$

□

**Remark** We have noted that a first-order linear differential equation of the type

$$\frac{dy}{dt} = g(t)\,y$$

is separable so that we can apply the technique of separation of variables:

$$\frac{1}{y}\frac{dy}{dt} = g(t)$$

⇒
$$\int \frac{1}{y}\frac{dy}{dt}\,dt = \int g(t)\,dt$$

⇒
$$\int \frac{1}{y}\,dy = \int g(t)\,dt$$

⇒
$$\ln(|y|) = \int g(t)\,dt + C$$

where $C$ is a constant. Thus

$$|y| = e^{\ln(|y|)} = e^{\int g(t)dt + C} = e^{\int g(t)dt} e^C.$$

If we replace $e^C$ by $C'$

$$|y| = C'e^{\int g(t)dt} \Rightarrow y = \pm C'e^{\int g(t)dt}.$$

If we replace $\pm C'$ simply by $C$

$$y = Ce^{\int g(t)dt}.$$

This expression agrees with the expression that we would have obtained if we had applied the technique of an integrating factor where the integrating factor is

$$e^{-\int g(t)dt}.$$

◊

**Example 2**

a) Determine the general solution of the differential equation

$$\frac{dy}{dt} = y^2(t).$$

b) Solve the initial-value problem,

$$\frac{dy}{dt} = y^2(t),\ y(0) = 1.$$

Determine the largest interval $I$ such that $0 \in J$ and $y(t)$ is defined for each $t \in I$.

c) Solve the initial-value problem,

$$\frac{dy}{dt} = y^2(t), \; y(0) = -1.$$

Determine the largest interval $J$ such that $0 \in J$ and $y(t)$ is defined for each $t \in J$.

**Solution**

a) The differential equation is separable. We can express the equation as

$$\frac{dy}{dt} = y^2 \Rightarrow \frac{1}{y^2}\frac{dy}{dt} = 1.$$

Therefore,

$$\int \frac{1}{y^2}\frac{dy}{dt} dt = \int 1 dt \Rightarrow \int \frac{1}{y^2} dy = t + C,$$

where $C$ is an arbitrary constant. Since

$$\int \frac{1}{y^2} dy = \int y^{-2} dy = -y^{-1} = -\frac{1}{y},$$

we obtain the relationship

$$-\frac{1}{y} = t + C.$$

Therefore,

$$y(t) = -\frac{1}{t+C}$$

is the general solution of the differential equation $y'(t) = y^2(t)$.

b) We have

$$y(0) = 1 \Leftrightarrow -\frac{1}{t+C}\bigg|_{t=0} = 1 \Leftrightarrow -\frac{1}{C} = 1 \Leftrightarrow C = -1.$$

Therefore, the solution of the initial-value problem

$$y'(t) = y^2(t), \; y(0) = 1$$

is

$$y(t) = -\frac{1}{t-1}.$$

Figure 2 shows the graph of $y(t)$.

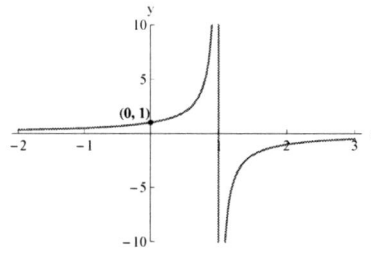

Figure 2

## 8.3. SEPARABLE DIFFERENTIAL EQUATIONS

The largest interval that contains 0 such that $y(t)$ is defined for each $t$ in the interval is $(-\infty, 1)$. Even though $y(t)$ is defined if $t > 1$, the values of $y(t)$ for $t > 1$ are irrelevant to the given initial value problem. If you imagine that $t$ is time, the initial value of $y$ is given at $t = 0$, and $y(t)$ evolves as $t$ increases towards 1. We have

$$\lim_{t \to 1-} y(t) = \lim_{t \to 1-} \left(-\frac{1}{t-1}\right) = +\infty,$$

i.e., $y(t)$ becomes arbitrarily large as $t$ approaches 1 from the left.

c) We have

$$y(0) = -1 \Leftrightarrow -\frac{1}{t+C}\bigg|_{t=0} = -1 \Leftrightarrow -\frac{1}{C} = -1 \Leftrightarrow C = 1.$$

Therefore, the solution of the initial-value problem

$$y'(t) = y^2(t),\ y(0) = -1$$

is

$$y(t) = -\frac{1}{t+1}.$$

The largest interval that contains 0 and $y(t)$ is defined for each $t$ in the interval is $(-1, +\infty)$. Figure 3 shows the graph of the solution. Note that

$$\lim_{t \to -1+} y(t) = -\infty.$$

□

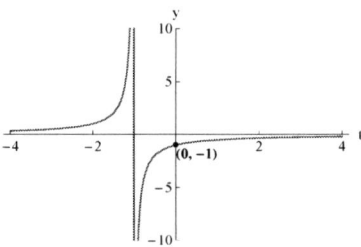

Figure 3

### Example 3

Consider the differential equation

$$\frac{dy}{dt} = y(t) - \frac{1}{10}y^2(t).$$

a) Determine the **steady-state solutions** (or **equilibrium solutions**), i.e., the constant solutions of the differential equations.
b) Determine the general solution of the differential equation.
c) Determine the solutions of the initial-value problems

$$\frac{dy(t)}{dt} = y(t) - \frac{1}{10}y^2(t),\ y(0) = y_0,$$

where $y_0 = 1, -10$ or $20$. For each solution, determine the largest interval that contains 0 such that the solution is defined on that interval

**Solution**

a) The constant $y$ is a solution of the differential equation if and only if
$$0 = \frac{dy}{dt} = y - \frac{1}{10}y^2 \Leftrightarrow 10y(10-y) = 0 \Leftrightarrow y = 0 \text{ or } y = 10.$$

Thus, the steady-state solutions are 0 and 10.

b) The given equation is separable. We will implement the technique of separation of variables. We write the differential equation as
$$\frac{dy}{dt} = y - \frac{1}{10}y^2.$$

Therefore,
$$\frac{1}{y - \frac{1}{10}y^2}\frac{dy}{dt} = 1 \Rightarrow \int \frac{1}{y - \frac{1}{10}y^2}\frac{dy}{dt}dt = \int 1 dt \Rightarrow \int \frac{1}{y - \frac{1}{10}y^2}dy = t + C.$$

The integrand on the left-hand side is a rational function and has a partial fraction decomposition:
$$\frac{1}{y - \frac{1}{10}y^2} = \frac{1}{y} - \frac{1}{y - 10}$$

(check). Therefore,
$$\int \frac{1}{y - \frac{1}{10}y^2}dy = \int \left(\frac{1}{y} - \frac{1}{y - 10}\right)dy == \ln(|y|) - \ln(|y - 10|) = \ln\left(\left|\frac{y}{y-10}\right|\right).$$

Thus,
$$\ln\left(\left|\frac{y}{y-10}\right|\right) = t + C.$$

We solve for $y$ by exponentiation:
$$\left|\frac{y}{y-10}\right| = e^{t+C} = e^C e^t \Leftrightarrow \frac{y}{y-10} = \pm e^C e^t.$$

Since $C$ is an arbitrary constant, we will relabel $\pm e^C$ as $C$, and express the relationship between $t$ and $y$ as
$$\frac{y}{y-10} = Ce^t.$$

We can solve for $y$:
$$y = Ce^t y - 10Ce^t \Leftrightarrow (1 - Ce^t)y = -10Ce^t \Rightarrow y = \frac{10Ce^t}{Ce^t - 1}.$$

Therefore, the general solution of the given differential equation can be expressed as
$$y(t) = \frac{10Ce^t}{Ce^t - 1},$$

where $C$ is an arbitrary constant.

c) We have $y(0) = 1$ if and only if
$$\left.\frac{10Ce^t}{Ce^t - 1}\right|_{t=0} = 1 \Leftrightarrow \frac{10C}{C-1} = 1 \Leftrightarrow C \Leftrightarrow -\frac{1}{9}.$$

## 8.3. SEPARABLE DIFFERENTIAL EQUATIONS

Therefore, the solution of the initial-value problem
$$\frac{dy}{dt} = y(t) - \frac{1}{10}y^2(t), \ y(0) = 1$$
is
$$y(t) = \frac{10\left(-\frac{1}{9}\right)e^t}{\left(-\frac{1}{9}\right)e^t - 1} = \frac{10e^t}{e^t + 9}.$$

We can express the solution as
$$y(t) = \frac{10e^t}{e^t\left(1 + \frac{9}{e^t}\right)} = \frac{10}{1 + 9e^{-t}}.$$

Since $e^{-t} > 0$ for each $t \in R$ we have $1 + 9e^{-t} > 1$. Therefore, $y(t)$ is defined for each $t \in R$. Note that
$$0 < \frac{10}{1 + 9e^{-t}} < \frac{10}{1} = 10$$
for each $t \in \mathbb{R}$, so that the values of the solution lie between the steady-states 0 and 10. We have
$$\lim_{t \to +\infty} y(t) = \lim_{t \to +\infty} \frac{10}{1 + 9e^{-t}} = 10,$$
since $\lim_{t \to +\infty} e^{-t} = 0$. Thus, $y(t)$ approaches the steady-state 10 as $t$ tends to infinity. As a matter of fact, $t$ does not have to be very large for $y(t)$ to be close to 10, since $e^{-t}$ approaches 0 very rapidly as $t$ increases.

As for the limit of $y(t)$ as $t$ tends to $-\infty$,
$$\lim_{t \to -\infty} y(t) = \lim_{t \to -\infty} \frac{10e^t}{e^t + 9} = 0,$$
since $\lim_{t \to -\infty} e^t = 0$.

Figure 4 shows the graph of the solution.

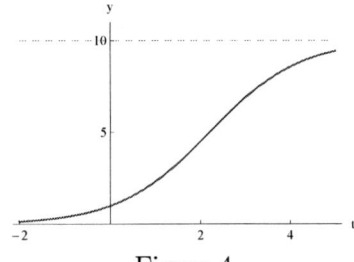

Figure 4

Now we will consider the initial-value problem
$$y'(t) = y(t) - \frac{1}{10}y^2(t), \ y(0) = -10.$$
Since the general solution is
$$y(t) = \frac{10Ce^t}{Ce^t - 1},$$

We will determine $C$ so that $y(0) = -10$:

$$-10 = \left.\frac{10Ce^t}{Ce^t - 1}\right|_{t=0} = \frac{10C}{C-1} \Leftrightarrow C = \frac{1}{2}$$

Therefore,

$$y(t) = \frac{10\left(\frac{1}{2}\right)e^t}{\frac{1}{2}e^t - 1} = \frac{10e^t}{e^t - 2}$$

is the solution that corresponds to the initial condition $y(0) = -10$. The value $y(t)$ is defined if and only if $e^t - 2 \neq 0$, i.e., $t \neq \ln(2)$. The interval $(-\infty, \ln(2))$ contains 0 and it is largest such interval on which the solution is defined. Figure 5 shows the graph of the solution on the interval $[-2, \ln(2))$.

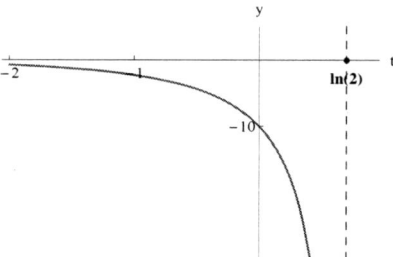

Figure 5

The picture indicates that $\lim_{t \to \ln(2)-} y(t) = -\infty$. Indeed, If $t \cong \ln(2)$,

$$y(t) = \frac{10e^t}{e^t - 2} \cong \frac{10e^{\ln(2)}}{e^t - 2} = \frac{20}{e^t - 2}.$$

If $t < \ln(2)$ then $e^t - 2 < 0$ and $\lim_{t \to \ln(2)} (e^t - 2) = 0$. Therefore,

$$\lim_{t \to \ln(2)-} y(t) = \lim_{t \to \ln(2)-} \frac{20}{e^t - 2} = -\infty$$

Finally, let us consider the initial condition $y(0) = 20$. We must have

$$y(t)|_{t=0} = \left.\frac{10Ce^t}{Ce^t - 1}\right|_{t=0} = \frac{10C}{C-1} = 20,$$

so that $C = 2$. Therefore,

$$y(t) = \frac{20e^t}{2e^t - 1}.$$

The solution is defined at $t$ if and only if $2e^t - 1 \neq 0$, i.e., $t \neq \ln(1/2) = -\ln(2)$. The interval $(-\ln(2), +\infty)$ contains 0 and it is largest such interval on which the solution is defined. Figure 6 shows the graph of the solution on the interval $(-\ln(2), 4]$. You should confirm that $\lim_{t \to -\ln(2)+} y(t) = +\infty$, as indicated by the picture. □

## 8.3. SEPARABLE DIFFERENTIAL EQUATIONS

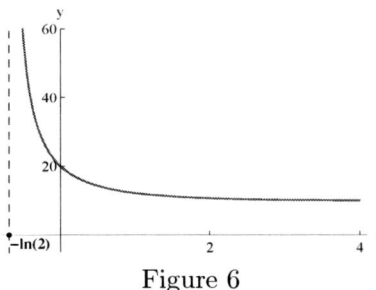

Figure 6

**Example 4**

a) Determine the steady-state solutions of the differential equation

$$\frac{dy(t)}{dt} = 1 - \frac{1}{9}y^2(t).$$

b) Determine the general solution of the differential equation

$$\frac{dy(t)}{dt} = 1 - \frac{1}{9}y^2(t).$$

c) Determine the solution $y(t)$ of the initial-value problem

$$\frac{dy}{dt} = 1 - \frac{1}{9}y^2, \ y(0) = 0,$$

and $\lim_{t \to \pm\infty} y(t)$. Are the limits related to the steady-state solutions?

**Solution**

a) The constant $y$ is a solution of the differential equation if and only if

$$1 - \frac{1}{9}y^2 = 0 \Leftrightarrow y = 3 \text{ or } y = -3.$$

Thus, the steady-state solutions are 3 and $-3$.

b) The differential equation is separable. We implement the technique of separation of variables:

$$\frac{dy}{dt} = 1 - \frac{1}{9}y^2 \Rightarrow \frac{1}{1 - \frac{1}{9}y^2}\frac{dy}{dt} = 1 \Rightarrow \int \frac{1}{1 - \frac{1}{9}y^2}\frac{dy}{dt}dt = \int 1 dt$$

$$\Rightarrow \int \frac{1}{1 - \frac{1}{9}y^2}dy = t + C,$$

where $C$ is an arbitrary constant. We have

$$\frac{1}{1 - \frac{1}{9}y^2} = -\frac{3}{2(y-3)} + \frac{3}{2(y+3)}$$

(check) so that

$$\int \frac{1}{1 - \frac{1}{9}y^2}dy = -\frac{3}{2}\ln(|y-3|) + \frac{3}{2}\ln(|y+3|) = \frac{3}{2}\ln\left(\left|\frac{y+3}{y-3}\right|\right).$$

Therefore,
$$\frac{3}{2}\ln\left(\left|\frac{y+3}{y-3}\right|\right) = t + C.$$

Thus,
$$\ln\left(\left|\frac{y+3}{y-3}\right|\right) = \frac{2}{3}t + \frac{2}{3}C.$$

We relabel $(2/3)C$ as $C$:
$$\ln\left(\left|\frac{y+3}{y-3}\right|\right) = \frac{2}{3}t + C.$$

Thus,
$$\left|\frac{y+3}{y-3}\right| = e^C e^{2t/3}.$$

Therefore,
$$\frac{y+3}{y-3} = \pm e^C e^{2t/3}.$$

We relabel $\pm e^C$ as $C$:
$$\frac{y+3}{y-3} = Ce^{2t/3}.$$

We solve the above relationship for $y$:
$$y + 3 = Ce^{2t/3}y - 3Ce^{2t/3} \Rightarrow \left(-Ce^{2t/3} + 1\right)y = -3\left(Ce^{2t/3} + 1\right) \Rightarrow y(t) = \frac{-3\left(Ce^{2t/3} + 1\right)}{-Ce^{2t/3} + 1}.$$

Therefore, the general solution of the given differential equation is
$$y(t) = \frac{3\left(Ce^{2t/3} + 1\right)}{Ce^{2t/3} - 1}.$$

b) We have
$$y(0) = 0 \Leftrightarrow \left.\frac{3\left(Ce^{2t/3} + 1\right)}{Ce^{2t/3} - 1}\right|_{t=0} = 0 \Leftrightarrow \frac{3(C+1)}{C-1} = 0 \Leftrightarrow C = -1.$$

Therefore, the solution of the initial-value problem
$$\frac{dy}{dt} = 1 - \frac{1}{9}y^2(t),\ y(0) = 0$$

is
$$y(t) = \frac{3\left(-e^{2t/3} + 1\right)}{-e^{2t/3} - 1} = \frac{3\left(e^{2t/3} - 1\right)}{e^{2t/3} + 1}$$

Figure 7 shows the graph of $f$.

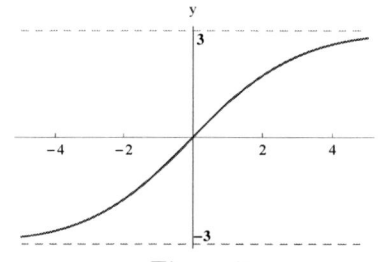

Figure 7

## 8.3. SEPARABLE DIFFERENTIAL EQUATIONS

We have
$$\lim_{t\to+\infty} y(t) = \lim_{t\to+\infty} \frac{3\left(e^{2t/3}-1\right)}{e^{2t/3}+1} = \lim_{t\to+\infty} \frac{3\left(1-e^{-2t/3}\right)}{1+e^{-2t/3}} = 3,$$
since $\lim_{t\to+\infty} e^{-2t/3} = 0$, and
$$\lim_{t\to-\infty} y(t) = \lim_{t\to-\infty} \frac{3\left(e^{2t/3}-1\right)}{e^{2t/3}+1} = -3,$$
since $\lim_{t\to-\infty} e^{2t/3} = 0$. Thus, $y(t)$ approaches the steady-state solution 3 as $t \to +\infty$, and the steady-state solution $-3$ as $t \to -\infty$.

Incidentally, the solution can be expressed in terms of hyperbolic tangent. Indeed,
$$\tanh(u) = \frac{\sinh(u)}{\cosh(u)} = \frac{e^u - e^{-u}}{e^u + e^u} = \frac{e^{2u}-1}{e^{2u}+1}.$$

Therefore,
$$y(t) = 3\left(\frac{e^{2t/3}-1}{e^{2t/3}+1}\right) = 3\tanh(t/3).$$

□

Here is an example where the technique of separation of variables leads to a relationship that cannot be solved for the unknown function explicitly:

**Example 5** Consider the separable differential equation
$$\frac{dy}{dt} = -\frac{y(t)}{y(t)+2}.$$

a) Use the technique of separation of variables to obtain a relationship that defines the solutions of the differential equation implicitly.
b) Make use of your computational/graphing utility to graph the relationship between $y$ and $t$ corresponding to the initial condition $y(0) = 1$.

**Solution**

a) We implement the technique of separation of variables:
$$\frac{y+2}{y}\frac{dy}{dt} = -1 \Rightarrow \int \frac{y+2}{y}\frac{dy}{dt}dt = -\int 1\,dt$$
$$\Rightarrow \int \left(1+\frac{2}{y}\right)dy = -t + C,$$

where $C$ is an arbitrary constant. Therefore,
$$y + 2\ln(|y|) = -t + C \Rightarrow y + \ln(y^2) = -t + C \Rightarrow e^{y+\ln(y^2)} = e^{-t+C} \Rightarrow e^y y^2 = Ce^{-t}.$$

Thus, we are led to the relationship
$$e^y y^2 = Ce^{-t}$$
between $y$ and $t$. For each $C$, the above relationship defines a solution of the differential equation
$$\frac{dy}{dt} = -\frac{y(t)}{y(t)+2}$$
implicitly.

b) We have $y(0) = 1$ if and only if
$$e^y y^2 = Ce^{-t}\big|_{t=0, y=1} \Leftrightarrow e = C.$$
Therefore, the relationship
$$e^y y^2 = ee^{-t}$$
defines the solution of the initial-value problem
$$\frac{dy}{dt} = -\frac{y(t)}{y(t)+2}, \ y(0) = 1$$
implicitly. Figure 8 shows the graph of the above relationship.

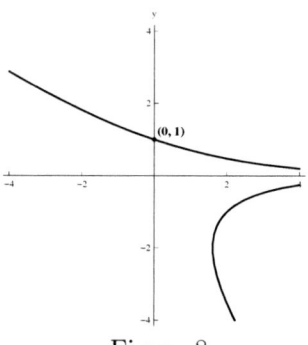

Figure 8

The picture indicates that there is a solution $y(t)$ that is defined implicitly by the relationship $e^y y^2 = ee^{-t}$ such that $y(0) = 1$ (the part of the graph shown in Figure 8 that lies above the $t$-axis is the graph of that solution). A computational/graphing utility such as the one that has produced Figure 8 enables us to compute approximations to $y(t)$ for desired values of $t$. □

In the next section we will discuss some interesting applications of nonlinear separable differential equations.

## Problems

In problems 1-9,
a) Determine the general solution of the given differential equation,
b) Determine the solution of the solution of the equation that corresponds to the given initial condition.

1.
$$\frac{dy}{dx} = -y^2, \ y(0) = 1.$$

2.
$$\frac{dy}{dx} = y^2 + 1, \ y\left(\frac{\pi}{4}\right) = 0.$$

3.
$$\frac{dy}{dx} = \sqrt{1+y^2}, \ y(0) = 2$$

4.
$$\frac{dy}{dx} + y^2 \sin(x) = 0, \ y\left(\frac{\pi}{3}\right) = \frac{1}{2}$$

5.
$$\frac{dy}{dx} = \frac{1+y^2}{1+x^2}, \ y(0) = 1$$

6.
$$\frac{dy}{dx} = \frac{y^2}{x}, \ y(e^2) = 2$$

7.
$$\frac{dy}{dx} = \frac{2xy^2}{1+x^2}, \ y(0) = -2$$

8.

9.

$$\frac{dy}{dx} = \frac{y^2}{1+x^2}, \; y(1) = -\frac{2}{\pi}$$

$$\frac{dy}{dt} = \sin(t)\, y^2, \; y(\pi) = 4$$

**10.**
a) Determine the steady-state solutions of the differential equation

$$\frac{dy}{dt} = \frac{1}{4} y(t) - \frac{1}{100} y^2(t).$$

b) Find the general solution and the solutions of the initial value problems for the above differential equation with $y(0) = 5$, $50$ and $-50$.
Specify the domains of the solutions. Determine the appropriate limits at $\pm\infty$ and at the vertical asymptotes of their graphs.
c) [C] Plot the graphs of the solutions with the help of your graphing utility. Are the pictures consistent with your responses to a) and b)?

**11.**
a) Determine the steady-state solutions of the differential equation

$$\frac{dy}{dt} = 4 - \frac{1}{100} y^2(t).$$

b) Find the general solution and solutions of the initial value problems for the above differential equation with $y(0) = 0$ and $40$.
Specify the domains of the solutions. Determine the appropriate limits at $\pm\infty$ and at the vertical asymptotes of their graphs.
c) [C] Plot the graphs of the solutions with the help of your graphing utility. Are the pictures consistent with your responses to a) and b)?

## 8.4 Applications of Separable Differential Equations

### Newtonian Damping

In Section 8.2 we discussed a falling body subject to viscous damping. The relevant differential equation is

$$mv'(t) = mg - \gamma v(t).$$

Here, $m$ is the mass of a falling object, $v$ is its velocity, $g$ is the gravitational acceleration, and $\gamma$ is a positive constant (you may assume that units are in the metric system). The term $\gamma v(t)$ represents the resistance of the air. The equation is a first-order linear differential equation. Experiments have shown that the model is realistic if the falling object is an object of low density, such as a feather or a snowflake. It does not appear to be realistic if the object is a dense body such as a raindrop, or a parachutist. In such a case, the modelling assumption that **the magnitude of the force due to air resistance is proportional to the square of the velocity** appears to lead to more credible results. This is referred to as **Newtonian damping**. Thus, if the positive direction is downward, $m, g$ and $v(t)$ have the same meaning as in the case of viscous damping, the net force acting on the body at the instant $t$ is $mg - \delta v^2(t)$, where $\delta$ is a positive constant. By **Newton's second law of motion**, the net force is equal to mass times acceleration $a(t)$, so that

$$mg - \delta v^2(t) = ma(t) = m\frac{dv}{dt}.$$

Therefore, the velocity $v(t)$ satisfies the differential equation,

$$\frac{dv}{dt} = g - \left(\frac{\delta}{m}\right)v^2(t).$$

If we set $k = \delta/m$, $k$ is a positive constant and is much smaller than $g$. We can rewrite the equation as

$$\frac{dv}{dt} = g - kv^2(t).$$

We will assume that the object is at rest at $t = 0$, so that the initial condition is that $v(0) = 0$. The equation is a first-order nonlinear separable differential equation. We will implement the technique of separation of variables. Thus, we write the equation as

$$\frac{dv}{dt} = g - kv^2 \Rightarrow \frac{1}{g - kv^2}\frac{dv}{dt} = 1,$$

and antidifferentiate:

$$\int \frac{1}{g - kv^2}\frac{dv}{dt}dt = \int 1 dt \Rightarrow \int \frac{1}{g - kv^2}dv = t + C,$$

where where $C$ is constant that will be determined in order to have $v(0) = 0$.
Since

$$g - kv^2 = -k\left(v^2 - \frac{g}{k}\right) = -k\left(v - \sqrt{\frac{g}{k}}\right)\left(v + \sqrt{\frac{g}{k}}\right),$$

we look for a partial fraction decomposition of the integrand as

$$\frac{1}{g - kv^2} \equiv \frac{A}{v - \sqrt{\frac{g}{k}}} + \frac{B}{v + \sqrt{\frac{g}{k}}},$$

where $A$ and $B$ are constants (depending on the parameters $g$ and $k$). You should check that

$$\frac{1}{g - kv^2} = -\frac{1}{2\sqrt{gk}}\left(\frac{1}{v - \sqrt{\frac{g}{k}}}\right) + \frac{1}{2\sqrt{gk}}\left(\frac{1}{v + \sqrt{\frac{g}{k}}}\right).$$

Therefore,

$$\int \frac{1}{g - kv^2}dv = -\frac{1}{2\sqrt{gk}}\int \frac{1}{v - \sqrt{\frac{g}{k}}}dv + \frac{1}{2\sqrt{gk}}\int \frac{1}{v + \sqrt{\frac{g}{k}}}dv$$

$$= -\frac{1}{2\sqrt{gk}}\ln\left(\left|v - \sqrt{\frac{g}{k}}\right|\right) + \frac{1}{2\sqrt{gk}}\ln\left(\left|v + \sqrt{\frac{g}{k}}\right|\right)$$

$$= \frac{1}{2\sqrt{gk}}\ln\left(\left|\frac{v + \sqrt{\frac{g}{k}}}{v - \sqrt{\frac{g}{k}}}\right|\right).$$

Thus,

$$\frac{1}{2\sqrt{gk}}\ln\left(\left|\frac{v + \sqrt{\frac{g}{k}}}{v - \sqrt{\frac{g}{k}}}\right|\right) = t + C \Rightarrow \ln\left(\left|\frac{v + \sqrt{\frac{g}{k}}}{v - \sqrt{\frac{g}{k}}}\right|\right) = \left(2\sqrt{gk}\right)t + \left(2\sqrt{gk}\right)C.$$

We will relabel $\left(2\sqrt{gk}\right)C$ as $C$ and exponentiate:

$$\left|\frac{\sqrt{\frac{g}{k}} + v}{\sqrt{\frac{g}{k}} - v}\right| = e^C e^{2\sqrt{gk}\,t}.$$

## 8.4. APPLICATIONS OF SEPARABLE DIFFERENTIAL EQUATIONS

Therefore,
$$\frac{\sqrt{\frac{g}{k}}+v}{\sqrt{\frac{g}{k}}-v}=\pm e^C e^{2\sqrt{gk}t}$$

We will relabel $\pm e^C$ as $C$:
$$\frac{\sqrt{\frac{g}{k}}+v}{\sqrt{\frac{g}{k}}-v}=Ce^{2\sqrt{gk}t}.$$

The initial condition is that $v(0)=0$. Therefore
$$\frac{\sqrt{\frac{g}{k}}}{\sqrt{\frac{g}{k}}}=C \Leftrightarrow C=1.$$

Thus,
$$\frac{\sqrt{\frac{g}{k}}+v}{\sqrt{\frac{g}{k}}-v}=e^{2\sqrt{gk}t}.$$

We solve for $v$:
$$v(t)=\sqrt{\frac{g}{k}}\left(\frac{1-e^{-2\sqrt{gk}t}}{1+e^{-2\sqrt{gk}t}}\right)=\sqrt{\frac{g}{k}}\left(\frac{e^{2\sqrt{gk}t}-1}{e^{2\sqrt{gk}t}+1}\right)=\sqrt{\frac{g}{k}}\tanh\left(\sqrt{gk}t\right)$$

(check). Since
$$0<\frac{1-e^{-2\sqrt{gk}t}}{1+e^{-2\sqrt{gk}t}}<1$$

if $t>0$, we have
$$0<v(t)<\sqrt{\frac{g}{k}}.$$

Figure 1 shows the graph of $v$ on a subinterval of $[0,+\infty)$.

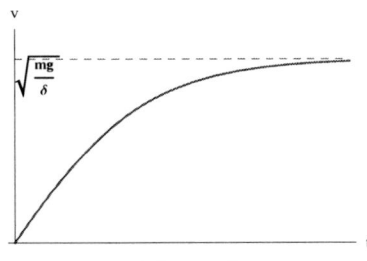

Figure 1

Note that
$$\lim_{t\to+\infty}v(t)=\lim_{t\to+\infty}\sqrt{\frac{g}{k}}\left(\frac{1-e^{-2\sqrt{gk}t}}{1+e^{-2\sqrt{gk}t}}\right)=\sqrt{\frac{g}{k}}=\sqrt{\frac{g}{\frac{\delta}{m}}}=\sqrt{\frac{mg}{\delta}}.$$

This model leads to the **terminal velocity**
$$\sqrt{\frac{mg}{\delta}}.$$

**Example 1** Assume that $g = 9.8$ meters/second$^2$, the mass of a sky diver and his equipment is 110 kilograms, and the Newtonian damping constant $\delta$ is 0.18. Then, the above model predicts that the terminal velocity of the sky diver is

$$\sqrt{\frac{mg}{\delta}} = \sqrt{\frac{110 \times 9.8}{0.18}} \cong 77.4 \text{ meters per second.}$$

□

## The Logistic Equation

Assume that $b > 0$, $c > 0$ and $b$ is much smaller than $c$. An equation of the form

$$\frac{dy}{dt} = cy - by^2(t),$$

is referred to as a **logistic equation**. This equation differs from the equation $y' = cy$ due to the term $-by^2$. It is no longer true that the rate of change of $y$ with respect to $t$ is proportional to $y(t)$. Even though $b$ is much smaller than $c$, the term $-by^2$ is significant if $y$ is large. Intuitively, that term inhibits the growth of $y$. Such an equation arises in **population models where factors which inhibit the growth of the population are taken into account.**

Let's begin by determining the steady-state solutions of such differential equations. The constant $y$ is the solution of the differential equation $y' = cy - by^2$ if and only if

$$0 = y' = cy - by^2 = y(c - by).$$

Therefore, the steady-state solutions are $0$ and $c/b$. We will determine the solution of the equation assuming that the initial condition is of the form $y(0) = y_0$, where $0 < y_0 < c/b$. The equation is separable, and we will apply the technique of separation of variables. We have

$$\frac{dy}{dt} = cy - by^2 \Rightarrow \frac{1}{cy - by^2}\frac{dy}{dt} = 1$$

so that

$$\int \frac{1}{cy - by^2}\frac{dy}{dt}dt = \int 1 dt \Rightarrow \int \frac{1}{cy - by^2}dy = t + C,$$

where $C$ is a constant. You can confirm that the integrand on the left-hand side has the following partial fraction decomposition:

$$\frac{1}{cy - by^2} = \frac{1}{cy} - \frac{1}{c\left(y - \frac{c}{b}\right)}.$$

Therefore,

$$\int \frac{1}{cy - by^2}dy = \frac{1}{c}\ln(|y|) - \frac{1}{c}\ln\left(\left|y - \frac{c}{b}\right|\right) = \frac{1}{c}\ln\left(\left|\frac{y}{y - \frac{c}{b}}\right|\right).$$

Thus,

$$\frac{1}{c}\ln\left(\left|\frac{y}{y - \frac{c}{b}}\right|\right) = t + C,$$

so that

$$\ln\left(\left|\frac{y}{y - \frac{c}{b}}\right|\right) = ct + cC.$$

## 8.4. APPLICATIONS OF SEPARABLE DIFFERENTIAL EQUATIONS

We relabel $cC$ as $C$:
$$\ln\left(\left|\frac{y}{y - \frac{c}{b}}\right|\right) = ct + C.$$

We exponentiate:
$$\left|\frac{y}{y - \frac{c}{b}}\right| = e^C e^{ct} \Rightarrow \frac{y}{y - \frac{c}{b}} = Ce^{ct}$$

(we relabeled $\pm e^C$ as $C$). The initial condition is that $y(0) = y_0$. Therefore,
$$\frac{y_0}{y_0 - \frac{c}{b}} = C.$$

Thus,
$$\frac{y}{y - \frac{c}{b}} = \frac{y_0}{y_0 - \frac{c}{b}} e^{ct}$$

We can solve for $y$. The result can be expressed as follows:
$$y(t) = \frac{c}{b + e^{-ct}\left(\frac{c - by_0}{y_0}\right)}$$

(check). Figure 2 shows the typical graph of such a solution.

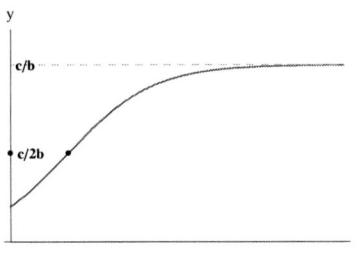

Figure 2

Since $0 < y_0 < c/b$, we have
$$\frac{c - by_0}{y_0} > 0,$$
so that $0 < y(t) < c/b$ for each $t$. Note that
$$\lim_{t \to +\infty} y(t) = \lim_{t \to +\infty} \frac{c}{e^{-ct}\left(\frac{c - by_0}{y_0}\right) + b} = \frac{c}{b}.$$

If $y(t)$ represents some population at time $t$, (e.g., the number of people in a country or the number of bears in a national park), that population approaches the steady state $c/b$, irrespective of the initial population $y_0$, as long as $0 < y_0 < c/b$.
We have
$$y''(t) = \frac{d}{dt}\left(cy(t) - by^2(t)\right) = \left(\frac{d}{dy}(cy - by^2)\bigg|_{y=y(t)}\right)\frac{dy}{dt} = (c - 2by(t))(cy(t) - by^2(t)),$$

We also have
$$cy(t) - by^2(t) = y(t)(c - by(t)) = by(t)\left(\frac{c}{b} - y(t)\right) > 0$$
since
$$0 < y(t) < \frac{c}{b}.$$
Thus,
$$y''(t) > 0 \text{ if } y(t) < \frac{c}{2b} \text{ and } y''(t) < 0 \text{ if } y(t) > \frac{c}{2b}.$$

The second derivative test for concavity tells us that there is an inflection point on the graph of the solution, and that the vertical coordinate of that inflection point is $c/2b$. This value is half of the steady-state solution to which $y(t)$ approaches as $t \to +\infty$. If $t_0$ is the corresponding time, the rate of increase of the population increases on the interval $(0, t_0)$ and decreases on the interval $(t_0, +\infty)$. The rate of increase of the population attains its maximum value at $t = t_0$. Thus, the inflection point of the graph of the solution has practical significance.

**Example 2** Assume that the population of a certain country was 3 million in 1930. If $t$ denotes time in years, $t = 0$ corresponds to the year 1930, and $y(t)$ denotes the population at time $t$, assume that the following model is valid:
$$y'(t) = 3 \times 10^{-2} y(t) - 3 \times 10^{-9} y^2(t)$$

With the notation of the discussion that preceded this example, $c = 3 \times 10^{-2}$ and $b = 3 \times 10^{-9}$. The steady-state population is
$$\frac{c}{b} = \frac{3 \times 10^{-2}}{3 \times 10^{-9}} = 10^7,$$

Thus, the model predicts the steady-state population of 10 million. If we had assumed a constant relative growth rate of $3 \times 10^{-2}$, we would have considered the linear model $y' = 3 \times 10^{-2} y$ that would have predicted unrestricted population growth. The solution of the differential equation with $y_0 = 3 \times 10^6$ is

$$y(t) = \frac{c}{e^{-ct}\left(\dfrac{c - by_0}{y_0}\right) + b}$$

$$= \frac{3 \times 10^{-2}}{\exp(-3 \times 10^{-2} t)\left(\dfrac{3 \times 10^{-2} - 3 \times 10^{-9} \times 3 \times 10^6}{3 \times 10^6}\right) + 3 \times 10^{-9}}$$

$$= \frac{3 \times 10^{-2}}{\exp(-3 \times 10^{-2} t)(7 \times 10^{-9}) + 3 \times 10^{-9}}$$

$$= \frac{3 \times 10^7}{7 \exp(-3 \times 10^{-2} t) + 3}$$

The growth rate of the population reaches its peak at $t_p$ such that $y(t_p) = c/2b = 10^7/2$. We have $t_p \cong 28.24$. Indeed,

$$\frac{3 \times 10^7}{7 \exp(-3 \times 10^{-2} t) + 3} = \frac{10^7}{2} \Leftrightarrow 6 = 7 \exp(-3 \times 10^{-2} t) + 3$$

$$\Leftrightarrow \exp(-3 \times 10^{-2} t) = \frac{3}{7}$$

$$\Leftrightarrow -3 \times 10^{-2} t = \ln\left(\frac{3}{7}\right)$$

$$\Leftrightarrow t = -\frac{\ln\left(\dfrac{3}{7}\right)}{3 \times 10^{-2}} \cong 28.24.$$

## The Hanging Cable

Assume that a cable is hanging freely, as shown in Figure 3, and that it has a mass density of $m$ kilograms/meter, so that its weight density is $mg$ newtons/meter, where $g$ is gravitational acceleration ($g$ may be assumed to be 9.8 meters/second$^2$).

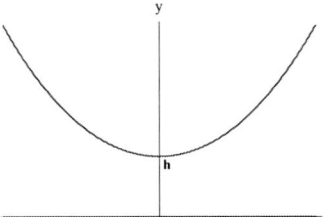

Figure 3: A hanging cable

Assume that the tension at the lowest point of the cable is $T_0$. If $f(x)$ is the height of the cable at $x$, then

$$\frac{d^2 f(x)}{dx^2} = \frac{mg}{T_0}\sqrt{1 + \left(\frac{df(x)}{dx}\right)^2}.$$

We will assume that $f$ is an even function, reflecting the symmetry of the cable with respect to the vertical axis, and that the slope of the cable at $x = 0$, so that $f'(0) = 0$.
We will set $w(x) = f'(x)$, so that $w(x)$ is the slope of the cable at $(x, f(x))$. We have

$$\frac{dw(x)}{dx} = \frac{mg}{T_0}\sqrt{1 + w^2(x)},$$

and $w(0) = f'(0) = 0$. The above equation is a nonlinear separable differential equation. We have

$$\frac{1}{\sqrt{1+w^2(x)}}\frac{dw(x)}{dx} = \frac{mg}{T_0},$$

so that

$$\int \frac{1}{\sqrt{1+w^2(x)}}\frac{dw(x)}{dx}dx = \int \frac{mg}{T_0}dx \to \int \frac{1}{\sqrt{1+w^2}}dw = \frac{mg}{T_0}x + C.$$

Therefore,

$$\operatorname{arcsinh}(w) = \frac{mg}{T_0}x + C \Leftrightarrow w(x) = \sinh\left(\frac{mg}{T_0}x + C\right),$$

We have

$$w(0) = 0 \Leftrightarrow \sinh(C) = 0 \Leftrightarrow C = 0.$$

Therefore,

$$w(x) = \sinh\left(\frac{mg}{T_0}x\right)$$

Since $w(x) = f'(x)$, we have

$$f(x) = \int w(x)\,dx = \int \sinh\left(\frac{mg}{T_0}x\right)dx = \frac{T_0}{mg}\cosh\left(\frac{mg}{T_0}x\right) + C.$$

If the height at $x = 0$ is $h$, we have

$$h = \frac{T_0}{mg} \cosh(0) + C.$$

Since $\cosh(0) = 1$,

$$h = \frac{T_0}{mg} + C \Rightarrow C = h - \frac{T_0}{mg}.$$

Therefore,

$$f(x) = \frac{T_0}{mg} \cosh\left(\frac{mg}{T_0} x\right) + h - \frac{T_0}{mg} = \frac{T_0}{mg} \left(\cosh\left(\frac{mg}{T_0} x\right) - 1\right) + h.$$

□

## Problems

**1** [C] Assume that the velocity $v(t)$ of a ball that was released from the top of the tower of Pisa is modeled under the assumption of Newtonian damping:

$$\frac{dv}{dt} = g - \left(\frac{\delta}{m}\right) v^2(t),$$

where $g = 9.8$ meters/sec$^2$, the mass $m = 10$ kg. and the friction constant $\delta$ is $0.2$.
a) Determine $v(t)$ and the terminal velocity of the ball.
b) Plot the graph of the velocity function. Indicate the asymptote at $+\infty$.

**2** [C] Assume that the velocity of a parachutist is modeled under the assumption of Newtonian damping:

$$\frac{dv}{dt} = g - \left(\frac{\delta}{m}\right) v^2(t),$$

where $g = 9.8$ meters/sec$^2$, the parachutist's mass $m = 75$ kg. and the friction constant $\delta$ is $0.25$.
a) Determine $v(t)$ and the terminal velocity of the parachutist
b) Plot the graph of the velocity function. Indicate the asymptote at $+\infty$.

**3** [C] The population of Istanbul is approximately 10 million, with its suburbs. If $t$ denotes time in years, $t = 0$ corresponds to the present (2010), and $y(t)$ denotes the population of Istanbul at time $t$, assume that the following model is valid:

$$y'(t) = 4 \times 10^{-2} y(t) - 10^{-9} y^2(t)$$

a) Determine $y(t)$ and the time at which the growth rate of the population peaks.
b) Determine the smallest upper limit for $y(t)$. Plot the graph of the population. Indicate the asymptote at $+\infty$.

**4** [C] Let $y(t)$ be the number of fish in a fish farm at time $t$, in months, that is modeled as

$$y'(t) = 5 \times 10^{-2} y(t) - 10^{-5} y^2(t),$$

without harvesting, and $y(0) = 1000$ (the present population).
a) Determine $y(t)$ and the time at which the growth rate of the population peaks.
b) Determine the smallest upper limit for $y(t)$. Plot the graph of the population. Indicate the asymptote at $+\infty$.

## 8.5 Approximate Solutions and Slope Fields

It is not always possible to determine the solutions of a differential equation explicitly or implicitly, as in the previous sections of this chapter. In such a case, we have to resort to an approximate differential equation solver that may be available on our computational utility. In this section we will examine the simplest approximation technique, namely **Euler's method**. You may study more sophisticated approximation schemes in post-calculus courses on differential equations and numerical analysis. **The slope field** that is associated with a differential equation is helpful in visualizing the solutions of the equation.

### Euler's Method

Assume that $y(t)$ is the solution of the initial-value problem

$$\frac{dy}{dt} = f(t, y(t)), \ y(0) = y_0.$$

Let **"the time step $\Delta t$"** be positive and small. Since

$$\frac{y(t + \Delta t) - y(t)}{\Delta t} \cong \frac{dy}{dt},$$

we have

$$\frac{y(t + \Delta t) - y(t)}{\Delta t} \cong f(t, y(t)).$$

**Euler's method** (or **the Euler difference scheme**) replaces the derivative by the difference quotient and the differential equation by a **system of difference equations**. Thus $Y_j$, $j = 0, 1, 2, \ldots$ are determined so that

$$\frac{Y_{j+1} - Y_j}{\Delta t} = f(j\Delta t, Y_j),$$

where $Y_0 = y_0$. We expect that $Y_j \cong y(j\Delta t)$ if $\Delta t$ is small.

**Example 1** The exact solution of the initial-value problem

$$\frac{dy}{dt} = -y(t) + 2, \ y(0) = 4$$

is

$$y(t) = 2 + 2e^{-t}$$

(derive the expression by using the technique of an integrating factor).

a) Determine the system of difference equations for the approximation of the solution in accordance with Euler's method.
b) With the notation in the preceding discussion, plot the points $(j\Delta t, Y_j)$, where $\Delta t = 0.1$ and $j = 0, 1, 2, \ldots, 40$, and the graph of $y(t) = 2 + 2e^{-t}$ on the interval $[0, 4]$. Does the picture support the expectation that $Y_j$ approximates $y(j\Delta t)$ if $\Delta t$ is small?
c) Compute $Y_{1/\Delta t}$ for $\Delta t = 0.1, 0.05$ and $0.01$. Do the numbers support the expectation that $\lim_{\Delta t \to 0} Y_{1/\Delta t} = y(1)$?

**Solution**

a) We have

$$\frac{dy}{dt} = f(y),$$

where $f(y) = -y + 2$. Therefore, the Euler method leads to the difference equations

$$\frac{Y_{j+1} - Y_j}{\Delta t} = f(Y_j) = -Y_j + 2, \ j = 0, 1, 2, \ldots,$$

where $Y_0 = 4$. Thus,

$$Y_{j+1} = Y_j + \Delta t (-Y_j + 2) = (1 - \Delta t) Y_j + 2\Delta t.$$

b) If $\Delta t = 0.1$,

$$Y_{j+1} = (1 - 0.1) Y_j + 0.2 = 0.9 Y_j + 0.2, \ j = 0, 1, 2, \ldots,$$

If we set $F(Y) = 0.9Y + 0.2$, then $Y_{j+1} = F(Y_j)$, $j = 0, 1, 2, \ldots$, where $Y_0 = 4$. Thus, the sequence $\{Y_j\}_{j=0}^{\infty}$ is generated recursively. Figure 1 shows the points $(j\Delta t, Y_j)$, where $\Delta t = 0.1$ and $j = 0, 1, 2, \ldots, 40$, and the graph of $y(t) = 2 + 2e^{-t}$ on the interval $[0, 4]$. The picture supports the expectation that $Y_j$ approximates $y(j\Delta t)$ if $\Delta t$ is small. $\square$

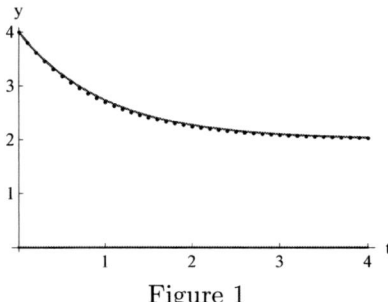

Figure 1

The solution of the system of difference equations

$$\frac{Y_{j+1} - Y_j}{\Delta t} = f(Y_j, j\Delta t), \ j = 0, 1, 2, \ldots,$$

where $Y_0 = y_0$, depends on the time step $\Delta t$. Let us indicate that dependence by denoting $Y_j$ as $Y_{\Delta t, j}$. Thus,

$$\frac{Y_{\Delta t, j+1} - Y_{\Delta t, j}}{\Delta t} = f(j\Delta t, Y_{\Delta t, j}).$$

If $y(t)$ is the solution of the initial-value problem

$$\frac{dy}{dt} = f(t, y(t)), \ y(0) = y_0,$$

we expect that

$$Y_{\Delta t, j} \cong y(j\Delta t)$$

if $\Delta t$ is small. If $t = j\Delta t$, then $j = t/\Delta t$, so that

$$Y_{\Delta t, t/\Delta t} \cong y(t),$$

and we expect that the approximation is as accurate as desired if $\Delta t$ is sufficiently small. Thus, we expect that

$$\lim_{\Delta t \to 0} Y_{\Delta t, t/\Delta t} = y(t)$$

for fixed $t$.

## 8.5. APPROXIMATE SOLUTIONS AND SLOPE FIELDS

**Example 2** Consider the initial-value problem

$$\frac{dy}{dt} = y - \frac{1}{10}y^2, \ y(0) = 1,$$

as in Example 3 of Section 8.3. The exact solution is

$$y(t) = \frac{10}{1 + 9e^{-t}}.$$

a) Determine the system of difference equations for the approximation of the solution in accordance with Euler's method.
b) If $\{Y_j\}$ denotes the sequence obtained as the solution of the difference equations of part a), where $\Delta t = 0.1$, plot the points $(j\Delta t, Y_j)$, $j = 0, 1, 2, \ldots, 60$, and the graph of

$$y(t) = \frac{10}{1 + 9e^{-t}}$$

on the interval $[0, 6]$. Does the picture support the expectation that $Y_j$ approximates $y(j\Delta t)$ if $\Delta t$ is small?
c) Calculate $Y_{\Delta t,2}$ for $\Delta t = 0.1, 0.05$ and $0.025$. Do the numbers support the expectation that we should have

$$\lim_{\Delta t \to 0} Y_{\Delta t, 2/\Delta t} = y(2)?$$

**Solution**

a) The Euler difference scheme for the given differential equation leads to the following system of difference equations:

$$Y_{j+1} = Y_j + \Delta t f(Y_j, j\Delta t) = Y_j + \Delta t \left( Y_j - \frac{1}{10} Y_j^2 \right), \ j = 0, 1, 2, \ldots,$$

where $Y_0 = 1$.
b) Figure 2 shows the points $(j\Delta t, Y_j)$, $j = 0, 1, 2, \ldots, 60$, where $\Delta t = 0.1$, and the graph of $y(t)$. The picture support the expectation that $Y_j$ approximates $y(j\Delta t)$ if $\Delta t$ is small.

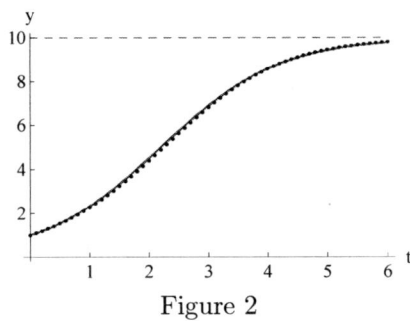

Figure 2

c) Since we expect that $Y_{\Delta t, 2/\Delta t} \cong t(2)$ if $\Delta t$ is small, we will calculate $Y_{\Delta t, 2/\Delta t}$ and $|Y_{\Delta t, 2/\Delta t} - y(2)|$ for $\Delta t = 0.1, 0.05$ and $0.025$, i.e., $Y_{0.1, 20}$, $Y_{0.05, 40}$, $Y_{0.01, 80}$ and the corresponding errors. We have

$$y(2) = \left.\frac{10}{1 + 9e^{-t}}\right|_{t=2} \cong 4.50853$$

Table 1 displays the relevant data. Notice that the absolute value is halved when the step size is halved. Indeed,
$$\left|Y_{\Delta t,t/\Delta t} - y(t)\right| \cong C\Delta t,$$
where $C$ is come constant that depends on $f$ and $t$. Thus, the Euler difference scheme is said to be **first-order accurate**. It is desirable to work with approximations which are more accurate. For example, a second-order accurate scheme involves an error which is comparable to $(\Delta t)^2$, if the step size is $\Delta t$. □

| $\Delta t$ | $Y_{\Delta t,2/\Delta t}$ | $\left|Y_{\Delta t,2} - y(2)\right|$ |
|---|---|---|
| 0.1 | 4.38414 | 0.12 |
| 0.05 | 4.44609 | 0.062 |
| 0.0025 | 4.47726 | 0.031 |

Table 1

**Example 3 (The Logistic Equation with Immigration)** Let us consider a differential equation of the form
$$\frac{dy}{dt} = cy - by^2 + s,$$
where $c > 0$, $b > 0$, $s > 0$ and $b$ is much smaller than $c$. If $y(t)$ represents a population at time $t$, the terms $s$ represents immigration (per year, if $t$ is measured in years).

For example,
$$\frac{dy}{dt} = y - \frac{1}{10}y^2 + 2, \ y(0) = 4.$$

The exact solution can be expressed as
$$y(t) = 3\sqrt{5}\tanh\left(\frac{3\sqrt{5}}{10}t + \frac{3}{10}\operatorname{arctanh}(\frac{1}{3\sqrt{5}})\right)$$

(you can determine the solution by the method of separation of variables after some hard work). Figure 3 shows the graph of the solution.

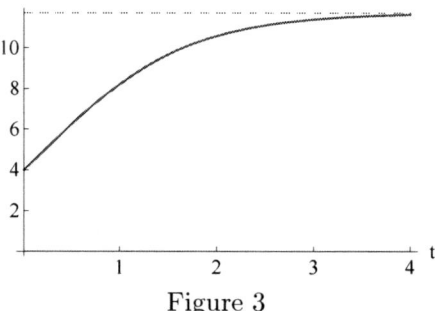

Figure 3

If we set
$$f(y) = y - \frac{1}{10}y^2 + 2,$$
the initial-value problem is
$$\frac{dy}{dt} = f(y), \ y(0) = 4.$$

## 8.5. APPROXIMATE SOLUTIONS AND SLOPE FIELDS

The Euler method leads to the difference equations

$$Y_{j+1} = f(Y_j)\Delta t + Y_j = \left(Y_j - \frac{1}{10}Y_j^2 + 2\right)\Delta t + Y_j,$$

where $Y_0 = 4$.

Figure 4 shows the points $(j\Delta t, Y_j)$, $j = 0, 1, 2, \ldots, 40$, where to $\Delta t = 0.1$, and the graph of $y(t)$ on the interval $[0, 4]$. The picture is consistent with the expectation that $Y_j \cong y(j\Delta t)$ if $\Delta t$ is small.

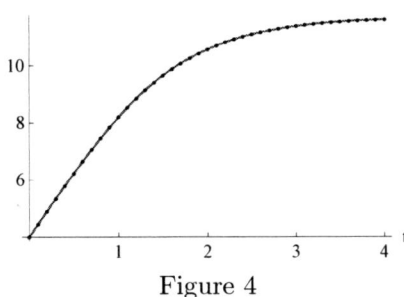

Figure 4

Table 2 displays $Y_{\Delta t, 1}$ and $|Y_{\Delta t, 1/\Delta t} - y(1)|$ for $\Delta t = 0.1, 0.05$ and $0.025$.
We have

$$y(1) = 3\sqrt{5}\tanh\left(\frac{3\sqrt{5}}{10}t + \frac{3}{10}\operatorname{arctanh}(\frac{1}{3\sqrt{5}})\right)\bigg|_{t=1} \cong 8.20777$$

The numbers in Table 2 support the expectation that $Y_{\Delta t, 1/\Delta t}$ should approximate $y(1)$ with increasing accuracy as $\Delta t$ gets smaller. □

| $\Delta t$ | $Y_{\Delta t, 1}$ | $|Y_{\Delta t, 1} - y(1)|$ |
|---|---|---|
| 0.1 | 8.24792 | 0.04 |
| 0.05 | 8.22811 | 0.02 |
| 0.0025 | 8.218 | 0.01 |

Table 2

## Slope Fields

Assume that the graph of a solution of the differential equation $y' = f(y, t)$ passes through the point $(y, t)$. The slope of the tangent line to the curve at $(t, y)$ is $f(y, t)$. Let's imagine that a small arrow that has its origin at $(t, y)$ and has slope $f(y, t)$ passes through passes through each point $(t, y)$. The collection of these arrows is **the slope field of the differential equation** $y' = f(y, t)$. In order to visualize the slope field, we can form a grid and plot the arrows that correspond to the points determined by the grid. If the grid is of sufficiently small mesh size, the picture gives some idea about the graphs of the solutions of the differential equations.

**Example 4** The general Solution of the differential equation

$$\frac{dy}{dt} = -y$$

is $Ce^{-t}$.

Therefore, the graphs of the solutions form a one-parameter family of curves. Figure 5 shows some members of this family of curves and the arrows indicate the slope field of the differential equation. □

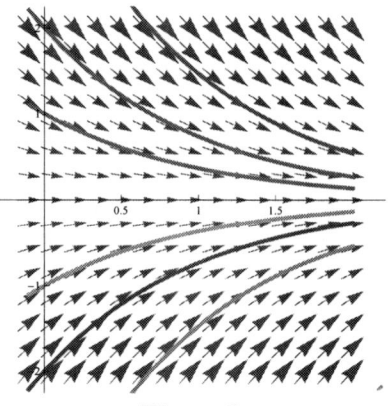

Figure 5

**Example 5** Figure 6 shows the slope field of the differential equation

$$\frac{dy}{dt} = y - \frac{1}{10}y^2,$$

and the solution of the initial value problem

$$\frac{dy}{dt} = y - \frac{1}{10}y^2, \; y(0) = 1,$$

i.e.

$$y(t) = \frac{10}{1 + 9e^{-t}}.$$

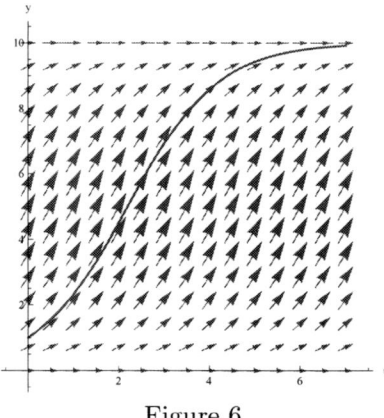

Figure 6

Slope fields were more popular in the past, since they could be obtained by hand calculations. These days, we can construct many approximate, but highly accurate solutions of a differential equation with the help of smart software and fast hardware, so that slope fields are not as important as they used to be in the old days.

## Problems

**1.** [C] Let
$$\frac{dy}{dt} = -\frac{1}{2}y(t) + 1, \; y(0) = 6$$

The exact solution is $y(t) = 2 + 4e^{-t/2}$.

a) With the notation that was used in the text, determine the system of difference equations for the approximation of the solution in accordance with Euler's method.

b) With the notation that was used in the text, plot the points $(j\Delta t, Y_j)$, where $\Delta t = 0.1$ and $j = 0, 1, 2, \ldots, 40$, and the graph of the solution on the interval $[0, 4]$. Does the picture support the expectation that $Y_j$ approximates $y(j\Delta t)$ if $\Delta t$ is small?

**2** [C] Let
$$\frac{dy}{dt} = 1 - \frac{1}{4}y^2(t), \; y(0) = 0.$$

The exact solution is
$$2\left(\frac{e^t - 1}{e^t + 1}\right).$$

a) With the notation that was used in the text, determine the system of difference equations for the approximation of the solution in accordance with Euler's method.

b) With the notation that was used in the text, plot the points $(j\Delta t, Y_j)$, where $\Delta t = 0.1$ and $j = 0, 1, 2, \ldots, 40$, and the graph of the solution on the interval $[0, 4]$. Does the picture support the expectation that $Y_j$ approximates $y(j\Delta t)$ if $\Delta t$ is small?

[C] In problems 3 and 4, the general solution of the differential equation is given. Make use of your computational and graphing utilities to visualize the associated slope field and several solution of the differential equation. Does the picture of the slope field give an indication of the behavior of the solutions?

**3.**
$$\frac{dy}{dt} = -\frac{1}{2}y + 1$$

The general solution is
$$y(t) = 2 + Ce^{-t/2}$$

**4.**
$$\frac{dy}{dt} = y - \frac{1}{4}y^2.$$

The general solution is
$$y(t) = \frac{4}{1 + Ce^{-t}}.$$

# Chapter 9

# Infinite Series

In this chapter we discuss **infinite series**, with special emphasis on **power series**. A section offers a glimpse of **Fourier series**.

## 9.1 Taylor Polynomials: Part 1

In the first two sections of this chapter we will consider the approximation of arbitrary functions by polynomials. Such approximations are useful since the values of a polynomial can be computed easily. It should be assumed that a given function has all the required derivatives even if such conditions are not spelled out explicitly.

### Taylor Polynomials based at 0

Let's begin by recalling some terminology. The polynomial

$$P(x) = a_0 + a_1 x + a_2 x^2 + \cdots + a_n x^n$$

is of **order** $n$. The **degree** of $P(x)$ is $k \leq n$ if $a_k \neq 0$ and $a_{k+1} = a_{k+2} = \cdots = a_n = 0$. For example, if

$$P(x) = 1 - \frac{1}{2}x^2 + \frac{1}{24}x^4.$$

then the order of $P(x)$ = degree of $P(x) = 4$. On the other hand, if

$$Q(x) = 1 - \frac{1}{2}x^2 + \frac{1}{24}x^4 + (0)x^5,$$

the order of $Q(x)$ is 5, whereas the degree of $Q(x)$ is 4.

Recall that **the linear approximation to** $f$ **based at 0** is

$$L_0(x) = f(0) + f'(0)x,$$

provided that $f$ is differentiable at 0. The graph of $y = L_0(x)$ is the tangent line to the graph of $y = f(x)$ at the point $(0, f(0))$.

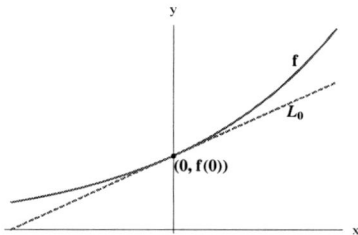

Figure 1: The linear approximation to $f$ based at 0

Note that $L_0(x)$ is a polynomial of order 1, and
$$L_0(0) = f(0),\ L_0'(x) = f'(0).$$

We have seen that $L_0(x)$ approximates $f(x)$ very well if $x$ is near the basepoint 0. We may expect even better accuracy near 0 if we approximate $f$ by a polynomial $P(x)$ of order $n > 1$ such that $P(0) = f(0)$ and $P^{(k)}(0) = f^{(k)}(0)$ for $k = 1, 2, \ldots, n$ (recall that $f^{(k)}$ denotes the $k$th derivative of $f$). The following expression for a polynomial of order $n$ shows us how to construct such a polynomial:

**Proposition 1** Assume that $P(x)$ is a polynomial of order $n$. Then,
$$P(x) = P(0) + P'(0)x + \frac{1}{2}P''(0)x^2 + \frac{1}{3!}P^{(3)}(0)x^3 + \cdots + \frac{1}{n!}P^{(n)}(0)x^n.$$

**We will confirm the statement of Proposition 1 for** $n = 4$ (it is possible to prove the proposition by mathematical induction):

We can express a polynomial of order 4 as
$$P(x) = a_0 + a_1 x + a_2 x^2 + a_3 x^3 + a_4 x^4.$$

Therefore,
$$P'(x) = a_1 + 2a_2 x + 3a_3 x^2 + 4a_4 x^3,$$
$$P''(x) = 2a_2 + (2)(3)a_3 x + (3)(4)a_4 x^2,$$
$$P^{(3)}(x) = (2)(3)a_3 + (2)(3)(4)a_4 x,$$
$$P^{(4)}(x) = (2)(3)(4)a_4.$$

Thus,
$$P(0) = a_0,$$
$$P'(0) = a_1,$$
$$P''(0) = 2a_2,$$
$$P^{(3)}(0) = (2)(3)a_3 = 3!a_3,$$
$$P^{(4)}(0) = (2)(3)(4)a_4 = 4!a_4.$$

Therefore,
$$P(x) = a_0 + a_1 x + a_2 x^2 + a_3 x^3 + a_4 x^4$$
$$= P(0) + P'(0)x + \frac{1}{2}P''(0)x^2 + \frac{1}{3!}P^{(3)}(0)x^3 + \frac{1}{4!}P^{(4)}(0)x^4.$$

■

## 9.1. TAYLOR POLYNOMIALS: PART 1

**Theorem 1** Assume that $f$ has derivatives up to order $n$ at 0. There exists a polynomial $P(x)$ of order $n$ such that

$$P(0) = f(0), \; P'(0) = f'(0), \; P''(0) = f''(0), \ldots, P^{(n)}(0) = f^{(n)}(0).$$

**We can express $P(x)$ as**

$$f(0) + f'(0)x + \frac{1}{2!}f''(0)x^2 + \cdots + \frac{1}{n!}f^{(n)}(0)x^n.$$

**Theorem 1 follows from Proposition 1:**

Since we can express a polynomial $P(x)$ of order $n$ as

$$P(0) + P'(0)x + \frac{1}{2}P''(0)x^2 + \frac{1}{3!}P^{(3)}(0)x^3 + \cdots + \frac{1}{n!}P^{(n)}(0)x^n,$$

we have

$$P(0) = f(0), \; P'(0) = f'(0), \; P''(0) = f''(0), \ldots, P^{(n)}(0) = f^{(n)}(0)$$

if and only if

$$P(x) = f(0) + f'(0)x + \frac{1}{2!}f''(0)x^2 + \cdots + \frac{1}{n!}f^{(n)}(0)x^n.$$

∎

**Definition 1** The polynomial

$$f(0) + f'(0)x + \frac{1}{2!}f''(0)x^2 + \cdots + \frac{1}{n!}f^{(n)}(0)x^n$$

is **the Taylor polynomial of order $n$ for $f$ based at 0.**

We will denote the Taylor polynomial of order $n$ for $f$ based at 0 by $P_n$. If we use the summation notation, we can express $P_n(x)$ as

$$\sum_{k=0}^{n} \frac{1}{k!} f^{(k)}(0) x^k$$

(recall that $0! = 1$ and $f^{(0)} = f$). The Taylor polynomial of order $n$ for $f$ based at 0 is also referred to as **the Maclaurin polynomial of order $n$ for $f$**.

**Remark** Taylor was Newton's student, and investigated the approximation of arbitrary functions by polynomials systematically. Later in this section we will discuss Taylor polynomials based at points other than 0. Taylor polynomials based at 0 are also referred to as **Maclaurin polynomials**. Maclaurin was the author of a book that included a discussion of approximation of arbitrary functions by polynomials.◊

Note that **the Maclaurin polynomial of order 1 for $f$ is the same as the linear approximation to $f$ based at 0**, since

$$P_1(x) = f(0) + f'(0)x = L_0(x).$$

The Maclaurin polynomial of order 2 for $f$ is

$$P_2(x) = f(0) + f'(0)x + \frac{1}{2}f''(0)x^2.$$

We may refer to $P_2$ as **the quadratic approximation to $f$ based at 0**.

**Example 1** Let
$$f(x) = \frac{1}{1-x}.$$

Construct the Maclaurin polynomials for $f$.

**Solution**

We have
$$f'(x) = \frac{d}{dx}(1-x)^{-1} = (1-x)^{-2},$$
$$f''(x) = \frac{d}{dx}(1-x)^{-2} = 2(1-x)^{-3},$$
$$f^{(3)}(x) = \frac{d}{dx}\left(2(1-x)^{-3}\right) = 3!(1-x)^{-4},$$
$$\vdots$$
$$f^{(n)}(x) = n!(1-x)^{-(n+1)}.$$

Therefore,
$$f(0) = 1, \ f'(0) = 1, \ f''(0) = 2, \ f^{(3)}(0) = 3!, \ldots, f^{(n)}(0) = n!$$

Thus,
$$P_n(x) = 1 + x + \frac{2}{2}x^2 + \frac{3!}{3!}x^3 + \cdots + \frac{n!}{n!}x^n$$
$$= 1 + x + x^2 + x^3 + \cdots + x^n = \sum_{k=0}^{n} x^k.$$

In particular, the linear approximation to $f$ based at 0 is
$$P_1(x) = 1 + x,$$

and the quadratic approximation to $f$ based at 0 is
$$P_2(x) = 1 + x + x^2.$$

The third-order Maclaurin polynomial for $f$ is
$$P_3(x) = 1 + x + x^2 + x^3.$$

Figure 2 shows the graph of $f$, $P_1$, $P_2$ and $P_3$ (the dashed curves indicate the graphs of the polynomials). The pictures indicate that $P_n(x)$ approximates $f(x)$ very well if $x$ is near the basepoint 0, and that the accuracy of the approximation increases as $n$ increases. □

## 9.1. TAYLOR POLYNOMIALS: PART 1

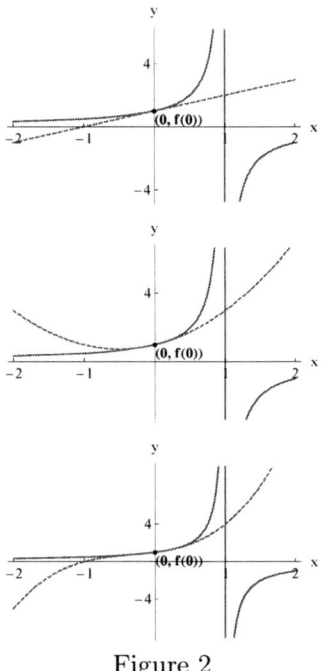

Figure 2

**Example 2** Let $f(x) = e^x$.
a) Construct the Maclaurin polynomials $P_n$ for $f$ based at 0.
b) Make use of your graphing utility to compare the graphs of $P_1$, $P_2$ and $P_3$ on $[-1, 1]$ with the graph of $f$. Do the pictures indicate that $P_n(x)$ should approximate $e^x$ better as $n$ increases if $x$ close to the basepoint 0?

**Solution**

a) We have $f^{(n)}(x) = e^x$, so that $f^{(n)}(0) = 1$, $n = 0, 1, 2, \ldots$. Therefore, the Maclaurin polynomial of order $n$ for the natural exponential function is

$$P_n(x) = f(0) + f'(0)x + \frac{1}{2!}f''(0)x^2 + \frac{1}{3!}f^{(3)}(0)x^3 + \cdots + \frac{1}{n!}f^{(n)}(0)x^n$$
$$= 1 + x + \frac{1}{2!}x^2 + \frac{1}{3!}x^3 + \cdots + \frac{1}{n!}x^n.$$

b) In particular, the linear approximation to $f$ based at 0 is $P_1(x) = 1 + x$, the quadratic approximation to $f$ based at 0 is

$$P_2(x) = 1 + x + \frac{1}{2}x^2.$$

The third-order Maclaurin polynomial for $f$ is

$$P_3(x) = 1 + x + \frac{1}{2}x^2 + \frac{1}{6}x^3.$$

Figure 3 compares the graph of the natural exponential function with the graphs of $P_1, P_2$ and $P_3$ on the interval $[-1, 1]$ (the dashed curves indicate the graphs of the polynomials). It becomes more and more difficult to distinguish between the graph of the function and the graph of $P_n$

near the basepoint 0 as $n$ increases. This indicates that $P_n(x)$ approximates $e^x$ better as $n$ increases if $x$ is near the basepoint 0. □

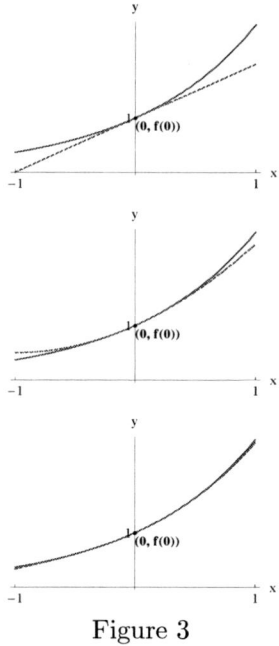

Figure 3

**Example 3** Let $f(x) = \sin(x)$.
a) Construct the Maclaurin polynomials for $f$.
b) Make use of your graphing utility to compare the graphs of $P_1$, $P_3$ and $P_5$ on $[-\pi, \pi]$ with the graph of $f$. Do the pictures indicate that $P_n(x)$ should approximate $\sin(x)$ better as $n$ increases if $x$ close to the basepoint 0?

**Solution**

a) We have

$$f'(x) = \frac{d}{dx}\sin(x) = \cos(x),$$

$$f''(x) = \frac{d^2}{dx^2}\sin(x) = -\sin(x),$$

$$f^{(3)}(x) = \frac{d^3}{dx^3}\sin(x) = -\cos(x),$$

$$f^{(4)}(x) = \frac{d^4}{dx^4}\sin(x) = \sin(x),$$

$$f^{(5)}(x) = \frac{d^5}{dx^5}\sin(x) = \cos(x),$$

$$\vdots$$

The general pattern can be expressed as follows:

$$f^{(2k)}(x) = (-1)^k \sin(x), \ k = 0, 1, 2, \ldots,$$
$$f^{(2k+1)}(x) = (-1)^k \cos(x), \ k = 0, 1, 2, \ldots.$$

## 9.1. TAYLOR POLYNOMIALS: PART 1

Therefore,
$$f^{(2k)}(0) = 0, \ k = 0, 1, 2, 3, \ldots,$$
$$f^{(2k+1)}(0) = (-1)^k, \ k = 0, 1, 2, 3, \ldots.$$

Thus,
$$P_{2k+2}(x) = P_{2k+1}(x) = x - \frac{1}{3!}x^3 + \frac{1}{5!}x^5 - \frac{1}{3!}x^7 + \cdots + (-1)^k \frac{1}{(2k+1)!}x^{2k+1},$$

$k = 0, 1, 2, 3, \ldots$.

b) $P_1(x) = x$ defines the linear approximation to sine based at 0, and $P_2(x) = P_1(x)$. The cubic Taylor polynomial based at 0 is

$$P_3(x) = x - \frac{1}{3!}x^3 = x - \frac{1}{6}x^3.$$

The Taylor polynomial of order 4 based at 0 is the same as $P_3$. The Taylor polynomial of order 5 based at 0 is
$$P_5(x) = x - \frac{1}{3!}x^3 + \frac{1}{5!}x^5 = 1 - \frac{1}{6}x^3 + \frac{1}{120}x^5,$$

and $P_6(x) = P_5(x)$.

Figure 4 compares the graph of sine with the graphs of $P_1, P_3$ and $P_5$ on the interval $[-\pi, \pi]$ (the dashed curves indicate the graphs of the polynomials). It becomes harder to distinguish between the graphs and sine and $P_n$ near the basepoint 0 as $n$ increases. This indicates that $P_n(x)$ approximates $\sin(x)$ with increasing accuracy as $n$ increases, at least for $x$ near the basepoint 0. □

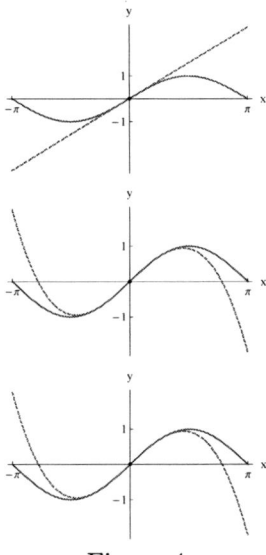

Figure 4

**Example 4** Let $f(x) = \cos(x)$. The Taylor polynomial of order $n$ for $f$ based at 0 is

$$P_{2n}(x) = 1 - \frac{1}{2}x^2 + \frac{1}{4!}x^4 - \frac{1}{6!}x^6 + \cdots (-1)^n \frac{1}{(2n)!}x^{2n} = \sum_{k=0}^{n} (-1)^k \frac{1}{(2k)!}x^{2k}$$

(Confirm as an exercise). In particular,

$$P_2(x) = 1 - \frac{1}{2}x^2, \ P_4(x) = 1 - \frac{1}{2}x^2 + \frac{1}{4!}x^4 \text{ and } P_6(x) = 1 - \frac{1}{2}x^2 + \frac{1}{4!}x^4 - \frac{1}{6!}x^6.$$

Figure 5 shows the graphs of $\cos(x)$, $P_2(x)$, $P_4(x)$ and $P_6(x)$. The pictures indicate that $P_{2n}(x)$ approximates $\cos(x)$ very well if $x$ is near the basepoint 0 and that the accuracy increases with increasing $n$. $\square$

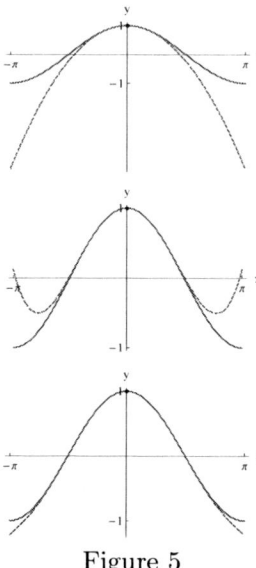

Figure 5

## Taylor Polynomials Based at an Arbitrary Point

We can construct Taylor polynomials for a given function based at a point other than 0:

**Definition 2 The Taylor polynomial of order $n$ for the function $f$ based at $c$ is the polynomial $P_{c,n}(x)$ of order $n$ such that**

$$P_{c,n}(c) = f(c), \ P'_{c,n}(c) = f'(c), \ P''_{c,n}(c) = f''(c), \ldots, P^{(n)}_{c,n} = f^{(n)}(c).$$

Thus, $P_{0,n}(x)$ coincides with $P_n$, the Maclaurin polynomial of order $n$ for the function $f$. We can express a polynomial of order $n$ in terms of powers of $x - c$:

**Proposition 2**

$$P(x) = P(c) + P'(c)(x-c) + \frac{1}{2}P''(c)(x-c)^2 + \frac{1}{3!}P^{(3)}(c)(x-c)^3 + \cdots$$
$$+ \frac{1}{n!}P^{(n)}(c)(x-c)^n.$$

Proposition 2 follows from Proposition 1 by setting $X = x - c$, so that $X = 0$ corresponds to $x = c$, and by applying Proposition 1 to the polynomial $Q(X) = P(X + c)$ (exercise). Proposition 2 leads to construction of Taylor polynomials:

## 9.1. TAYLOR POLYNOMIALS: PART 1

**Theorem 2** The Taylor polynomial of order $n$ for $f$ based at $c$ is

$$P_{c,n}(x) = f(c) + f'(c)(x-c) + \frac{1}{2!}f''(c)(x-c)^n + \cdots + \frac{1}{n!}f^{(n)}(c)(x-c)^n = \sum_{k=0}^{n} \frac{1}{k!}f^{(k)}(c)(x-c)^k.$$

**Example 5** Let $f(x) = \ln(x)$. Determine the Taylor polynomials for $f$ based at 1.

**Solution**

We have

$$f'(x) = \frac{d}{dx}\ln(x) = \frac{1}{x} = x^{-1},$$

$$f''(x) = \frac{d}{dx}\left(x^{-1}\right) = -x^{-2},$$

$$f^{(3)}(x) = \frac{d}{dx}\left(-x^{-2}\right) = 2x^{-3},$$

$$f^{(4)}(x) = \frac{d}{dx}\left(2x^{-3}\right) = -3!x^{-4},$$

$$f^{(5)}(x) = 4!x^{-5},$$

$$f^{(6)}(x) = -5!x^6,$$

$$\vdots$$

$$f^{(n)}(x) = (-1)^{n-1}(n-1)!x^{-n}.$$

Therefore,

$$f(1) = \ln(1) = 0,$$
$$f'(1) = 1,$$
$$f''(1) = -1,$$
$$f^{(3)}(1) = 2,$$
$$f^{(4)}(1) = -3!,$$
$$f^{(5)}(1) = 4!,$$
$$f^{(6)}(1) = -5!$$

Indeed,

$$f(1) = 0 \text{ and } f^{(n)} = (-1)^{n-1}(n-1)! \text{ for } n = 1, 2, 3, \ldots$$

(recall that $0! = 1$). Therefore, the Taylor polynomial of order $n$ for $f(x) = \ln(x)$ based at 1 is

$$P_{1,n}(x) = f(1) + f'(1)(x-1) + \frac{1}{2!}f''(1)(x-1)^2 + \frac{1}{3!}f^{(3)}(1)(x-1)^3 + \cdots + \frac{1}{n!}f^{(n)}(1)(x-1)^n$$

$$= (x-1) - \frac{1}{2}(x-1)^2 + \frac{2}{3!}(x-1)^3 - \frac{3!}{4!}(x-1)^4 + \cdots + (-1)^{n-1}\frac{(n-1)}{n!}(x-1)^n$$

$$= (x-1) - \frac{1}{2}(x-1)^2 + \frac{1}{3}(x-1)^3 - \frac{1}{4}(x-1)^4 + \cdots + (-1)^{n-1}\frac{1}{n}(x-1)^n$$

$$= \sum_{k=1}^{n}(-1)^{k-1}\frac{1}{k}(x-1)^k.$$

In particular, the linear approximation to $f$ based at 1 is

$$P_{1,1}(x) = x - 1,$$

and the quadratic approximation to $f$ based at 1 is

$$P_{1,2}(x) = (x-1) - \frac{1}{2}(x-1)^2.$$

The Taylor polynomial of order 3 for $f$ based at 1 is

$$P_{1,3}(x) = (x-1) - \frac{1}{2}(x-1)^2 + \frac{1}{3}(x-1)^3.$$

Figure 6 shows the graph of $f$, $P_{1,1}$, $P_{1,2}$ and $P_{1,3}$. The pictures indicate that $P_{1,n}(x)$ approximates $f(x)$ well if $x$ is near the basepoint 1, and that the accuracy of the approximation increases as $n$ increases. □

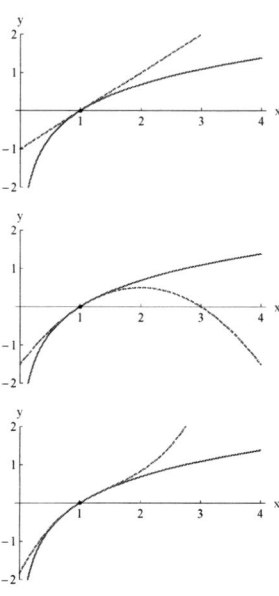

Figure 6

## Problems

1.
a) Derive the expression for the Maclaurin polynomial $P_n$ of order $n = 2, 4$ and 6 for $\cos(x)$,
b) Derive the general expression for $P_n$,
c) [C] Make use of your graphing utility in order to plot the graph of $\cos(x)$ and $P_6$ on the interval $[-\pi, \pi]$. Does the picture indicate that the approximation of $\cos(x)$ by $P_6(x)$ is accurate if $x$ is near the basepoint 0?

2. Derive the expression for the Maclaurin polynomial $P_n$ of order $n = 1, 3$ and 5 for $\sinh(x)$,
b) Derive the general expression for $P_n$,
c) [C] Make use of your graphing utility in order to plot the graph of $\sinh(x)$ and $P_5$ on the interval $[-4, 4]$. Does the picture indicate that the approximation of $\sinh(x)$ by $P_5(x)$ is accurate if $x$ is near the basepoint 0?

In problems 3 and 4,
a) Derive the expressions for the Maclaurin polynomials $P_n$ of the indicated orders for $f(x)$,
b) Derive the general expression for $P_n$.

3. $f(x) = \ln(x+1)$, $n = 1, 2, 3, 4$   4. $f(x) = \cosh(x)$, $n = 2, 4, 6$

In problems 5 and 6,
a) Derive the expression of the Taylor polynomial of the indicated order for $f(x)$ based at $a$,
b) [C] Make use of your graphing utility in order to plot the graph of $f(x)$ and the Taylor polynomial that has been constructed in part a). Does the pictures indicate that the approximation of $f(x)$ by the Taylor polynomial is accurate if $x$ is near the relevant basepoint ?

5. $f(x) = \arctan(x)$, $a = 0$, $n = 3$ (plot the graphs on the interval $[-2, 2]$.

6. $f(x) = \sqrt{x}$, $a = 1$, $n = 4$ (plot the graphs on the interval $[-2, 2]$.

In problems 7 and 8 derive the expression of the Taylor polynomial of the indicated order for $f(x)$ based at $a$.

7. $f(x) = \arcsin(x)$, $a = 0$, $n = 3$   8. $f(x) = x^{1/3}$, $a = 1$, $n = 4$

## 9.2 Taylor Polynomials: Part 2

### The Error in the Approximation by a Taylor Polynomial

Let $P_n$ be the Maclaurin polynomial of order $n$ for $f$. We will denote the difference $f(x) - P_n(x)$ as $R_n(x)$ and refer to $R_n(x)$ as **the error** or **the remainder in the approximation of $f(x)$ by $P_n(x)$**. We would like to compare $|R_n(x)|$ with $|x|$, the distance of $x$ from the basepoint 0. Let's begin with a case where we can express $R_n(x)$ explicitly.

**Example 1** Let
$$f(x) = \frac{1}{1-x} \text{ and } P_n(x) = 1 + x + x^2 + \cdots + x^n,$$

as in Example 1 of Section 9.1. If $f(x) = P_n(x) + R_n(x)$, compare $|R_n(x)|$ with $|x|^n$ if $x$ is near the basepoint 0.

**Solution**

We have

$$\begin{aligned}f(x) - P_n(x) &= \frac{1}{1-x} - \left(1 + x + x^2 + \cdots + x^n\right) \\ &= \frac{1 - (1-x)\left(1 + x + x^2 + \cdots + x^n\right)}{1-x} \\ &= \frac{1 - \left(1 + x + x^2 + \cdots + x^n\right) + \left(x + x^2 + \cdots + x^n + x^{n+1}\right)}{1-x} \\ &= \frac{x^{n+1}}{1-x}.\end{aligned}$$

Thus, $f(x) = P_n(x) + R_n(x)$, where

$$R_n(x) = \frac{x^{n+1}}{1-x}.$$

If $x$ is close to 0, $|R_n(x)|$ is comparable to $|x|^{n+1}$. In particular, $|R_n(x)|$ is much smaller than $|x|$ if $x$ is near 0. For example, if $|x| = 10^{-1}$, $|R_2(x)| \cong \left(10^{-1}\right)^3 = 10^{-3}$, and $|R_3(x)| \cong \left(10^{-1}\right)^4 = 10^{-4}$. Thus, the accuracy of the approximation of $f(x)$ by $P_n(x)$ improves as $n$ increases. $\square$

In many case, it is not feasible to obtain an expression for the remainder directly as in the above example. The following Theorem is helpful is assessing the magnitude of the remainder in such cases:

**Theorem 1 (Taylor's Formula for the Remainder)** Assume that $f$ has continuous derivatives up to order $n+1$ in an open interval $J$ containing 0. If $x \in J$ there exists a point $c_n(x)$ between $x$ and 0 such that

$$f(x) = P_n(x) + R_n(x) = f(0) + f'(0)x + \frac{1}{2!}f''(0)x^2 + \cdots + \frac{1}{n!}f^{(n)}(0)x^n + R_n(x),$$

**where**

$$R_n(x) = \frac{1}{(n+1)!}f^{(n+1)}(c_n(x))x^{n+1}.$$

You can find the proof of Taylor's Formula for the Remainder in Appendix F. We have used the notation "$c_n(x)$" to indicate the dependence of $c_n(x)$ both on $n$ and $x$.

If $x$ is near 0,

$$|R_n(x)| = \frac{1}{(n+1)!}\left|f^{(n+1)}(c_n(x))\right||x|^{n+1} \cong \frac{1}{(n+1)!}\left|f^{(n+1)}(0)\right||x|^{n+1}.$$

Therefore, $|R_n(x)|$ is comparable to $|x|^{n+1}$ if $x$ is near 0. In particular,

$$f(x) = P_1(x) + R_1(x) = f(0) + f'(0)x + \frac{1}{2}f''(c_1(x))x^2,$$

where $c_1(x)$ is between $x$ and 0. Thus, the magnitude of the error in the approximation of $f(x)$ by the linear approximation based at 0 is comparable to $x^2$ if $x$ is near 0.

**Example 2** Let $f(x) = e^x$. As in Example 2 of Section 9.1,

$$P_n(x) = 1 + x + \frac{1}{2!}x^2 + \cdots + \frac{1}{n!}x^n.$$

If $f(x) = P_n(x) + R_n(x)$, compare $|R_n(x)|$ with $|x|^{n+1}$ if $x$ is near the basepoint 0.

**Solution**

$$f(x) - P_n(x) = R_n(x) = \frac{1}{(n+1)!}e^{c_n(x)}x^{n+1},$$

where $c_n(x)$ is between $x$ and 0. Therefore, if $x$ is near 0,

$$|R_n(x)| \cong \frac{1}{(n+1)!}e^0|x|^{n+1} = \frac{1}{(n+1)!}|x|^{n+1}.$$

Thus, $|R_n(x)|$ is much smaller than $|x|$ if $x$ is near 0 and the accuracy of the approximation of $e^x$ by $P_n(x)$ improves as $n$ increases. □

**Example 3** Let $f(x) = \sin(x)$. As in Example 3 of Section 91,

$$f^{(2k)}(x) = (-1)^k \sin(x), \ k = 0, 1, 2, \ldots,$$
$$f^{(2k+1)}(x) = (-1)^k \cos(x), \ k = 0, 1, 2, \ldots.$$

and

$$P_{2k+2}(x) = P_{2k+1}(x) = x - \frac{1}{3!}x^3 + \frac{1}{5!}x^5 - \frac{1}{3!}x^7 + \cdots + (-1)^k \frac{1}{(2k+1)!}x^{2k+1}.$$

## 9.2. TAYLOR POLYNOMIALS: PART 2

Therefore,
$$\sin(x) - P_{2k+1}(x) = \sin(x) - P_{2k+2}(x) = R_{2k+2}(x),$$
where
$$R_{2k+2}(x) = \frac{1}{(2k+3)!} f^{(2k+3)}(c_{2k+2}(x)) x^{2k+3} = \frac{1}{(2k+3)!} (-1)^{k+1} \cos(c_{2k+2}(x)) x^{2k+3}.$$

Thus,
$$|R_{2k+2}(x)| \cong |\cos(0)| \frac{|x|^{2k+3}}{(2k+3)!} = \frac{|x|^{2k+3}}{(2k+3)!}$$

if $|x|$ is small. Thus, $|R_{2k+2}(x)|$ is much smaller than $|x|$ if $x$ is near 0 and the accuracy of the approximation of $e^x$ by $P_{2k}(x)$ improves as $k$ increases. □

Taylor's formula for the remainder can be generalized to an arbitrary basepoint:

**Theorem 2 (Taylor's formula for the remainder: The general case)** Assume that $f$ has continuous derivatives up to order $n+1$ in an open interval $J$ containing the point $c$. If $x \in J$ there exists a point $c_n(x)$ between $x$ and $c$ such that

$$f(x) = P_{c,n}(x) + R_{c,n}(x)$$
$$= f(c) + f'(c)(x-c) + \frac{1}{2!} f''(c)(x-c)^2 + \cdots + \frac{1}{n!} f^{(n)}(c)(x-c)^n + R_{c,n}(x),$$

where
$$R_{c,n}(x) = \frac{1}{(n+1)!} f^{(n+1)}(c_n(x))(x-c)^{n+1}.$$

You can find the proof of Theorem 2 in Appendix F.

**Example 4** Let $f(x) = \ln(x)$ and

$$P_{1,n}(x) = \sum_{k=1}^{n} (-1)^{k-1} \frac{1}{k} (x-1)^k$$
$$= (x-1) - \frac{1}{2}(x-1)^2 + \frac{1}{3}(x-1)^3 - \frac{1}{4}(x-1)^4 + \cdots + (-1)^{n-1} \frac{1}{n}(x-1)^n,$$

as in Example 5 of Section 9.1. Compare $|R_{1,n}(x)|$ with $|x-1|^{n+1}$ if $x$ is near the basepoint 1.

**Solution**

By Taylor's formula for the remainder,
$$R_{1,n}(x) = \frac{1}{(n+1)!} f^{(n+1)}(c_n(x))(x-1)^{n+1},$$
where $c_n(x)$ is between $x$ and the basepoint 1. In Example 5 of Section 9.1 we showed that
$$f^{(n)}(x) = (-1)^{n-1} (n-1)! x^{-n}$$
Therefore,
$$f^{(n+1)}(x) = (-1)^n n! x^{-(n+1)}.$$
Thus,
$$R_{1,n}(x) = \frac{1}{(n+1)!} (-1)^n n! (c_n(x))^{-(n+1)} (x-1)^{n+1} = (-1)^n \frac{(x-1)^{n+1}}{(n+1)(c_n(x))^{n+1}}.$$

If $x$ is near 1, then $c_n(x) \cong 1$, since $c_n(x)$ is between $x$ and 1. Therefore,

$$|R_{1,n}(x)| \cong \frac{|x-c|^{n+1}}{(n+1)}$$

if $x$ is near 1. In particular, the magnitude of the remainder is much smaller than $|x-1|^{n+1}$ if $|x-1|$ is small, and the accuracy of the approximation of $\ln(x)$ by $P_{1,n}(x)$ improves as $n$ increases. □

## The Limit as the order of Taylor polynomial increases

Until now we discussed the error in the approximation of $f(x)$ by $P_{c,n}(x)$, where $P_{c,n}$ is the Taylor polynomial of order $n$ for $f$ based at $c$. We saw that the magnitude of the error is comparable to $|x-c|^{n+1}$ if $x$ is near the basepoint $c$. Now we will discuss whether the error is as small as desired for a given $x$ if $n$ is large enough, irrespective of the distance of $x$ from the basepoint.

**Proposition 1**

$$\lim_{n \to \infty} \left(1 + x + x^2 + \cdots + x^n\right) = \frac{1}{1-x}$$

**if and only if** $-1 < x < 1$.

**Proof**

If we set

$$f(x) = \frac{1}{1-x} \text{ and } P_n(x) = 1 + x + x^2 + \cdots + x^n,$$

we have

$$\left(1 + x + x^2 + \cdots + x^n\right) - \frac{1}{1-x} = f(x) - P_n(x) = \frac{x^{n+1}}{1-x},$$

as in Example 1 of Section 9.1. Therefore,

$$|f(x) - P_n(x)| = \frac{|x|^{n+1}}{|1-x|}.$$

If $-1 < x < 1$, we have $|x| < 1$. Thus,

$$\lim_{n \to \infty} |f(x) - P_n(x)| = \lim_{n \to \infty} \frac{|x|^{n+1}}{|1-x|} = \frac{1}{|1-x|} \lim_{n \to \infty} |x|^{n+1} = 0,$$

Therefore,

$$\lim_{n \to \infty} P_n(x) = \lim_{n \to \infty} (P_n(x) - f(x) + f(x)) = \lim_{n \to \infty} (P_n(x) - f(x)) + f(x) = f(x)$$

if $-1 < x < 1$.
If $|x| > 1$,

$$\lim_{n \to \infty} |f(x) - P_n(x)| = \lim_{n \to \infty} \frac{|x|^{n+1}}{|1-x|} = +\infty,$$

so that $P_n(x)$ does not converge to $f(x)$.
If $x = 1$, $f(1)$ is not defined, so that the convergence of $P_n(1)$ to $f(1)$ is out of the question. In any case, note that

$$P_n(1) = 1 + 1 + \cdots + 1 = n + 1,$$

## 9.2. TAYLOR POLYNOMIALS: PART 2

since there are $n+1$ terms, so that
$$\lim_{n \to \infty} P_n(1) = \lim_{n \to \infty} (n+1) = +\infty.$$

If $x = -1$,
$$P_n(-1) = 1 - 1 + 1 - 1 + \cdots + (-1)^n = \begin{cases} 1 & \text{if } n = 0, 2, 4, \ldots, \\ 0 & \text{if } n = 1, 3, 5, \ldots \end{cases}$$

The sequence
$$1, 0, 1, 0, 1, 0, \ldots$$
does not have a limit. ∎

**Example 5** With the notation of Proposition 1, compute $P_n(1/2)$ and $|P_n(1/2) - f(1/2)|$ for $n = 4, 6, 8, 10$. Are the numbers consistent with the fact
$$\lim_{n \to \infty} P_n\left(\frac{1}{2}\right) = f\left(\frac{1}{2}\right),$$
i.e.,
$$\lim_{n \to \infty} \left(1 + \frac{1}{2} + \frac{1}{2^2} + \cdots + \frac{1}{2^n}\right) = \frac{1}{1 - \frac{1}{2}} = 2?$$

**Solution**

Table 1 displays the required data. The numbers in Table 1 are consistent with the fact that $\lim_{n \to \infty} P_n(1/2) = f(1/2) = 2$. □

| $n$ | $P_n(1/2)$ | $\|P_n(1/2) - f(1/2)\|$ |
|---|---|---|
| 4 | 1.9375 | $6.3 \times 10^{-2}$ |
| 6 | 1.98438 | $1.6 \times 10^{-2}$ |
| 8 | 1.99609 | $3.9 \times 10^{-3}$ |
| 10 | 1.99902 | $9.8 \times 10^{-4}$ |

Table 1

Taylor's formula for the remainder will enable us to show that it is possible to approximate $e^x$, $\sin(x)$ and $\cos(x)$ by Taylor polynomials with desired accuracy for any $x \in \mathbb{R}$. Since
$$f(x) = P_{c,n}(x) + R_{c,n}(x),$$
we have
$$\lim_{n \to \infty} P_{c,n}(x) = f(x) \text{ if and only if } \lim_{n \to \infty} R_{c,n}(x) = 0.$$

The following fact will play an important role in establishing the relevant propositions:

**Proposition 2** We have
$$\lim_{n \to \infty} \frac{x^n}{n!} = 0$$
for each $x \in \mathbb{R}$.

Note that the statement of Proposition 2 is far from obvious: If $|x| > 1$, $\lim_{n \to \infty} |x^n| = \lim_{n \to \infty} |x|^n = +\infty$ and $\lim_{n \to \infty} n! = +\infty$, since $n! > n$. A naive attempt to evaluate the limit by applying the quotient rule for limits leads to the indeterminate expression $\infty/\infty$. You can find the proof of Proposition 2 at the end of this section.

**Proposition 3** We have

$$\lim_{n \to \infty} \left( 1 + x + \frac{1}{2!}x^2 + \frac{1}{3!}x^3 + \cdots + \frac{1}{n!}x^n \right) = e^x$$

**for each real number** $x$.
**Proof**

For each $x \in \mathbb{R}$ there exists $c_n(x)$ between $0$ and $x$ such that

$$R_n(x) = \frac{1}{(n+1)!} e^{c_n(x)} x^{n+1},$$

as in Example 2.
If $x > 0$, we have $0 \leq c_n(x) \leq x$. Since the natural exponential function is an increasing function on the entire number line, we have

$$0 < R_n(x) \leq \frac{e^x}{(n+1)!} x^{n+1}.$$

If $x < 0$, we have $x \leq c_n(x) \leq 0$, so that

$$|R_n(x)| = \frac{\exp(c_n(x))}{(n+1)!} |x|^{n+1} \leq \frac{\exp(0)}{(n+1)!} |x|^{n+1} \leq \frac{1}{(n+1)!} |x|^{n+1}.$$

Thus,

$$|R_n(x)| \leq \begin{cases} \dfrac{e^x}{(n+1)!} x^{n+1} & \text{if } x \geq 0, \\ \dfrac{1}{(n+1)!} |x|^{n+1} & \text{if } x \leq 0. \end{cases}$$

By Proposition 2,

$$\lim_{n \to \infty} \frac{|x|^{n+1}}{(n+1)!} = 0.$$

By the above inequalities,

$$\lim_{n \to \infty} R_n(x) = 0$$

as well. Therefore,

$$\lim_{n \to \infty} P_n(x) = e^x,$$

i.e.,

$$\lim_{n \to \infty} \left( 1 + x + \frac{1}{2!}x^2 + \frac{1}{3!}x^3 + \cdots + \frac{1}{n!}x^n \right) = e^x$$

for each $x \in \mathbb{R}$. ∎

**Example 6** If we set $x = 1$ in the statement of Proposition 3 we obtain the fact that

$$\lim_{n \to \infty} \left( 1 + 1 + \frac{1}{2!} + \frac{1}{3!} + \cdots + \frac{1}{n!} \right) = e.$$

## 9.2. TAYLOR POLYNOMIALS: PART 2

The above expression for $e$ as a limit provides an efficient method for the approximation of $e$. As in the proof of Proposition 3, we have

$$\left| e - \left(1 + 1 + \frac{1}{2!} + \frac{1}{3!} + \cdots + \frac{1}{n!}\right) \right| = |R_n(1)|$$
$$\leq \frac{e^1}{(n+1)!} 1^{n+1} < \frac{3}{(n+1)!}.$$

Since $(n+1)!$ grows very rapidly as $n$ increases, we can obtain an approximation to $e$ with desired accuracy with a moderately large $n$. For example, if it is desired to approximate $e$ with an absolute error less than $10^{-4}$, it is sufficient to have

$$\frac{3}{(n+1)!} < 10^{-4}.$$

You can check that

$$\frac{3}{8!} < 10^{-4},$$

so that it is sufficient to set $n = 7$. Indeed, we have

$$\sum_{k=0}^{7} \frac{1}{k!} \cong 2.718\,25,$$

and

$$\left| e - \sum_{k=0}^{7} \frac{1}{k!} \right| \cong 2.8 \times 10^{-5} < 10^{-4}.$$

$\square$

**Proposition 4** We have

$$\lim_{n \to \infty} \left( x - \frac{1}{3!} x^3 + \cdots + (-1)^n \frac{1}{(2n+1)!} x^{2n+1} \right) = \sin(x)$$

for each $x \in \mathbb{R}$.

**Proof**

As in Example 3,

$$\sin(x) - P_{2n+1}(x) = \sin(x) - P_{2n+2}(x) = R_{2n+2}(x) = \frac{1}{(2n+3)!} \cos(c_{2n+2}(x)) x^{2n+3},$$

where $f(x) = \sin(x)$ and $c_{2n+2}(x)$ is a point between the basepoint 0 and $x$. Therefore,

$$|R_{2n+2}(x)| = \frac{1}{(2n+3)!} |\cos(c_{2n+2}(x))| |x|^{2n+3}.$$

Since $|\cos(\theta)| \leq 1$ for each $\theta \in \mathbb{R}$,

$$|R_{2n+2}(x)| \leq \frac{1}{(2n+3)!} |x|^{2n+3}.$$

Thanks to Proposition 2,

$$\lim_{n \to \infty} \frac{1}{(2n+3)!} |x|^{2n+3} = 0.$$

Therefore $\lim_{n \to \infty} R_{2n+2}(x) = 0$. Therefore,

$$\lim_{n \to \infty} P_{2n+1}(x) = \lim_{n \to \infty} \left( x - \frac{1}{3!} x^3 + \cdots + (-1)^k \frac{1}{(2k+1)!} x^{2k+1} \right) = \sin(x).$$

for each $x \in \mathbb{R}$. ∎

**Example 7** With the notation of Proposition 4 calculate $P_{2n+1}(\pi/6)$ (round to 10 significant digits) and $|P_{2n+1}(\pi/6) - \sin(\pi/6)|$ (round to 2 significant digits) for $n = 1, 2, 3, 4$. Are the numbers consistent with the fact that

$$\lim_{n\to\infty} P_{2n+1}(\pi/6) = \lim_{n\to\infty} \left(\frac{\pi}{6} - \frac{1}{3!}\left(\frac{\pi}{6}\right)^3 + \cdots + (-1)^n \frac{1}{(2n+1)!}\left(\frac{\pi}{6}\right)^{2n+1}\right) = \sin(\pi/6) = \frac{1}{2}?$$

**Solution**

Let

$$S_n = P_{2n+1}(\pi/6) = \frac{\pi}{6} - \frac{1}{3!}\left(\frac{\pi}{6}\right)^3 + \cdots + (-1)^n \frac{1}{(2n+1)!}\left(\frac{\pi}{6}\right)^{2n+1}$$

Table 2 displays $S_n$ and $|S_n - 1/2|$ for $n = 1, 2, 3, 4$. The numbers in Table 2 indicate that $S_n$ approaches $1/2$ very rapidly as $n$ increases. Note that $S_4 \cong 0.5 = \sin(\pi/6)$, rounded to 10 significant digits. $\square$

| $n$ | $S_n$ | $|S_n - 1/2|$ |
| --- | --- | --- |
| 1 | 0.499 674 1794 | $3.3 \times 10^{-4}$ |
| 2 | 0.500 002 1326 | $2.1 \times 10^{-6}$ |
| 3 | 0.499 999 9919 | $8.1 \times 10^{-9}$ |
| 4 | 0.5 | $2 \times 10^{-11}$ |

Table 2

**Proposition 5** We have

$$\lim_{n\to\infty} P_{2n}(x) = \lim_{n\to\infty} \left(1 - \frac{1}{2}x^2 + \frac{1}{4!}x^4 + \cdots + (-1)^n \frac{x^{2n}}{(2n)!}\right) = \cos(x)$$

**for each** $x \in \mathbb{R}$.

The proof of Proposition 5 is similar to the proof of Proposition 4 (exercise).

**Proposition 6** We have

$$\lim_{n\to\infty} \left((x-1) - \frac{1}{2}(x-1)^2 + \frac{1}{3}(x-1)^3 - \frac{1}{4}(x-1)^4 + \cdots + (-1)^{n-1}\frac{1}{n}(x-1)^n\right) = \ln(x)$$

if $x \in [1, 2]$.

**Proof**

As in Example 4, The Taylor polynomial of order $n$ for $f(x) = \ln(x)$ based at 1 is

$$P_n(x) = (x-1) - \frac{1}{2}(x-1)^2 + \frac{1}{3}(x-1)^3 - \frac{1}{4}(x-1)^4 + \cdots + (-1)^{n-1}\frac{1}{n}(x-1)^n,$$

and

$$R_n(x) = \frac{(-1)^n c^{-(n+1)} n!}{(n+1)!}(x-1)^{n+1} = (-1)^n \frac{(x-1)^{n+1}}{(n+1)c^{n+1}},$$

where $c$ is between $x$ and 1. Therefore,

$$|R_n(x)| = \frac{|x-1|^{n+1}}{(n+1)|c|^{n+1}}.$$

## 9.2. TAYLOR POLYNOMIALS: PART 2

If $1 \leq x \leq 2$ then $|x-1| \leq 1$ and $1 \leq c \leq 2$, so that $1/|c| \leq 1$. Thus,

$$|R_n(x)| \leq \frac{|x-1|^{n+1}}{n+1} \leq \frac{1}{n+1}$$

Therefore, $\lim_{n \to \infty} R_n(x) = 1 \leq x \leq 2$. ∎

**Remark 1** In Section 9.6 we will show that

$$\lim_{n \to \infty} \left( (x-1) - \frac{1}{2}(x-1)^2 + \frac{1}{3}(x-1)^3 - \frac{1}{4}(x-1)^4 + \cdots + (-1)^{n-1} \frac{1}{n}(x-1)^n \right) = \ln(x)$$

for each $x \in (0, 2]$. ◊

**Example 8** If we set $x = 2$ in the statement of Proposition 6, we obtain the fact that

$$\lim_{n \to \infty} \left( 1 - \frac{1}{2} + \frac{1}{3} - \frac{1}{4} + \cdots + (-1)^{n-1} \frac{1}{n} \right) = \ln(2).$$

□

## The Proof of Proposition 2

Since

$$\lim_{n \to \infty} \frac{x^n}{n!} = 0 \Leftrightarrow \lim_{n \to \infty} \frac{|x|^n}{n!} = 0,$$

and the limit of the constant sequence 0 is 0, it is enough to show that

$$\lim_{n \to \infty} \frac{r^n}{n!} = 0$$

for any $r > 0$.

Given $r > 0$, there exists a positive integer $N$ such that $N > 2r$, no matter how large $r$ may be. If $n \geq N$ then $n \geq 2r$, so that

$$\frac{r}{n} \leq \frac{1}{2}.$$

Therefore, if $n \geq N$,

$$\frac{r^n}{n!} = \frac{r^N r^{n-N}}{N!(N+1)(N+2)\cdots(n)} = \frac{r^N}{N!} \left( \frac{r}{N+1} \right) \left( \frac{r}{N+2} \right) \cdots \left( \frac{r}{n} \right)$$

$$\leq \frac{r^N}{N!} \underbrace{\left( \frac{1}{2} \right) \left( \frac{1}{2} \right) \cdots \left( \frac{1}{2} \right)}_{(n-N) \text{ factors}} = \frac{r^N}{N!} \left( \frac{1}{2^{n-N}} \right) = \frac{(2r)^N}{N!} \left( \frac{1}{2^n} \right).$$

Thus,

$$0 \leq \frac{r^n}{n!} \leq \frac{(2r)^N}{N!} \left( \frac{1}{2^n} \right)$$

if $n \geq N$. We keep $N$ fixed, once it is chosen so that $N \geq 2r$, and let $n \to \infty$. Since

$$\lim_{n \to \infty} \frac{1}{2^n} = 0,$$

the above inequality shows that

$$\lim_{n \to \infty} \frac{r^n}{n!} = 0$$

also. ∎

## Problems

**1 [C].** The Maclaurin polynomial $P_n(x)$ for

$$f(x) = \frac{1}{1-x}$$

is

$$1 + x + x^2 + \cdots + x^n.$$

We showed that

$$|f(x) - P_n(x)| = \frac{1 - x^{n+1}}{1 - x} \text{ if } x \neq 1.$$

Determine the smallest $n$ such that

$$|f(1/3) - P_n(1/3)| < 10^{-3}.$$

**2 [C].** The Maclaurin polynomial $P_n(x)$ for $e^x$ is

$$1 + x + \frac{1}{2}x^2 + \frac{1}{3!}x^3 + \cdots + \frac{1}{n!}x^n.$$

We showed that

$$|e^x - P_n(x)| = |R_n(x)| \leq \begin{cases} \dfrac{e^x}{(n+1)!} x^{n+1} & \text{if } x \geq 0, \\ \dfrac{1}{(n+1)!} |x|^{n+1} & \text{if } x \leq 0. \end{cases}$$

Make use of the above estimate in the search for the smallest $n$ such that the absolute error in the approximation of $e^2$ by $P_n(2)$ is less than $10^{-4}$. Determine the actual value of $n$ by calculating the actual absolute error (you will have to consider the value that is given by your computational utility to be the exact value).

**3 [C].** The Maclaurin polynomial $P_{2n+1}(x)$ for $\sin(x)$ is

$$x - \frac{1}{3!}x^3 + \cdots + (-1)^n \frac{1}{(2n+1)!} x^{2n+1}.$$

We showed that

$$|\sin(x) - P_{2n+1}(x)| = |\sin(x) - P_{2n+2}(x)| = |R_{2n+2}(x)| \leq \frac{1}{(2n+3)!} |x|^{2n+3}.$$

Make use of the above estimate in the search for the smallest $n$ such that the absolute error in the approximation of $\sin(\pi/3)$ by $P_{2n+1}(\pi/3)$ is less than $10^{-6}$. Determine the actual value of $n$ by calculating the actual absolute error (you will have to consider the value that is given by your computational utility to be the exact value).

**4 [C].** The Taylor polynomial for $\ln(x)$ based at 1 is

$$P_n(x) = (x-1) - \frac{1}{2}(x-1)^2 + \frac{1}{3}(x-1)^3 - \frac{1}{4}(x-1)^4 + \cdots + (-1)^{n-1} \frac{1}{n}(x-1)^n.$$

We showed that

$$|\ln(x) - P_n(x)| = |R_n(x)| \leq \frac{|x-1|^{n+1}}{n+1}$$

if $1 \leq x \leq 2$. Make use of the above estimate in the search for the smallest $n$ such that the absolute error in the approximation of $\ln(1.5)$ by $P_2(3/2)$ is less than $10^{-4}$. Determine the actual value of $n$ by calculating the actual absolute error (you will have to consider the value that is given by your computational utility to be the exact value).

5 [C]. We showed that
$$\lim_{n \to \infty} \left(1 + x + x^2 + \cdots + x^n\right) = \frac{1}{1-x}$$
if $-1 < x < 1$. Confirm this statement for the special case $x = -2/3$.

6 [C]. We showed that
$$\lim_{n \to \infty} \left(1 + x + \frac{1}{2!}x^2 + \frac{1}{3!}x^3 + \cdots + \frac{1}{n!}x^n\right) = e^x$$
for each real number $x$. Provide numerical evidence for this assertion if $x = 3$ by calculating the sums corresponding to $n = 2, 4, 8, 16$ and the absolute errors in the approximation of the limit by such sums.

7 [C]. We have
$$\lim_{n \to \infty} \left(1 - \frac{1}{2}x^2 + \frac{1}{4!}x^4 + \cdots + (-1)^n \frac{x^{2n}}{(2n)!}\right) = \cos(x)$$
for each $x \in \mathbb{R}$. Provide numerical evidence for this assertion if $x = \pi/4$ by calculating the sums corresponding to $n = 2, 3, 4, 5$ and the absolute errors in the approximation of the limit by such sums.

8 [C]. It is known that
$$\lim_{n \to \infty} \left((x-1) - \frac{1}{2}(x-1)^2 + \frac{1}{3}(x-1)^3 - \frac{1}{4}(x-1)^4 + \cdots + (-1)^{n-1} \frac{1}{n}(x-1)^n\right) = \ln(x)$$
if $x \in (0, 2]$. Provide numerical evidence for this assertion if $x = 2/3$ by calculating the sums corresponding to $n = 4, 8, 16, 32$ and the absolute errors in the approximation of the limit by such sums.

## 9.3 The Concept of an Infinite Series

Let's summarize some of the facts that were established in Section 9.2:
$$\lim_{n \to \infty} \left(1 + \frac{1}{2} + \frac{1}{2^2} + \cdots \frac{1}{2^n}\right) = 2 \text{ (Example 7 of Section 9.2)},$$
$$\lim_{n \to \infty} \left(1 + 1 + \frac{1}{2!} + \frac{1}{3!} + \cdots + \frac{1}{n!}\right) = e \text{ (Example 8 of Section 9.2)},$$
$$\lim_{n \to \infty} \left(1 - \frac{1}{2} + \frac{1}{3} - \frac{1}{4} + \cdots + (-1)^{n-1} \frac{1}{n}\right) = \ln(2) \text{ (Example 8 of Section 9.2)}.$$

There is a common theme in all of the above examples: A sequence is formed by adding more and more terms according to a specific rule, and that sequence has a certain limit. The concept of an **infinite series** captures that common theme.

Given the sequence $c_1, c_2, c_3, \ldots, c_n, \ldots$, consider the sequence

$$S_1 = c_1,$$
$$S_2 = c_1 + c_2,$$
$$S_3 = c_1 + c_2 + c_3,$$
$$\vdots$$
$$S_n = c_1 + c_2 + c_3 + \cdots + c_n,$$
$$\vdots$$

If the sequence $S_n$, $n = 1, 2, 3, \ldots$, has a limit and that limit is $S$, i.e., if

$$\lim_{n \to \infty} S_n = \lim_{n \to \infty} (c_1 + c_2 + c_3 + \cdots + c_n) = S,$$

we will write

$$c_1 + c_2 + c_3 + \cdots + c_n + \cdots = S.$$

For example,

$$1 + \frac{1}{2} + \frac{1}{2^2} + \cdots \frac{1}{2^n} + \cdots = 2$$

and

$$1 + 1 + \frac{1}{2!} + \frac{1}{3!} + \cdots + \frac{1}{n!} + \cdots = e.$$

As in the above examples, there is nothing mysterious or mystical about an expression of the form

$$c_1 + c_2 + c_3 + \cdots + c_n + \cdots = S.$$

It just means that

$$\lim_{n \to \infty} (c_1 + c_2 + c_3 + \cdots + c_n) = S.$$

Now we will take more liberty with the above expression, and we will write

$$c_1 + c_2 + c_3 + \cdots + c_n + \cdots$$

even if

$$\lim_{n \to \infty} (c_1 + c_2 + c_3 + \cdots + c_n)$$

does not exist.

**Definition 1** Given a sequence $c_1, c_2, c_3, \ldots, c_n, \ldots$, we define the corresponding **infinite series** as the formal expression

$$c_1 + c_2 + c_3 + \cdots + c_n + \cdots.$$

We say "formal expression", since we do not require that

$$\lim_{n \to \infty} (c_1 + c_2 + c_3 + \cdots + c_n)$$

exists in order to speak of the infinite series $c_1 + c_2 + c_3 + \cdots + c_n + \cdots$. For example, we may refer to the infinite series

$$1 + 2 + 2^2 + \cdots + 2^{n-1} + \cdots,$$

even though

$$\lim_{n \to \infty} \left(1 + 2 + 2^2 + \cdots + 2^{n-1}\right)$$

## 9.3. THE CONCEPT OF AN INFINITE SERIES

does not exist. Indeed, we have

$$1 + x + x^2 + \cdots + x^{n-1} = \frac{1 - x^n}{1 - x} \text{ if } x \neq 1,$$

as in Proposition 1 of Section 9.2, so that

$$1 + 2 + 2^2 + \cdots + 2^{n-1} = \frac{1 - 2^n}{1 - 2} = 2^n - 1$$

Therefore,

$$\lim_{n \to \infty} \left(1 + 2 + 2^2 + \cdots + 2^{n-1}\right) = \lim_{n \to \infty} \left(\frac{1 - 2^n}{1 - 2}\right) = \lim_{n \to \infty} (2^n - 1) = +\infty.$$

This is merely shorthand for the fact that the sequence $S_n$ exceeds any given number provided that $n$ is sufficiently large. Thus, the sequence $S_n = 1 + 2 + 2^2 + \cdots + 2^{n-1}$ does not have a finite limit.

**Definition 2** We say that **the infinite series** $c_1 + c_2 + \cdots + c_n + \cdots$ **converges** if

$$\lim_{n \to \infty} S_n = \lim_{n \to \infty} (c_1 + c_2 + \cdots + c_n)$$

exists. In this case, we define **the sum of the infinite series** $c_1 + c_2 + \cdots + c_n + \cdots$ to be $\lim_{n \to \infty} S_n$.

**Terminology and Notation:** The sequence $S_1, S_2, \ldots, S_n, \ldots$ is referred to as the sequence of **partial sums** of the given infinite series. Thus, a series converges if and only if the corresponding sequence of partial sums converges. The sum $S$ of the convergent infinite series $c_1 + c_2 + \cdots + c_n + \cdots$ is the limit of the corresponding sequence of partial sums. We write

$$S = c_1 + c_2 + \cdots + c_n + \cdots.$$

The infinite series $c_1 + c_2 + \cdots + c_n + \cdots$ is said to **diverge** if the corresponding sequence of partial sums does not have a (finite) limit.

We may refer to an infinite series simply as a **series**. Given the series $c_1 + c_2 + \cdots + c_n + \cdots$, $c_n$ is referred to as the $n$**th term of the series**.

**Remark (Caution)** The expression $c_1 + c_2 + \cdots + c_n + \cdots$ plays a dual role: It expresses a series, even though the series may not converge, and it expresses the sum of a series, provided that the series converges. You will have to get used to this mathematical "doublespeak". ◊

**Example 1** The infinite series

$$1 + 1 + \frac{1}{2!} + \frac{1}{3!} + \cdots + \frac{1}{(n-1)!} + \frac{1}{n!} + \cdots$$

converges and its sum is $e$:

$$1 + 1 + \frac{1}{2!} + \frac{1}{3!} + \cdots + \frac{1}{(n-1)!} + \frac{1}{n!} + \cdots = e.$$

The $n$th term of the series is

$$\frac{1}{(n-1)!}.$$

The $n$th partial sum of the series is

$$1 + 1 + \frac{1}{2!} + \frac{1}{3!} + \cdots + \frac{1}{(n-1)!}.$$

□

**Example 2** The infinite series
$$1 + 2 + 2^2 + \cdots + 2^{n-1} + \cdots$$
diverges, since
$$\lim_{n \to \infty} \left(1 + 2 + 2^2 + \cdots + 2^{n-1}\right) = +\infty.$$
□

We may use **the summation notation** to refer to a series or the sum of a series. Just as we may denote $c_1 + c_2 + \cdots + c_n$ as $\sum_{k=1}^{n} c_k$, the infinite series $c_1 + c_2 + \cdots + c_n + \cdots$ and its sum can be denoted as $\sum_{k=1}^{\infty} c_k$. The index is a dummy index, just as in the case of finite sums, and can be replaced by any convenient letter. Thus,

$$\sum_{k=0}^{\infty} \frac{1}{k!} = 1 + 1 + \frac{1}{2!} + \frac{1}{3!} + \cdots + \frac{1}{n!} + \cdots = e$$

(recall that $0! = 1$).

The first value of the summation index can be any integer. For example, we may refer to the series
$$\sum_{n=4}^{\infty} \frac{1}{2^n} = \frac{1}{2^4} + \frac{1}{2^5} + \cdots + \frac{1}{2^n} + \cdots.$$

**Definition 3** **The geometric series** corresponding to $x \in \mathbb{R}$ is the series

$$\sum_{n=1}^{\infty} x^{n-1} = 1 + x + x^2 + \cdots + x^{n-1} + \cdots.$$

The $n$th term of the series is $x^{n-1}$. If $x = 0$ the series is not very interesting:
$$1 + 0 + 0 + \cdots.$$

If $x \neq 0$, the ratio of consecutive terms is constant and is equal to $x$:
$$\frac{x^n}{x^{n-1}} = x, \; n = 1, 2, 3, \cdots.$$

Let's restate Proposition 1 of Section 9.2, using the language of infinite series:

**Proposition 1** The geometric series
$$\sum_{n=1}^{\infty} x^{n-1} = 1 + x + x^2 + \cdots + x^{n-1} + \cdots$$

**converges if $|x| < 1$, i.e., if $-1 < x < 1$, and has the sum**
$$\frac{1}{1-x}.$$

**The geometric series diverges if $|x| \geq 1$**, i.e., if $x \leq -1$ or $x \geq 1$.

Note that the $n$th partial sum of the geometric series
$$1 + x + x^2 + \cdots + x^{n-1} + x^n + \cdots$$

## 9.3. THE CONCEPT OF AN INFINITE SERIES

is
$$P_{n-1}(x) = 1 + x + x^2 + \cdots + x^{n-1} = \frac{1-x^n}{1-x}$$
if $x \neq 1$, as in the proof of Proposition 1 of Section 9.2. $P_{n-1}(x)$ is the Maclaurin polynomial of order $n-1$ for
$$f(x) = \frac{1}{1-x}.$$
We don't have to rewrite the proof of Proposition 1 of Section 9.2, since $\lim_{n \to +\infty} P_{n-1}(x)$ exists if and only if $\lim_{n \to \infty} P_n(x)$ exists and $\lim_{n \to \infty} P_{n-1}(x) = \lim_{n \to \infty} P_n(x)$. The geometric series corresponding to $x=1$ is
$$1 + 1 + 1 + \cdots + 1 + \cdots.$$
The $n$th partial sum is $S_n(1) = n$, so that $\lim_{n \to \infty} S_n(1) = +\infty$. The geometric series corresponding to $x = -1$ is
$$1 - 1 + 1 - 1 + \cdots + (-1)^{n-1} + \cdots.$$
Therefore, the $n$th partial sum of the series is
$$S_n(-1) = \begin{cases} 0 & \text{if } n \text{ is even,} \\ 1 & \text{if } n \text{ is odd.} \end{cases}$$
The sequence
$$1, 0, 1, 0, \ldots$$
does not have a limit. Therefore, the series
$$1 - 1 + 1 - 1 + \cdots + (-1)^{n-1} + \cdots.$$
diverges, even though $|S_n(-1)|$ is not unbounded as $n$ becomes large.

**Example 3** The geometric series
$$\sum_{n=1}^{\infty} \left(-\frac{1}{3}\right)^{n-1} = 1 - \frac{1}{3} + \frac{1}{3^2} - \frac{1}{3^3} + \cdots + \left(-\frac{1}{3}\right)^{n-1} + \cdots$$
converges and has the sum
$$\frac{1}{1 - \left(-\frac{1}{3}\right)} = \frac{1}{1 + \frac{1}{3}} = \frac{3}{4}.$$
The geometric series
$$\sum_{n=1}^{\infty} \left(\frac{4}{3}\right)^{n-1} = 1 + \frac{4}{3} + \left(\frac{4}{3}\right)^2 + \cdots$$
diverges, since
$$\frac{4}{3} > 1.$$
□

**Definition 4** If $c$ is a constant we define **the multiplication of a series by** $c$ via term-by-term multiplication:
$$c \sum_{n=1}^{\infty} a_n = \sum_{n=1}^{\infty} c a_n = c a_1 + c a_2 + c a_3 + \cdots + c a_n + \cdots.$$

**Proposition 2** If the series $\sum_{n=1}^{\infty} a_n$ converges, so does $\sum_{n=1}^{\infty} ca_n$, and we have

$$\sum_{n=1}^{\infty} ca_n = c \sum_{n=1}^{\infty} a_n,$$

where the summation sign refers to the sum of a series.

**Proof**

If

$$S_n = a_1 + a_2 + \cdots + a_n$$

is the $n$th partial sum of the series $\sum_{n=1}^{\infty} a_n$, then

$$cS_n = ca_1 + ca_2 + \cdots + ca_n$$

is the $n$th partial sum of the series $\sum_{n=1}^{\infty} ca_n$. By the constant multiple rule for limits,

$$\lim_{n \to \infty} cS_n = c \lim_{n \to \infty} S_n = c \sum_{n=1}^{\infty} a_n.$$

Therefore, the series $\sum_{n=1}^{\infty} ca_n$ converges, and we have

$$\sum_{n=1}^{\infty} ca_n = \lim_{n \to \infty} cS_n = c \sum_{n=1}^{\infty} a_n,$$

where the summation sign refers to the sum of a series. ∎

**Definition 5** We define **the addition** of the series $\sum_{n=1}^{\infty} a_n$ and $\sum_{n=1}^{\infty} b_n$ by adding the corresponding terms:

$$\sum_{n=1}^{\infty} a_n + \sum_{n=1}^{\infty} b_n = \sum_{n=1}^{\infty} (a_n + b_n).$$

Thus,

$$(a_1 + a_2 + \cdots + a_n + \cdots) + (b_1 + b_2 + \cdots b_n + \cdots)$$
$$= (a_1 + b_1) + (a_2 + b_2) + \cdots + (a_n + b_n) + \cdots$$

**Proposition 3** Assume that the series $\sum_{n=1}^{\infty} a_n$ and $\sum_{n=1}^{\infty} b_n$ converge. Then the addition of the series also converges and we have

$$\sum_{n=1}^{\infty} (a_n + b_n) = \sum_{n=1}^{\infty} a_n + \sum_{n=1}^{\infty} b_n,$$

where each summation sign denotes the sum of the corresponding series.

**Proof**

We have

$$(a_1 + b_1) + (a_2 + b_2) + \cdots + (a_n + b_n) = (a_1 + a_2 + \cdots + a_n) + (b_1 + b_2 + \cdots + b_n).$$

## 9.3. THE CONCEPT OF AN INFINITE SERIES

Therefore,
$$\lim_{n \to \infty} ((a_1 + b_1) + (a_2 + b_2) + \cdots + (a_n + b_n))$$
$$= \lim_{n \to \infty} ((a_1 + a_2 + \cdots + a_n) + (b_1 + b_2 + \cdots + b_n))$$
$$= \lim_{n \to \infty} (a_1 + a_2 + \cdots + a_n) + \lim_{n \to \infty} (b_1 + b_2 + \cdots + b_n)$$
$$= \sum_{n=1}^{\infty} a_n + \sum_{n=1}^{\infty} b_n.$$

Thus, $\sum_{n=1}^{\infty} (a_n + b_n)$ converges, and we have
$$\sum_{n=1}^{\infty} (a_n + b_n) = \sum_{n=1}^{\infty} a_n + \sum_{n=1}^{\infty} b_n,$$
where the summation sign refers to the sum of a series. ∎

**Example 4** We have
$$\sum_{n=1}^{\infty} \left( \frac{1}{(n-1)!} + (-1)^{n-1} \frac{1}{n} \right) = \sum_{n=1}^{\infty} \frac{1}{(n-1)!} + \sum_{n=1}^{\infty} (-1)^{n-1} \frac{1}{n}.$$

As in Example 8 of Section 9.2,
$$\sum_{n=1}^{\infty} \frac{1}{(n-1)!} = \lim_{n \to \infty} \left( 1 + 1 + \frac{1}{2!} + \frac{1}{3!} + \cdots + \frac{1}{(n-1)!} \right) = e.$$

As in Example 10 of Section 9.2,
$$\sum_{n=1}^{\infty} (-1)^{n-1} \frac{1}{n} = \lim_{n \to \infty} \left( 1 - \frac{1}{2} + \frac{1}{3} - \frac{1}{4} + \cdots + (-1)^{n-1} \frac{1}{n} \right) = \ln(2).$$

Therefore,
$$\sum_{n=1}^{\infty} \left( \frac{1}{(n-1)!} + (-1)^{n-1} \frac{1}{n} \right) = e + \ln(2).$$

□

There is a useful necessary condition for the convergence of an infinite series:

**Theorem 1** If the infinite series $\sum_{n=1}^{\infty} c_n$ converges, we must have
$$\lim_{n \to \infty} c_n = 0.$$

**Thus, in order for a series to converge, the $n$th term of the series must converge to 0 as $n \to \infty$.**

**Proof**

Assume that the series $\sum_{n=1}^{\infty} c_n$ converges. This means that the corresponding sequence of partial sums has a limit which is the sum $S$ of the series:
$$S = \lim_{n \to \infty} S_n = \lim_{n \to \infty} (c_1 + c_2 + \cdots + c_n)$$

We also have
$$\lim_{n\to\infty} S_{n+1} = \lim_{n\to\infty} (c_1 + c_2 + \cdots + c_n + c_{n+1}) = S,$$
since the only difference between the sequences $S_1, S_2, S_3, \ldots$ and $S_2, S_3, S_4, \ldots$ is a shifting of the index. Therefore,
$$\lim_{n\to\infty} (S_{n+1} - S_n) = \lim_{n\to\infty} S_{n+1} - \lim_{n\to\infty} S_n = S - S = 0.$$
But $S_{n+1} - S_n = c_{n+1}$. Therefore, $\lim_{n\to\infty} c_{n+1} = 0$. This is equivalent to the statement that $\lim_{n\to\infty} c_n = 0$. ■

**Remark 1 Caution)** Even though the condition
$$\lim_{n\to\infty} c_n = 0$$
is necessary for the convergence of the infinite series $c_1 + c_2 + c_3 + \cdots$, **the condition is not sufficient for the convergence of the series.**◊

**Proposition 4 The infinite series**
$$\sum_{n=1}^{\infty} \frac{1}{n} = 1 + \frac{1}{2} + \frac{1}{3} + \frac{1}{4} + \cdots + \frac{1}{n} + \cdots$$
**diverges, even though $\lim_{n\to\infty} 1/n = 0$.**

**Proof**

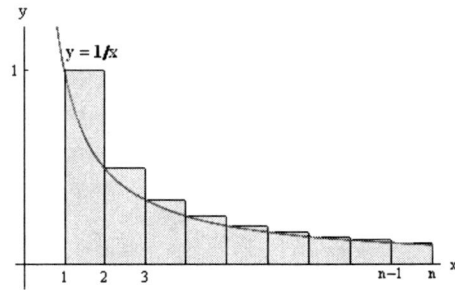

Figure 1

With reference to Figure 1, the area between the graph of $y = 1/x$ and the interval $[1, n]$ is less than the sum of the areas of the rectangles:
$$\int_1^n \frac{1}{x} dx < (1)(1) + (1)\left(\frac{1}{2}\right) + (1)\left(\frac{1}{3}\right) + \cdots + (1)\left(\frac{1}{n-1}\right)$$
$$= 1 + \frac{1}{2} + \frac{1}{3} + \cdots + \frac{1}{n-1}.$$

We have
$$\int_1^n \frac{1}{x} dx = \ln(n).$$

Therefore,
$$1 + \frac{1}{2} + \frac{1}{3} + \cdots + \frac{1}{n-1} > \ln(n), \quad n = 2, 3, \ldots.$$

## 9.3. THE CONCEPT OF AN INFINITE SERIES

The above inequality shows that

$$\lim_{n\to\infty}\left(1+\frac{1}{2}+\frac{1}{3}+\cdots+\frac{1}{n-1}\right)=+\infty,$$

since $\lim_{n\to\infty}\ln(n)=+\infty$. Thus, the series

$$\sum_{n=1}^{\infty}\frac{1}{n}$$

diverges. ∎

We will refer to the infinite series

$$\sum_{n=1}^{\infty}\frac{1}{n}=1+\frac{1}{2}+\frac{1}{3}+\cdots+\frac{1}{n}+\cdots$$

as **the harmonic series**. Thus, the harmonic series is an example of an infinite series that diverges even though the $n$th term tends to 0 as $n$ tends to infinity. The sequence of partial sums corresponding to the harmonic series grows very slowly, though. Therefore, it is difficult to detect the divergence of the harmonic series numerically.

**Remark** The convergence or divergence of an infinite series is not changed if finitely many terms of the series are modified. Indeed, if we have the series $c_1+c_2+\cdots+c_N+c_{N+1}+c_{N+2}+\cdots+c_{N+k}+\cdots$, and modify the first $N$ terms of the series and form the new series $d_1+d_2+\cdots+d_N+c_{N+1}+c_{N+2}+\cdots+c_{N+k}+\cdots$, either both series converge, or both series diverge. If we denote the partial sums of the first series as

$$S_{N+k}=c_1+c_2+\cdots+c_N+c_{N+1}+c_{N+2}+\cdots+c_{N+k},\ k=1,2,3,\ldots,$$

and the partial sums of the second series as

$$T_{N+k}=d_1+d_2+\cdots+d_N+c_{N+1}+c_{N+2}+\cdots+c_{N+k},\ k=1,2,3,\ldots,$$

we have

$$\begin{aligned}T_{N+k}&=(T_{N+k}-S_{N+k})+S_{N+k}\\&=((d_1-c_1)+(d_2-c_2)+\cdots+(d_N-c_N))+S_{N+k}.\end{aligned}$$

Therefore, $\lim_{k\to\infty}T_{N+1}$ exists if and only if $\lim_{k\to\infty}S_{N+k}$ exists. In the case of convergence, we have

$$\begin{aligned}&d_1+d_2+\cdots+d_N+c_{N+1}+c_{N+2}+\cdots+c_{N+k}+\cdots\\&=((d_1-c_1)+(d_2-c_2)+\cdots+(d_N-c_N))\\&\quad+(c_1+c_2+\cdots+c_N+c_{N+1}+c_{N+2}+\cdots+c_{N+k}+\cdots),\end{aligned}$$

so that the sums of the series differ by the sum of the differences of the first $N$ terms. ◊

If the initial value of the index $n$ is not important in a given context, we may denote a series such as $\sum_{n=1}^{\infty}c_n$ is $\sum_{n=4}^{\infty}c_n$ simply as $\sum c_n$.

## Problems

In problems 1-6,
a) Determine $S_n$, the $n$th partial sum of the infinite series,
b) Determine $S$, the sum of the infinite series as $\lim_{n\to\infty}S_n$.

1.
$$\sum_{k=0}^{\infty} \frac{1}{4^k}$$

2.
$$\sum_{k=0}^{\infty} \frac{2^k}{3^k}$$

3.
$$\sum_{k=1}^{\infty} \frac{3^k}{4^k}$$

4.
$$\sum_{k=0}^{\infty} (-1)^k \frac{4^k}{5^k}$$

5.
$$\sum_{k=1}^{\infty} \left( \frac{1}{k} - \frac{1}{k+1} \right)$$

6.
$$\sum_{k=1}^{\infty} \left( \frac{1}{k^2 + 3k + 2} \right)$$

Hint: Partial fraction decomposition.

**7** [C] Let $S_n$ be the $n$th partial sum of the series
$$\sum_{n=1}^{\infty} \frac{n}{2^n}$$
that has the sum $S = 2$. Make use of your computational utility to compute $S_n$ for $n = 4, 8, 16, 32$ and the absolute errors $|2 - S_n|$. Do the numbers support the claim that the sum of the series is 2?

**8.** [C] Let $S_n$ be the $n$th partial sum of the series
$$\sum_{n=1}^{\infty} \frac{1}{n^2}$$
that has the sum $S = \pi^2/6$. Make use of your computational utility to compute $S_n$ for $n = 10^k$, where $k = 1, 2, 3, 4$, and the absolute errors $|S - S_n|$. Do the numbers support the claim that the sum of the series is $\pi^2/6$ ?

In problems 9-12, show that the given infinite series diverges (Hint: Consider the limit of the $n$th term):

9.
$$\sum_{n=0}^{\infty} \left( \frac{4}{3} \right)^n$$

10.
$$\sum_{n=1}^{\infty} (-1)^{n-1} \frac{n}{\ln^2 (n)}$$

11.
$$\sum_{n=1}^{\infty} \frac{2^n}{n^2}$$

12.
$$\sum_{n=1}^{\infty} \frac{\sqrt{n}}{\ln (n)}$$

**13.** Show that the series
$$\sum_{n=1}^{\infty} \sin \left( \frac{n\pi}{2} \right)$$
diverges by displaying the sequence of partial sums.

**14.** Show that the series
$$\sum_{n=1}^{\infty} \cos (n\pi)$$
diverges by displaying the sequence of partial sums.

## 9.4 The Ratio Test and the Root Test

In this section we will introduce the concept of **absolute convergence** and two tests that are frequently used to establish absolute convergence: The **ratio test** and the **root test**.

### The Monotone Convergence Principle and Absolute Convergence

Let's begin by introducing some terminology: We will say that a sequence $\{a_n\}_{n=1}^{\infty}$ is **increasing** if $a_{n+1} \geq a_n$ for each $n$ (strictly speaking, we should say that the sequence is nondecreasing, but it is easier to refer to an "increasing sequence"). A sequence $\{a_n\}_{n=1}^{\infty}$ is said to be **bounded above** if there is a number $M$ such that $a_n \leq M$ for each $n$. Such a number $M$ is referred to as **an upper bound** for the sequence $\{a_n\}_{n=1}^{\infty}$.

The following fact about increasing sequences will play an important role in our discussion of series with nonnegative terms:

**Theorem 1 (The Monotone Convergence Principle) Assume that $\{a_n\}_{n=1}^{\infty}$ is an increasing sequence. Either the sequence is bounded above, in which case $\lim_{n \to \infty} a_n$ exists and $a_k \leq \lim_{n \to \infty} a_n$ for each $k$, or there is no upper bound for the sequence, in which case $\lim_{n \to \infty} a_n = +\infty$.**

We will leave the proof of the Monotone Convergence Principle to a course in advanced calculus. The principle is intuitively plausible, though: As illustrated in Figure 1, if each $a_n \leq M$ and $\{a_n\}_{n=1}^{\infty}$ is increasing, the sequence should converge to a limit $L \leq M$. Otherwise, the sequence should increase beyond all bounds, i.e., $\lim_{n \to \infty} a_n = +\infty$.

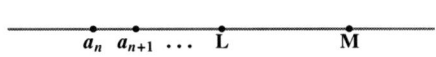

Figure 1

The Monotone Convergence Principle leads to a very useful criterion for the convergence of a series with nonnegative terms:

**Theorem 2 Assume that $c_n \geq 0$ for $n = 1, 2, 3, \ldots$. The infinite series $\sum c_n$ converges if the corresponding sequence of partial sums is bounded from above. In this case we have**

$$c_1 + c_2 + \cdots + c_n \leq c_1 + c_2 + \cdots + c_n + \cdots$$

**for each $n$. If the sequence of partial sums is not bounded above, the infinite series $\sum c_n$ diverges and we have $\lim_{n \to \infty} (c_1 + c_2 + \cdots + c_n) = +\infty$.**

**Proof**

Let $S_n = c_1 + c_2 + \cdots + c_n$ be the $n$th partial sum of the series $\sum c_n$. We have

$$S_{n+1} = c_1 + c_2 + \cdots + c_n + c_{n+1} \geq c_1 + c_2 + \cdots + c_n = S_n,$$

since $c_{n+1} \geq 0$. Thus, the sequence of partial sums $\{S_n\}_{n=1}^{\infty}$ that corresponds to the series $\sum c_n$ is an increasing sequence. By the Monotone Convergence Principle, if the sequence $\{S_n\}_{n=1}^{\infty}$ is bounded from above, $\lim_{n \to \infty} S_n$ exists, i.e., the infinite series $\sum c_n$ converges. We have $S_n \leq \lim_{k \to \infty} S_k$ for each $n$. Since $\lim_{k \to \infty} S_k$ is the sum of the series, we can express this fact by writing
$$c_1 + c_2 + \cdots + c_n \leq c_1 + c_2 + \cdots + c_n + \cdots.$$

If the monotone increasing sequence of partial sums is not bounded from above, we have $\lim_{n \to \infty} S_n = \lim_{n \to \infty} (c_1 + c_2 + \cdots + c_n) = +\infty$. Therefore, the infinite series $\sum c_n$ diverges. Theorem 2 leads to the fact that a series has to converge if the series that is formed by the absolute values of its terms converges:

**Theorem 3** *If $\sum |c_n|$ converges then $\sum c_n$ converges as well.*

You can find the proof of Theorem 3 at the end of this section.

**Definition 1** *The series $\sum a_n$ converges absolutely if $\sum |a_n|$ converges.*

By Theorem 3, a series that converges absolutely is convergent. On the other hand, a series may converge, even though it does not converge absolutely:

**Example 1** As in Example 8 of Section 9.2, the series
$$1 - \frac{1}{2} + \frac{1}{3} - \frac{1}{4} + \cdots + (-1)^{n-1}\frac{1}{n} + \cdots$$
converges (and its sum is $\ln(2)$). The series that is formed by the absolute values of the terms of the above series is the harmonic series
$$1 + \frac{1}{2} + \frac{1}{3} + \cdots + \frac{1}{n} + \cdots,$$
and we have shown that the harmonic series diverges (Proposition 4 of Section 9.3). Thus, the series
$$\sum_{n=1}^{\infty} (-1)^{n-1}\frac{1}{n}$$
is convergent, even if it does not converge absolutely. $\square$

**Definition 2** *We say that the series $\sum a_n$ converges conditionally if $\sum a_n$ converges, but $\sum a_n$ does not converge absolutely, i.e., $\sum |a_n|$ diverges.*

Thus, the series
$$1 - \frac{1}{2} + \frac{1}{3} - \frac{1}{4} + \cdots + (-1)^{n-1}\frac{1}{n} + \cdots$$
of Example 1 converges conditionally.

In the rest of this section we will discuss useful criteria for absolute convergence. Since absolute convergence implies convergence, these criteria will enable us to test many series for convergence. We will take up the issue of conditional convergence in Section 9.8.

## 9.4. THE RATIO TEST AND THE ROOT TEST

## The Ratio Test

We begin by observing that a geometric series either converges absolutely or diverges:

**Proposition 1** **The geometric series $\sum x^{n-1}$ converges absolutely if $|x| < 1$, i.e., if $-1 < x < 1$, and diverges if $|x| \geq 1$, i.e., if $x \leq -1$ or $x \geq 1$.**

**Proof**

In Proposition 1 of Section 9.3 we established that $\sum x^{n-1}$ converges if $|x| < 1$ and diverges if $|x| \geq 1$. Therefore, all we need to show is that $\sum x^{n-1}$ converges absolutely if $|x| < 1$. As in the proof of Proposition 1 of Section 9.2,

$$1 + |x| + |x|^2 + |x|^3 + \cdots + |x|^{n-1} = \frac{1}{1-|x|} - \frac{|x|^n}{1-|x|}.$$

Therefore,

$$\lim_{n \to \infty} \left(1 + |x| + |x|^2 + |x|^3 + \cdots + |x|^{n-1}\right) = \frac{1}{1-|x|} - \frac{1}{1-|x|} \lim_{n \to \infty} |x|^n = \frac{1}{1-|x|},$$

since $\lim_{n \to \infty} |x|^n = 0$ if $|x| < 1$. ∎

Note that the ratio of the absolute values of the consecutive terms in the geometric series $\sum x^{n-1}$ is $|x|$ for each $n$:

$$\frac{|x^n|}{|x^{n-1}|} = |x| \text{ for all } n.$$

The ratio test is about series whose convergence behavior is comparable to a geometric series:

**Theorem 4 (The Ratio Test)**

**Assume that**

$$\lim_{n \to \infty} \frac{|c_{n+1}|}{|c_n|} = L \text{ or } \lim_{n \to \infty} \frac{|c_{n+1}|}{|c_n|} = +\infty.$$

**a) If $L < 1$, the series $\sum c_n$ converges absolutely.**
**b) If $L > 1$ or $\lim_{n \to \infty} |c_{n+1}|/|c_n| = +\infty$, the series $\sum c_n$ diverges.**

You can find the proof of the ratio test at the end of this section.

**Remark** **The ratio test is inconclusive if**

$$\lim_{n \to \infty} \frac{|c_{n+1}|}{|c_n|} = 1,$$

i.e., the series $\sum c_n$ may converge or diverge. Here is an example:

We showed that the harmonic series $\sum_{n=1}^{\infty} 1/n$ diverges (Proposition 4 of Section 9.3). By using a similar technique, we can show that the series $\sum_{n=1}^{\infty} 1/n^2$ converges: We set $f(x) = 1/x^2$, so that $\sum 1/n^2 = \sum f(n)$. With reference to Figure 2, the sum of the areas of the rectangles is less than the area of the region between the graph of $f$ and the interval $[1, n]$.

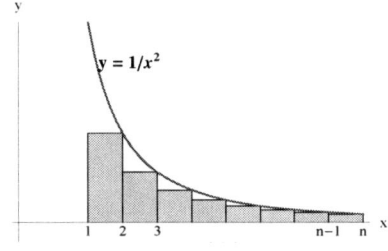

Figure 2

Therefore,
$$\frac{1}{2^2} + \frac{1}{3^2} + \cdots + \frac{1}{n^2} < \int_1^n \frac{1}{x^2} dx = 1 - \frac{1}{n} < 1.$$
Thus,
$$1 + \frac{1}{2^2} + \frac{1}{3^2} + \cdots + \frac{1}{n^2} < 2$$
for each $n$. This shows that the sequence of partial sums that corresponds to the series $\sum 1/n^2$ is bounded above. We also have $1/n^2 > 0$ for each $n$. Therefore $\sum 1/n^2$ converges, by Theorem 2.

Now,
$$\lim_{n \to \infty} \frac{\frac{1}{n+1}}{\frac{1}{n}} = \lim_{n \to \infty} \frac{n}{n+1} = \lim_{n \to \infty} \frac{n}{n\left(1 + \frac{1}{n}\right)} = \lim_{n \to \infty} \frac{1}{1 + \frac{1}{n}} = 1,$$

and
$$\lim_{n \to \infty} \frac{\frac{1}{(n+1)^2}}{\frac{1}{n^2}} = \lim_{n \to \infty} \frac{n^2}{(n+1)^2} = \lim_{n \to \infty} \frac{n^2}{n^2\left(1 + \frac{1}{n}\right)^2} = \lim_{n \to \infty} \frac{1}{\left(1 + \frac{1}{n}\right)^2} = 1.$$

Thus, in both cases the limit of the ratio of successive terms of the series is 1, even though the harmonic series $\sum 1/n$ diverges, and the series $\sum 1/n^2$ converges. ◊

**Example 2** Use the ratio test to determine whether the series
$$\sum_{n=1}^{\infty} (-1)^{n-1} \frac{3^{2n-2}}{n!}$$
converges absolutely or diverges.

**Solution**

We compute the limit that is required by the ratio test:
$$\lim_{n \to \infty} \frac{\left|(-1)^n \frac{3^{2(n+1)-2}}{(n+1)!}\right|}{\left|(-1)^{n-1} \frac{3^{2n-2}}{n!}\right|} = \lim_{n \to \infty} \frac{\frac{3^{2n+2-2}}{(n+1)!}}{\frac{3^{2n-2}}{n!}} = \lim_{n \to \infty} \left(\frac{3^{2n}}{3^{2n-2}} \frac{n!}{(n+1)!}\right) = \lim_{n \to \infty} \frac{9}{n+1} = 0 < 1.$$

Therefore the series converges absolutely. □

**Example 3** Use the ratio test to determine whether the series
$$\sum_{n=1}^{\infty} (-1)^{n-1} \frac{n!}{10^n}.$$
converges absolutely or diverges.

**Solution**

## 9.4. THE RATIO TEST AND THE ROOT TEST

We have

$$\lim_{n\to\infty} \frac{\left|(-1)^n \frac{(n+1)!}{10^{n+1}}\right|}{\left|(-1)^{n=1} \frac{n!}{10^n}\right|} = \lim_{n\to\infty}\left(\frac{(n+1)!}{n!}\cdot\frac{10^n}{10^{n+1}}\right) = \lim_{n\to\infty}\left(\frac{n+1}{10}\right) = +\infty.$$

Therefore, the ratio test predicts that the series diverges.

Note that we could have predicted the divergence of the given series by observing that the limit of the $n$th term is not 0. Indeed,

$$\lim_{n\to\infty}\left|(-1)^{n-1}\frac{n!}{10^n}\right| = \lim_{n\to\infty}\frac{n!}{10^n} = +\infty,$$

since $n!/10^n > 0$ and

$$\lim_{n\to\infty}\frac{1}{\frac{n!}{10^n}} = \lim_{n\to\infty}\frac{10^n}{n!} = 0$$

by Proposition 2 of Section 9.2. $\square$

### The Root Test

Just as the ratio test, the root test is based on the geometric series. Given the geometric series $\sum x^{n-1}$, we have

$$|x^n|^{1/n} = |x|^{n/n} = |x| \text{ for all } n.$$

The series converges absolutely if $|x| < 1$ and diverges if $|x| > 1$. The root test makes predictions about series with similar behavior:

**Theorem 5 (The Root Test)** Assume that

$$\lim_{n\to\infty}|c_n|^{1/n} = L \text{ or } \lim_{n\to\infty}|c_n|^{1/n} = +\infty.$$

**a) If $L < 1$, the series $\sum_{n=1}^{\infty} c_n$ converges absolutely.**
**b) If $L > 1$ or $\lim_{n\to\infty}|c_n|^{\frac{1}{n}} = +\infty$, the series $\sum_{n=1}^{\infty} c_n$ diverges.**

You can find the proof of the root test at the end of this section.

**Remark** The series $\sum c_n$ may converge or diverge if $\lim_{n\to\infty}|c_n|^{1/n} = 1$. Indeed, the harmonic series $\sum_{n=1}^{\infty} 1/n$ diverges and the series $\sum_{n=1}^{\infty} 1/n^2$ converges. In both cases the relevant limit is 1. Indeed,

$$\lim_{n\to\infty}\left(\frac{1}{n}\right)^{1/n} = \frac{1}{\lim_{n\to\infty} n^{1/n}} = 1,$$

and

$$\lim_{n\to\infty}\left(\frac{1}{n^2}\right)^{1/n} = \frac{1}{\lim_{n\to\infty} n^{2/n}} = 1,$$

since $\lim_{n\to\infty} n^{1/n} = 1$ (as in Example 10 of Section 4.8 on L'Hôpital's rule). $\Diamond$

**Remark** It can be shown that

$$\lim_{n\to\infty}\frac{|c_{n+1}|}{|c_n|} = L \Rightarrow \lim_{n\to\infty}|c_n|^{1/n} = L.$$

Thus, if the ratio test is applicable, so is the root test. In general the ratio test is easier to implement, even though in some cases you may find that the root test is more convenient. $\Diamond$

**Example 4** Use the root test in order to determine whether the series

$$\sum_{n=1}^{\infty} \frac{n}{2^{n-1}}.$$

converges absolutely or diverges.

**Solution**

We have

$$\lim_{n \to \infty} \left( \frac{n}{2^{n-1}} \right)^{1/n} = \lim_{n \to \infty} \frac{n^{1/n}}{2^{1-1/n}} = \frac{\lim_{n \to \infty} n^{1/n}}{2} = \frac{1}{2} < 1.$$

By the root test, the series converges (absolutely, since the terms are positive anyway). □

**Example 5** Use the root test to determine whether the series

$$\sum_{n=1}^{\infty} (-1)^{n-1} \frac{2^{2n-1}}{2n-1}$$

converges absolutely or diverges.

**Solution**

We have

$$\lim_{n \to \infty} \left| (-1)^{n-1} \frac{2^{2n-1}}{2n-1} \right|^{1/n} = \lim_{n \to \infty} \left( \frac{2^{2n-1}}{2n-1} \right)^{1/n} = \lim_{n \to \infty} \frac{2^{2-1/n}}{(2n-1)^{1/n}} = \frac{4}{\lim_{n \to \infty} (2n-1)^{1/n}}.$$

Now,

$$\lim_{n \to \infty} (2n-1)^{1/n} = \lim_{n \to \infty} \exp \left( \frac{1}{n} \ln (2n-1) \right) = \exp \left( \lim_{n \to \infty} \frac{\ln (2n-1)}{n} \right)$$

$$= \exp \left( \lim_{n \to \infty} \frac{\frac{2}{2n-1}}{1} \right) = \exp (0) = 1.$$

Therefore,

$$\lim_{n \to \infty} \left| (-1)^{n-1} \frac{2^{2n-1}}{(2n-1)} \right|^{1/n} = \frac{4}{1} > 1,$$

so that the root test predicts that the series diverges.
Note that we could have predicted the divergence of the given series by observing that the limit of the $n$th term is not 0:

$$\lim_{n \to \infty} \left| (-1)^{n-1} \frac{2^{2n-1}}{2n-1} \right| = \lim_{n \to \infty} \frac{2^{2n-1}}{2n-1} = \lim_{n \to \infty} \frac{e^{(2n-1) \ln(2)}}{2n-1} = \lim_{x \to \infty} \frac{e^{x \ln(2)}}{x} = +\infty,$$

since $\ln (2) > 0$. □

## The Proof of Theorem 3

We will make use of the following fact:

**Lemma** Given any real number $c$,

$$c = \frac{|c| + c}{2} - \frac{|c| - c}{2}.$$

## 9.4. THE RATIO TEST AND THE ROOT TEST

**We have**
$$\frac{|c|+c}{2} = \begin{cases} |c| & \text{if } c \geq 0, \\ 0 & \text{if } c \leq 0, \end{cases} \text{ and } \frac{|c|-c}{2} = \begin{cases} 0 & \text{if } c \leq 0 \\ |c| & \text{if } c \geq 0. \end{cases}$$

**In particular,**
$$0 \leq \frac{|c|+c}{2} \leq |c| \text{ and } 0 \leq \frac{|c|-c}{2} \leq |c|.$$

**Proof**

The first equality is confirmed by simple algebra. Let us confirm that
$$\frac{|c|+c}{2} = \begin{cases} |c| & \text{if } c \geq 0, \\ 0 & \text{if } c \leq 0, \end{cases}$$

Indeed, if $c \geq 0$, we have $|c| = c$. Therefore,
$$\frac{|c|+c}{2} = \frac{|c|+|c|}{2} = |c|.$$

If $c \leq 0$, we have $|c| = -c$. Therefore,
$$\frac{|c|+c}{2} = \frac{-c+c}{2} = 0.$$

Now let us confirm that
$$\frac{|c|-c}{2} = \begin{cases} 0 & \text{if } c \leq 0 \\ |c| & \text{if } c \geq 0. \end{cases}$$

If $c \geq 0$, we have $|c| = c$, so that
$$\frac{|c|-c}{2} = \frac{c-c}{2} = 0.$$

If $c \geq 0$, we have $|c| = -c$, so that
$$\frac{|c|-c}{2} = \frac{|c|+|c|}{2} = |c|.$$

∎

Now we are ready to prove Theorem 3:

By the Lemma we have
$$c_n = \frac{|c_n|+c_n}{2} - \frac{|c_n|-c_n}{2}.$$

Thus, in order to show that $\sum c_n$ converges, it is sufficient to show that the series
$$\sum_{n=1}^{\infty} \frac{|c_n|+c_n}{2} \text{ and } \sum_{n=1}^{\infty} \frac{|c_n|-c_n}{2}$$

converge.
Since
$$\frac{|c_n|+c_n}{2} \leq |c_n|,$$

we have
$$\frac{|c_1|+c_1}{2} + \frac{|c_2|+c_2}{2} + \cdots + \frac{|c_n|+c_n}{2} \leq |c_1| + |c_2| + \cdots + |c_n|$$

Since the series $\sum |c_n|$ converges, there is a number $M$ such that $|c_1| + |c_2| + \cdots + |c_n| \leq M$ for each $n$. By the above inequality,
$$\frac{|c_1|+c_1}{2} + \frac{|c_2|+c_2}{2} + \cdots + \frac{|c_n|+c_n}{2} \leq M$$

for each $n$, as well. Thus, the sequence of partial sums that corresponds to the series
$$\sum_{n=1}^{\infty} \frac{|c_n| + c_n}{2}$$
is bounded above. We also have
$$\frac{|c_n| + c_n}{2} \geq 0,$$
for each $n$. Therefore, Theorem 2 is applicable, and we conclude that
$$\sum_{n=1}^{\infty} \frac{|c_n| + c_n}{2}$$
converges.
Similarly, the series
$$\sum_{n=1}^{\infty} \frac{|c_n| - c_n}{2}$$
converges as well, since
$$0 \leq \frac{|c_n| - c_n}{2} \leq |c_n|$$
for each $n$.
Therefore,
$$\sum_{n=1}^{\infty} c_n = \sum_{n=1}^{\infty} \left( \frac{|c_n| + c_n}{2} - \frac{|c_n| - c_n}{2} \right)$$
converges, as a sum of convergent series. ■

## The Proofs of the Ratio Test and the Root Test

### The Proof of the Ratio Test

a) We have
$$\lim_{n \to \infty} \frac{|c_{n+1}|}{|c_n|} = L < 1.$$
By the definition of the limit of a sequence, $|c_{n+1}|/|c_n|$ is arbitrarily close to $L < 1$ if $n$ is sufficiently large. Therefore, there exists a real number $r$ such that $L < r < 1$ and a positive integer $N$ such that
$$\frac{|c_{n+1}|}{|c_n|} < r < 1$$
if $n \geq N$. Thus,
$$|c_{n+1}| < r |c_n|, \ n = N, N+1, N+2, \ldots$$
Therefore,
$$|c_{N+1}| \leq r |c_N|,$$
$$|c_{N+2}| \leq r |c_{N+1}| \leq r \left( r |c_N| \right) = r^2 |c_N|,$$
$$|c_{N+3}| \leq r |c_{N+2}| \leq r \left( r^2 |c_N| \right) = r^3 |c_N|,$$
$$\vdots,$$
so that
$$|c_{N+j}| \leq r^j |c_N|, \ j = 0, 1, 2, 3, \ldots$$

## 9.4. THE RATIO TEST AND THE ROOT TEST

Therefore,

$$|c_N|+|c_{N+1}|+|c_{N+2}|+\cdots|c_{N+k}| \leq |c_N|+r|c_N|+r^2|c_N|+\cdots+r^k|c_N| = |c_N|\left(1+r+r^2+\cdots+r^k\right)$$

for $k = 0, 1, 2, \ldots$. Since $0 < r < 1$, the geometric series $1 + r + r^2 + \cdots + r^k + \cdots$ converges, and we have

$$1 + r + r^2 + \cdots + r^k < 1 + r + r^2 + \cdots + r^k + \cdots = \frac{1}{1-r}.$$

Therefore,

$$|c_N| + |c_{N+1}| + |c_{N+2}| + \cdots |c_{N+k}| \leq |c_N|\left(1 + r + r^2 + \cdots + r^k\right) \leq |c_N|\left(\frac{1}{1-r}\right),$$

so that

$$|c_1| + |c_2| + \cdots + |c_{N-1}| + |c_N| + |c_{N+1}| + \cdots |c_{N+k}| \leq |c_1| + |c_2| + \cdots + |c_{N-1}| + |c_N|\left(\frac{1}{1-r}\right)$$

for $k = 0, 1, 2, 3, \ldots$. If we set

$$M = |c_1| + |c_2| + \cdots + |c_{N-1}| + |c_N|\left(\frac{1}{1-r}\right),$$

we have

$$|c_1| + |c_2| + \cdots + |c_{N-1}| + |c_N| + |c_{N+1}| + \cdots |c_{N+k}| \leq M$$

for each $k$. Thus, the sequence of partial sums that corresponds to the infinite series $\sum |c_n|$ is bounded from above. By Theorem 2, the series $\sum |c_n|$ converges, i.e., the series $\sum c_n$ converges absolutely.

b) Now let us prove "the divergence clause" of the ratio test. Thus, we assume that

$$\lim_{n\to\infty} \frac{|c_{n+1}|}{|c_n|} = L > 1 \text{ or } \lim_{n\to\infty} \frac{|c_{n+1}|}{|c_n|} = +\infty.$$

In either case, there exists $r > 1$ and positive integer $N$ such that

$$\frac{|c_{n+1}|}{|c_n|} \geq r$$

if $n \geq N$. Therefore,

$$|c_{N+1}| \geq r|c_N|,$$
$$|c_{N+2}| \geq r|c_{N+1}| \geq r^2|c_N|,$$
$$\vdots$$
$$|c_{N+k}| \geq r^k|c_N|.$$

Therefore, if $n \geq N$, we have

$$|c_n| \geq r^{n-N}|c_N|.$$

Since $r > 1$, $\lim_{n\to\infty} r^{n-N}|c_N| = +\infty$. By the above inequality, we have $\lim_{n\to\infty} |c_n| = +\infty$ as well. This implies that the series $\sum c_n$ diverges, since a necessary condition for the convergence of the series is that $\lim_{n\to\infty} c_n = 0$. ∎

**The Proof of the Root Test**

a) If
$$\lim_{n\to\infty} |c_n|^{1/n} = L < 1,$$
$|c_n|^{1/n}$ is arbitrarily close to $L$ if $n$ is sufficiently large. Therefore, there exists a number $r$ such that $0 \leq r < 1$ and a positive integer $N$ such that
$$0 \leq |c_n|^{1/n} \leq r$$
if $n \geq N$. Thus,
$$|c_n| \leq r^n, \; n = N, N+1, N+2, \ldots.$$
Therefore,
$$\begin{aligned}|c_N| + |c_{N+1}| + |c_{N+2}| + \cdots + |c_{N+k}| &\leq r^N + r^{N+1} + r^{N+2} + \cdots + r^{N+k}\\ &= r^N\left(1 + r + r^2 + \cdots + r^k\right)\\ &\leq r^N\left(1 + r + r^2 + \cdots + r^k + \cdots\right)\\ &= r^N\left(\frac{1}{1-r}\right).\end{aligned}$$

$k = 0, 1, 2, \ldots$. Thus,
$$|c_1| + |c_2| + \cdots + |c_{N-1}| + |c_N| + |c_{N+1}| + \cdots + |c_{N+k}| \leq |c_1| + |c_2| + \cdots + |c_{N-1}| + \frac{r^N}{1-r},$$

$k = 0, 1, 2, \ldots$. This implies that the sequence of partial sums for the infinite series $\sum |c_n|$ is bounded from above. By Theorem 2, $\sum |c_n|$ converges, i.e., the series $\sum c_n$ converges absolutely.

b) If
$$\lim_{n\to\infty} |c_n|^{1/n} = L > 1 \;\text{ or }\; \lim_{n\to\infty} |c_n|^{1/n} = +\infty$$
there exists a number $r > 1$ and integer $N$ such that
$$|c_n|^{1/n} \geq r, \; n = N, N+1, N+2, \ldots.$$
Therefore,
$$|c_n| \geq r^n, \; n = N, N+1, N+2, \ldots.$$
Since $r > 1$, $\lim_{n\to\infty} r^n = +\infty$. By the above inequalities, we have $\lim_{n\to\infty} |c_n| = +\infty$ as well. Therefore, $\sum c_n$ does not converge. ∎

## Problems

In problems 1 - 18 use **the ratio test** in order to determine whether the given series converges absolutely or whether it diverges, provided that the test is applicable. You need not investigate the convergence of the given series if the test is inconclusive:

1.
$$\sum_{n=0}^{\infty} e^{-n}$$

2.
$$\sum_{n=0}^{\infty} (-1.1)^n.$$

3.
$$\sum_{n=0}^{\infty} 3^n$$

4.
$$\sum_{n=0}^{\infty} (-1)^n \frac{1}{2^n}.$$

## 9.4. THE RATIO TEST AND THE ROOT TEST

5. 
$$\sum_{n=0}^{\infty} \frac{3^n}{n!}$$

6. 
$$\sum_{n=0}^{\infty} \frac{n!}{3^n}$$

7. 
$$\sum_{n=0}^{\infty} (-1)^n \frac{2^n}{n!}.$$

8. 
$$\sum_{n=1}^{\infty} \frac{n^2}{4^n}$$

9. 
$$\sum_{n=1}^{\infty} (-1)^n \frac{4^n}{n^2}$$

10. 
$$\sum_{n=2}^{\infty} \frac{\ln(n)}{n^2}$$

11. 
$$\sum_{n=1}^{\infty} \frac{n}{n^2+1}$$

12. 
$$\sum_{k=0}^{\infty} (-1)^k \frac{\pi^{2k+1}}{2^{2k+1}(2k+1)!}.$$

13. 
$$\sum_{n=1}^{\infty} (-1)^{n-1} \frac{1}{n2^n}.$$

14. 
$$\sum_{n=1}^{\infty} (-1)^{n-1} \frac{2^n}{n}.$$

15. 
$$\sum_{k=0}^{\infty} (-1)^k \frac{1}{2^{2k+1}(2k+1)}.$$

16. 
$$\sum_{k=0}^{\infty} (-1)^k \frac{4^{2k+1}}{2k+1}.$$

17. 
$$\sum_{n=1}^{\infty} (-1)^n \frac{1}{\sqrt{n}\ln(n)}$$

18. 
$$\sum_{k=0}^{\infty} (-1)^{k-1} \frac{(1.1)^k}{k}.$$

If the ratio test predicts that an infinite series converges, the convergence of the sequence of partial sums to the sum of the series is usually fast, i.e., a partial sum approximates the sum accurately without having to add a large number of terms. In problems 19 and 20,
a) Show that the given series converges by using the ratio test,
b) [CAS] Make use of your computer algebra system in order to find the exact value of the sum $S$ of the series, calculate the partial sums $S_n$ and $|S_n - S|$ for $n = 2, 4, 8, 16, 32$. Do the numbers support the claim of fast convergence?

19. 
$$\sum_{n=1}^{\infty} \frac{n}{2^n}$$

20. 
$$\sum_{n=0}^{\infty} \frac{3^n}{n!}$$

If the ratio test predicts that a series diverges, the magnitude of the $n$th term tends to infinity and usually grows very rapidly as $n$ increases. In problems 21 and 22,
a) Show that the series $\sum a_n$ diverges by the ratio test,
b) [C] Calculate $|a_n|$ for $n = 2, 4, 8, 16, 32$. Do the numbers support the claim that $|a_n|$ grows rapidly as $n$ increases?

21. 
$$\sum_{n=1}^{\infty} \frac{n!}{4^n}$$

22. 
$$\sum_{n=1}^{\infty} (-1)^{n-1} \frac{3^n}{n}$$

In problems 23 - 30 use **the root test** in order to determine whether the given series converges absolutely or whether it diverges, provided that the test is applicable. You need not investigate the convergence of the given series if the test is inconclusive:

23.
$$\sum_{n=1}^{\infty} \frac{10^n}{n^n}$$

24.
$$\sum_{n=1}^{\infty} \frac{n^2}{2^n}$$

25.
$$\sum_{n=2}^{\infty} \frac{\ln(n)}{n^3}$$

26.
$$\sum_{n=1}^{\infty} n^4 e^{-n}$$

27.
$$\sum_{n=0}^{\infty} (-1)^n \frac{1}{3^n}.$$

28.
$$\sum_{n=0}^{\infty} (-1.5)^n.$$

29.
$$\sum_{n=1}^{\infty} (-1)^{n-1} \frac{1}{n4^n}.$$

30.
$$\sum_{n=1}^{\infty} (-1)^{n-1} \frac{(1.2)^n}{n}.$$

If the root test predicts that an infinite series converges, the convergence of the sequence of partial sums to the sum of the series is usually fast, i.e., a partial sum approximates the sum accurately without having to add a large number of terms. In problems 31 and 32,
a) Show that the given series converges by using the root test,
b) [CAS] Make use of your computer algebra system in order to find the exact value of the sum $S$ of the series, calculate the partial sums $S_n$ and $|S_n - S|$ for $n = 2, 4, 8, 16, 32$. Do the numbers support the claim of fast convergence?

31.
$$\sum_{n=1}^{\infty} \frac{n}{3^n}$$

32.
$$\sum_{n=1}^{\infty} (-1)^{n-1} \frac{1}{n2^n}$$

If the root test predicts that a series diverges, the magnitude of the $n$th term tends to infinity and usually grows very rapidly as $n$ increases. In problems 21 and 22,
a) Show that the series $\sum a_n$ diverges by the root test,
b) [C] Calculate $|a_n|$ for $n = 2, 4, 8, 16, 32$. Do the numbers support the claim that $|a_n|$ grows rapidly as $n$ increases?

33.
$$\sum_{n=1}^{\infty} \frac{3^n}{n}$$

34.
$$\sum_{n=1}^{\infty} (-1)^{n-1} \frac{2^n}{n}$$

## 9.5 Power Series: Part 1

### The Definitions

In sections 9.1 and 9.2 we discussed Taylor polynomials. Recall that the Taylor polynomial of order $n$ based at $c$ for the function $f$ is

$$\sum_{k=0}^{n} \frac{1}{n!} f^{(k)}(c)(x-c)^k = f(c) + f'(c)(x-c) + \frac{1}{2!} f^{(2)}(c)(x-c)^2 + \cdots + \frac{1}{n!} f^{(n)}(c)(x-c)^n.$$

We saw examples of $f$ and $x$ such that

$$\lim_{n \to \infty} \left( c + f'(c)(x-c) + \frac{1}{2!} f^{(2)}(c)(x-c)^2 + \cdots + \frac{1}{n!} f^{(n)}(c)(x-c)^n \right) = f(x),$$

## 9.5. POWER SERIES: PART 1

i.e., examples of infinite series of the form

$$f(c) + f'(c)(x-c) + \frac{1}{2!}f^{(2)}(c)(x-c)^2 + \cdots + \frac{1}{n!}f^{(n)}(c)(x-c)^n + \cdots$$

such that the sum of the series for certain values of $x$ is $f(x)$:

$$f(c) + f'(c)(x-c) + \frac{1}{2!}f^{(2)}(c)(x-c)^2 + \cdots + \frac{1}{n!}f^{(n)}(c)(x-c)^n + \cdots = f(x).$$

**Definition 1** **The Taylor series for the function $f$ based at $c$ is the infinite series**

$$f(c) + f'(c)(x-c) + \frac{1}{2!}f''(c)(x-c)^n + \cdots + \frac{1}{n!}f^{(n)}(c)(x-c)^n + \cdots$$

Note that the $(n+1)$st partial sum of the Taylor series for $f$ based at $c$ is the Taylor polynomial of order $n$ for $f$ based at $c$. Just as in the case of Taylor polynomials, almost all the Taylor series that we will consider will be based at 0, so that they will be in the form

$$f(0) + f'(0)x + \frac{1}{2}f''(0)x^2 + \frac{1}{3!}f^{(3)}(0)x^3 + \cdots \frac{1}{n!}f^{(n)}(0)x^n + \cdots$$

The Taylor series for a function $f$ based at 0 is referred to as **the Maclaurin series for $f$.**

**Example 1** As we saw in Example 2 of Section 9.1, the Maclaurin polynomial of order $n$ for the natural exponential function is

$$P_n(x) = 1 + x + \frac{1}{2}x^2 + \frac{1}{3!}x^3 + \cdots + \frac{1}{n!}x^n.$$

$P_n(x)$ is the $(n+1)$st partial sum of the Maclaurin series for the natural exponential function:

$$1 + x + \frac{1}{2}x^2 + \frac{1}{3!}x^3 + \cdots + \frac{1}{n!}x^n + \cdots$$

In Section 9.2, we used Taylor's formula for the remainder to show that

$$\lim_{n \to \infty} \left(1 + x + \frac{1}{2}x^2 + \frac{1}{3!}x^3 + \cdots + \frac{1}{n!}x^n\right) = e^x$$

for each $x \in \mathbb{R}$ (Proposition 3 of Section 9.2). Thus,

$$1 + x + \frac{1}{2}x^2 + \frac{1}{3!}x^3 + \cdots + \frac{1}{n!}x^n + \cdots = e^x$$

for each $x \in \mathbb{R}$. $\square$

**Example 2** As we saw in Example 8 of Section 9.1, the Taylor polynomial of order $n$ for the natural logarithm based at 1 is

$$(x-1) - \frac{1}{2}(x-1)^2 + \frac{1}{3}(x-1)^3 - \frac{1}{4}(x-1)^4 + \cdots + (-1)^{n-1}\frac{1}{n}(x-1)^n$$

By Proposition 6 of Section 9.2,

$$\lim_{n \to \infty} \left((x-1) - \frac{1}{2}(x-1)^2 + \frac{1}{3}(x-1)^3 - \frac{1}{4}(x-1)^4 + \cdots + (-1)^{n-1}\frac{1}{n}(x-1)^n\right) = \ln(x)$$

if $1 \leq x \leq 2$. In Section 9.6 we will see that the above statement is valid for each $x \in (0, 2]$. Thus, the sum of the power series

$$(x-1) - \frac{1}{2}(x-1)^2 + \frac{1}{3}(x-1)^3 - \frac{1}{4}(x-1)^4 + \cdots + (-1)^{n-1}\frac{1}{n}(x-1)^n + \cdots$$

is $\ln(x)$ for each $x \in (0, 2]$. □

If we set

$$a_n = \frac{1}{n!}f^{(n)}(c), \; n = 0, 1, 2, \ldots,$$

we can write the Taylor series for $f$ based at $c$ as

$$a_0 + a_1(x-c) + a_2(x-c)^2 + \cdots + a_n(x-c)^n + \cdots = \sum_{n=0}^{\infty} a_n(x-c)^n.$$

**Definition 2** A **power series in powers of** $(x - c)$ is a series of the form

$$\sum_{n=0}^{\infty} a_n(x-c)^n,$$

where $a_0, a_1, a_2, \ldots$ are given constants. The number $a_n$ is the coefficient of $(x-c)^n$.

Thus, a Taylor series based at $c$ is a power series in powers of $x - c$. In this section and in Section 9.6 we will examine the basic properties of power series and functions that are defined by power series. As you will see in post-calculus courses, many special functions of Mathematics are defined via power series.

## Convergence Properties of a Power Series

Let's begin by stating a theorem that provides a guideline as to what to expect in terms of the convergence of a power series:

**Theorem 1** Given a power series $\sum_{n=0}^{\infty} a_n(x-c)^n$, one of the following cases is valid:

1. There is a number $r > 0$ such that $\sum_{n=0}^{\infty} a_n(x-c)^n$ **converges absolutely if** $|x - c| < r$. **i.e., if** $c - r < x < c + r$, **and diverges if** $|x - c| > r$. **i.e., if** $x < c - r$ **or** $x > c + r$, **or**
2. $\sum_{n=0}^{\infty} a_n(x-c)^n$ **converges for all** $x \in r$, **or**
3. $\sum_{n=0}^{\infty} a_n(x-c)^n$ **converges if and only if** $x = c$ **(in which case the series is reduced to the single term** $a_0$**).**

The proof of Theorem 1 is left to a course in advanced calculus.

**Remark** If case 1 of Theorem 1 is valid, we will refer to the interval $(c - r, c + r)$ as **the open interval of convergence** of the power series $\sum_{n=0}^{\infty} a_n(x-c)^n$, and to the number $r$ as its **radius of convergence**. The power series may converge or diverge if $x$ is an endpoint of the open interval of convergence. If we wish to determine the interval of convergence of the power series, we need to investigate the convergence of the series at the endpoints of the open interval of convergence. If case 2 of Theorem 1 is valid, we will refer to $(-\infty, +\infty)$ as the interval of convergence of the given power series, and declare the radius of convergence of the series to be $+\infty$. In case 3, you may imagine that the interval of convergence has degenerated to the singleton $\{c\}$, and declare the radius of convergence of the series to be 0. In the examples that we will discuss, we will be able to determine the open interval of convergence of the given power series by the ratio test or the root test. We will have to resort to some of the other tests that have been developed in the previous sections of this chapter if we wish to determine the behavior of the series at the endpoints of the interval of convergence. ◊

## 9.5. POWER SERIES: PART 1

**Example 3** The geometric series

$$1 + x + x^2 + \cdots + x^n + \cdots$$

is a power series in powers of $x$. This power series is the Taylor series for $1/(1-x)$ based at 0. As in Proposition 1 of Section 9.4, the series converges absolutely if $-1 < x < 1$, and diverges if $x \leq -1$ or $x \geq 1$. Thus, the open interval of convergence of the geometric series is $(-1, 1)$. The radius of convergence of the power series is 1. The series diverges at both endpoints $-1$ and $1$ of the open interval of convergence. □

**Example 4** The Taylor series for the natural exponential function, i.e.,

$$1 + x + \frac{1}{2!}x^2 + \frac{1}{3!}x^3 + \cdots + \frac{1}{n!}x^n + \cdots$$

converges absolutely for all $x \in \mathbb{R}$.

Indeed,

$$\lim_{n \to \infty} \frac{\left|\frac{1}{(n+1)!}x^{n+1}\right|}{\left|\frac{1}{n!}x^n\right|} = \lim_{n \to \infty} \left(\frac{n!\,|x^{n+1}|}{(n+1)!\,|x^n|}\right) = \lim_{n \to \infty} \left(\frac{1}{n+1}\frac{|x|^{n+1}}{|x|^n}\right)$$

$$= \lim_{n \to \infty} \left(\frac{1}{n+1}|x|\right) = |x| \lim_{n \to \infty} \frac{1}{n+1} = 0 < 1$$

for each $x \in \mathbb{R}$. Therefore, the series converges absolutely for each $x \in \mathbb{R}$, by the ratio test. Thus, the interval of convergence of the series is $(-\infty, +\infty)$, and the radius of convergence is $+\infty$. □

**Example 5** The power series

$$\sum_{n=1}^{\infty} n!x^n = x + 2!x^2 + 3!x^3 + \cdots$$

converges if and only if $x = 0$.

Indeed, we can apply the ratio test:

$$\lim_{n \to \infty} \frac{|(n+1)!x^{n+1}|}{|n!x^n|} = \lim_{n \to \infty} (n+1)|x| = +\infty$$

if $x \neq 0$. Therefore, the series diverges if $x \neq 0$. If $x = 0$ the series is reduced to $0 + 0 + \cdots$, so that convergence is trivial. The radius of convergence of this power series is 0. □

**Example 6** Consider the power series

$$\sum_{n=0}^{\infty} \frac{1}{2n+1}x^n = 1 + \frac{1}{3}x + \frac{1}{5}x^2 + \frac{1}{7}x^3 + \cdots.$$

Find the radius of convergence and the open interval of convergence of the power series.

**Solution**

We will apply the ratio test:

$$\lim_{n\to\infty} \left| \frac{\frac{1}{2n+3} x^{n+1}}{\frac{1}{2n+1} x^n} \right| = \lim_{n\to\infty} \left( \frac{2n+1}{2n+3} |x| \right) = |x| \lim_{n\to\infty} \frac{2n+1}{2n+3} = |x|.$$

Therefore, the given power series converges absolutely if $|x| < 1$, and diverges if $|x| > 1$. Thus, the radius of convergence of the series is 1, and the open interval of the series is $(-1, 1)$.. $\square$

**Example 7** Consider the power series

$$\sum_{n=1}^{\infty} \frac{1}{n^2} (x-2)^n.$$

Find the radius of convergence and the open interval of convergence of the power series.

**Solution**

We will apply the root test:

$$\lim_{n\to\infty} \left| \frac{1}{n^2} (x-2)^n \right|^{1/n} = \lim_{n\to\infty} \frac{1}{n^{2/n}} |x-2| = |x-2| \lim_{n\to\infty} \frac{1}{\left(n^{1/n}\right)^2} = |x-2|.$$

Therefore, the power series converges absolutely if $|x - 2| < 1$, and diverges if $|x - 2| > 1$. The open interval of convergence of the power series is

$$\{x : |x-2| < 1\} = (1, 3).$$

$\square$

## Differentiation of Functions defined by Power Series

A power series defines an infinitely differentiable function in its open interval of convergence:

**Theorem 2** Assume that the power series

$$\sum_{n=0}^{\infty} a_n (x-c)^n$$

has the nonempty open interval of convergence $J$, and that

$$f(x) = \sum_{n=0}^{\infty} a_n (x-c)^n = a_0 + a_1 (x-c) + a_2 (x-c)^2 + a_3 (x-c)^3 + a_4 (x-c)^4 + \cdots + a_n (x-c)^n + \cdots$$

for each $x \in J$. Then, $f$ has derivatives of all orders in the interval $J$, and its derivatives can be computed by differentiating the power series termwise:

$$f'(x) = a_1 + 2a_2 (x-c) + 3a_3 (x-c)^2 + 4a_4 (x-c)^3 + \cdots + na_n (x-c)^{n-1} + \cdots,$$

$$f''(x) = 2a_2 + (2)(3)a_3 (x-c) + (3)(4) a_4 (x-c)^2 + \cdots + (n-1)na_n (x-c)^{n-2} + \cdots,$$

$$f^{(3)}(x) = (2)(3)a_3 + (2)(3)(4) a_4 (x-c) + \cdots + (n-2)(n-1) na_n (x-c)^{n-3} + \cdots,$$

$$\vdots$$

The power series corresponding to the derivative of $f$ of any order has the same open interval of convergence as the power series corresponding to $f$.

## 9.5. POWER SERIES: PART 1

We will leave the proof of Theorem 2 to a course in advanced calculus.

**Example 8** We know that
$$f(x) = \frac{1}{1-x} = 1 + x + x^2 + x^3 + \cdots + x^n + \cdots \text{ if } -1 < x < 1.$$

a) Determine the power series in powers of $x$ for
$$f'(x) = \frac{1}{(1-x)^2}$$
by differentiating the power series for $f$ termwise. Confirm that the power series has the same open interval as the geometric series, i.e., the interval $(-1, 1)$.

b) Determine the power series in powers of $x$ for
$$f''(x) = \frac{d^2}{dx^2}\left(\frac{1}{1-x}\right) = \frac{2}{(1-x)^3}$$
by differentiating the power series for $f'$ termwise. Confirm that the power series has the same open interval as the geometric series, i.e., the interval $(-1, 1)$.

**Solution**

a) By Theorem 2,
$$f'(x) = \frac{d}{dx}\left(\frac{1}{1-x}\right) = \frac{1}{(1-x)^2} = \frac{d}{dx}\left(1 + x + x^2 + x^3 + \cdots + x^n + \cdots\right)$$
$$= 1 + 2x + 3x^2 + \cdots + nx^{n-1} + \cdots.$$

Theorem 2 predicts that the open interval of convergence of the above series is the same as the open interval of convergence of the original series, i.e., $(-1, 1)$. Let's confirm this by the ratio test:
$$\lim_{n\to\infty} \frac{|(n+1)x^n|}{|nx^{n-1}|} = \lim_{n\to\infty}\left(\frac{n+1}{n}\right)|x| = |x|\lim_{n\to\infty}\frac{n+1}{n} = |x|(1) = |x|.$$

Therefore, the series converges absolutely if $|x| < 1$ and diverges if $|x| > 1$.

b) Again, by Theorem 2,
$$f''(x) = \frac{d^2}{dx^2}\left(\frac{1}{1-x}\right) = \frac{d}{dx}\left(\frac{1}{(1-x)^2}\right)$$
$$= \frac{d}{dx}\left(1 + 2x + 3x^2 + \cdots + nx^{n-1} + \cdots\right)$$
$$= 2 + (2)(3)x + \cdots + (n-1)(n)x^{n-2} + \cdots, \; x \in (-1, 1).$$

Let's apply the ratio test the above power series:
$$\lim_{n\to\infty} \frac{n(n+1)|x^{n+1}|}{(n-1)n|x^n|} = |x|\lim_{n\to\infty}\frac{n+1}{n-1} = |x|.$$

Therefore, the open interval of convergence of the series is $(-1, 1)$, as predicted by Theorem 2.□

**Example 9** We know that
$$\sin(x) = x - \frac{1}{3!}x^3 + \frac{1}{5!}x^5 - \frac{1}{7!}x^7 + \cdots.$$

Determine a power series that has as its sum $\cos(x)$ for each $x \in \mathbb{R}$ by termwise differentiation.

**Solution**

Since the Taylor series for sine has the sum $\sin(x)$ for each $x$ in $(-\infty, +\infty)$, we can apply Theorem 2:

$$\cos(x) = \frac{d}{dx}\sin(x) = \frac{d}{dx}\left(x - \frac{1}{3!}x^3 + \frac{1}{5!}x^5 - \frac{1}{7!}x^7 + \cdots\right)$$

$$= 1 - \frac{1}{2!}x^2 + \frac{1}{4!}x^4 - \frac{1}{6!}x^6 + \cdots = \sum_{k=0}^{\infty}(-1)^k \frac{1}{(2k)!}x^{2k}.$$

□

The following theorem says that if a function is defined via a power series in powers of $(x - c)$, that power series is the Taylor series of the function based at $c$:

**Theorem 3** Assume that

$$f(x) = a_0 + a_1(x - c) + a_2(x - c)^2 + \cdots + a_n(x - c)^n + \cdots$$

**for all $x \in J$, where $J$ is the non-empty open interval of convergence of the power series. Then,**

$$a_n = \frac{1}{n!}f^{(n)}(c), \ n = 0, 1, 2, 3, \ldots.$$

**Proof**

By Theorem 2, $f$ has derivatives of all orders in the interval $J$ and

$$f'(x) = a_1 + 2a_2(x - c) + 3a_3(x - c)^2 + 4a_4(x - c)^3 + 5a_5(x - c)^4 + \cdots,$$
$$f''(x) = 2a_2 + (2)(3)a_3(x - c) + (3)(4)a_4(x - c)^2 + (4)(5)a_5(x - c)^3 + \cdots,$$
$$f^{(3)}(x) = (2)(3)a_3 + (2)(3)(4)a_4(x - c) + (3)(4)(5)a_5(x - c)^2 + \cdots,$$
$$f^{(4)}(x) = (2)(3)(4)a_4 + (2)(3)(4)(5)a_5(x - c) + \cdots,$$
$$\vdots$$

Therefore,

$$f(c) = a_0,$$
$$f'(c) = a_1,$$
$$f''(c) = 2a_2,$$
$$f^{(3)}(c) = (2)(3)a_3,$$
$$f^{(4)}(c) = (2)(3)(4)a_4,$$
$$\vdots$$

The trend should be clear:

$$f^{(n)}(c) = n!a_n, \ n = 0, 1, 2, 3, \ldots$$

Thus,

$$a_n = \frac{1}{n!}f^{(n)}(c), \ n = 0, 1, 2, 3, \ldots,$$

## 9.5. POWER SERIES: PART 1

so that

$$f(x) = f(c) + f'(c)(x-c) + \frac{1}{2}f''(c)(x-c)^2 + \frac{1}{3!}f^{(3)}(c)(x-c)^3 + \cdots$$
$$= \sum_{n=0}^{\infty} \frac{1}{n!} f^{(n)}(c)(x-c)^n.$$

Therefore, the power series which defines $f$ is the Taylor series of $f$. ■

Even though a function which is defined by a power series is infinitely differentiable, not every infinitely differentiable function can be expressed as a power series in an interval that does not degenerate to a single point:

**Example 10** Let

$$f(x) = \begin{cases} \exp(-1/x^2) & \text{if } x \neq 0, \\ 0 & \text{if } x = 0. \end{cases}$$

Figure 1 shows the graph of $f$. The picture suggests that

$$\lim_{x \to 0} f(x) = \lim_{x \to 0} \exp\left(-\frac{1}{x^2}\right) = 0.$$

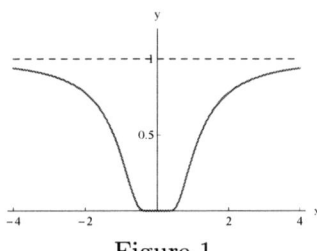

Figure 1

Indeed, if we set $z = 1/x^2$, we have $z \to +\infty$ as $x \to 0$. Therefore,

$$\lim_{x \to 0} \exp\left(-\frac{1}{x^2}\right) = \lim_{z \to +\infty} \exp(-z) = 0.$$

Thus, $f$ is continuous at 0.

Actually, $f$ has derivatives of all orders at 0 and $f^{(n)}(0) = 0$, $n = 1, 2, 3, \ldots$ (notice that the graph of $f$ is very flat near the origin). We will only show that

$$f'(0) = 0 \text{ and } f''(0) = 0,$$

and leave the technically tedious proof as an exercise. In order to show that $f'(0) = 0$, we must establish that

$$\lim_{h \to 0} \frac{f(h) - f(0)}{h} = \lim_{h \to 0} \frac{\exp\left(-\frac{1}{h^2}\right)}{h} = 0.$$

Indeed, if we set $z = 1/h^2$, $z \to +\infty$ as $h \to 0$, and $h = \pm 1/\sqrt{z}$. Therefore,

$$\lim_{h \to 0+} \frac{\exp\left(-\frac{1}{h^2}\right)}{h} = \lim_{z \to +\infty} \frac{\exp(-z)}{\frac{1}{\sqrt{z}}} = \lim_{z \to +\infty} \sqrt{z} e^{-z} = 0.$$

Similarly,

$$\lim_{h \to 0-} \frac{\exp\left(-\frac{1}{h^2}\right)}{h} = 0,$$

so that
$$\lim_{h\to 0} \frac{\exp\left(-\frac{1}{h^2}\right)}{h} = 0,$$
as claimed.

Now let us show that $f''(0) = 0$. We have
$$\frac{f'(h) - f'(0)}{h} = \frac{f'(h)}{h} = \frac{1}{h}\left(\frac{d}{dh}\exp\left(-h^{-2}\right)\right) = \frac{1}{h}\left(2h^{-3}\exp\left(-h^{-2}\right)\right) = 2\frac{\exp(-h^{-2})}{h^4}$$

If we set $z = 1/h^2$,
$$\lim_{h\to 0}\left(2\frac{\exp(-h^{-2})}{h^4}\right) = 2\lim_{z\to +\infty}\left(z^2 \exp(-z)\right) = 2(0) = 0.$$

Therefore, $f''(0) = 0$.

The Taylor series of $f$ is very simple:
$$0 + (0)x + (0)x^2 + \cdots + (0)x^n + \cdots.$$

Since $f(x) \neq 0$ if $x \neq 0$, the only point at which the above series has the same value as $f$ is 0. □

You may wonder why we bother to discuss power series in a more general context than Taylor series, if any power series is the Taylor series of a function. The reason is that new useful functions can be defined via power series, as in the following example:

**Example 11** Let's define the function $J_0$ by the expression
$$J_0(x) = \sum_{n=0}^{\infty} (-1)^n \frac{1}{(n!)^2 2^{2n}} x^{2n} = 1 - \frac{1}{2^2}x^2 + \frac{1}{(2!)^2 2^4}x^4 - \frac{1}{(3!)^2 2^6}x^6 + \cdots.$$

As an exercise, you can check that the power series converges on $\mathbb{R}$ by the ratio test. Therefore $J_0$ is infinitely differentiable on $\mathbb{R}$. Even though the power series that defines $J$ may seem strange, $J_0$ is a very respectable and useful function. It is referred to as the **Bessel function** of the first kind of order 0, and shows up as a solution of an important differential equation. In fact, $J_0$ is so important in certain applications that it is a built-in function in computer algebra systems such as Mathematica or Maple. Figure 2 shows the graph of $J_0$ on the interval $[-10, 10]$. □

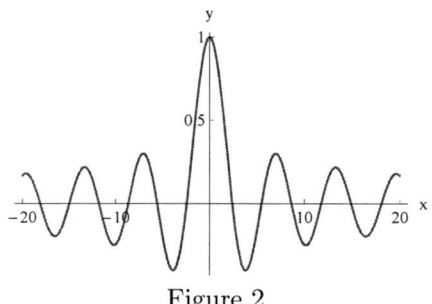

Figure 2

The following theorem on the uniqueness of the power series representation of a function has practical implications in the determination of new Taylor series by making use of known results:

## 9.5. POWER SERIES: PART 1

**Theorem 4 (The Uniqueness of Power Series Representations)** Assume that

$$a_0+a_1(x-c)+a_2(x-c)^2+\cdots+a_n(x-c)^n+\cdots = b_0+b_1(x-c)+b_2(x-c)^2+\cdots+b_n(x-c)^n+\cdots$$

**for all $x$ in an open interval $J$ which contains $c$. Then, $a_n = b_n$, $n = 0, 1, 2, \ldots$.**

**Proof**

We showed that the power series which represents a function on its open interval of convergence $J$ is the Taylor series of that function, i.e., if

$$f(x) = a_0 + a_1(x-c) + a_2(x-c)^2 + \cdots + a_n(x-c)^n + \cdots, \quad x \in J,$$

then

$$a_0 = f(c), \ a_1 = f'(c), \ a_2 = \frac{1}{2!}f''(c), \ldots, a_n = \frac{1}{n!}f^{(n)}(c), \ldots.$$

Since

$$f(x) = a_0+a_1(x-c)+a_2(x-c)^2+\cdots+a_n(x-c)^n+\cdots = b_0+b_1(x-c)+b_2(x-c)^2+\cdots+b_n(x-c)^n+\cdots,$$

we have

$$b_n = \frac{1}{n!}f^{(n)}(c), \ n = 0, 1, 2, \ldots,$$

also. Therefore,

$$a_n = b_n, \ n = 0, 1, 2, \ldots,$$

as claimed. ∎

**Example 12** Let

$$f(x) = \frac{1}{1+x}, \ x \neq -1.$$

Determine the Taylor series for $f$.

**Solution**

Thanks to our previous discussion about geometric series,

$$\frac{1}{1+x} = \frac{1}{1-(-x)} = 1 + (-x) + (-x)^2 + (-x)^3 + \cdots + (-x)^n + \cdots$$

$$= 1 - x + x^2 - x^3 + \cdots + (-1)^n x^n + \cdots$$

if $|-x| = |x| < 1$. By the uniqueness of the power series corresponding to a given function, we have computed the Taylor series for $f$ in powers of $x$. □

**Example 13** Let $f(x) = e^{-x^2}$.

a) Determine the Taylor series for $f$.
b) Let $P_{12}(x)$ be the Taylor polynomial of order 8 for $f$. Make use of your calculator to compare the graph of $P_8$ with the graph of $f$. Does the picture indicate that $P_8(x)$ approximates $e^{-x^2}$ very well if $x$ is not too large?

**Solution**

a) We know that

$$e^u = 1 + u + \frac{1}{2!}u^2 + \frac{1}{3!}u^3 + \cdots + \frac{1}{n!}u^n + \cdots$$

for any $u \in R$. Therefore, if we set $u = -x^2$, we obtain

$$e^{-x^2} = 1 + (-x^2) + \frac{1}{2!}(-x^2)^2 + \frac{1}{3!}(-x^2)^3 + \cdots + \frac{1}{n!}(-x^2)^n + \cdots$$
$$= 1 - x^2 + \frac{1}{2!}x^4 - \frac{1}{3!}x^6 + \cdots + (-1)^n \frac{1}{n!}x^{2n} + \cdots.$$

b) By part a),
$$P_{12}(x) = 1 - x^2 + \frac{1}{2!}x^4 - \frac{1}{3!}x^6 + \frac{1}{4!}x^8 - \frac{1}{5!}x^{10} + \frac{1}{6!}x^{12}.$$

Figure 3 shows the graphs of $f$ and $P_{12}$ (the graph of $P_{12}$ is the dashed curve). The picture indicates that $P_{12}(x)$ approximates $e^{-x^2}$ very well if $x \in [-1.4, 1.4]$. $\square$

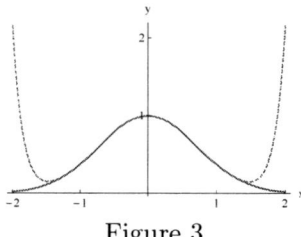

Figure 3

The theorems on the termwise differentiation of power series and the uniqueness of power series representations can be used to determine solutions of differentiable equations that are represented by power series, as in the following example:

**Example 14** Let's consider the differential equation
$$y''(x) = -y(x),$$

and seek a solution in the form
$$y(x) = a_0 + a_1 x + a_2 x^2 + a_3 x^3 + a_4 x^4 + \cdots + a_n x^n + a_{n+1} x^{n+1} + a_{n+2} x^{n+2} + \cdots.$$

We have
$$y'(x) = a_1 + 2a_2 x + 3a_3 x^2 + 4a_4 x^3 + \cdots + n a_n x^{n-1} + (n+1) a_{n+1} x^n + (n+2) a_{n+2} x^{n+1} + \cdots,$$

and
$$y''(x) = 2a_2 + (2)(3)a_3 x + (3)(4)a_4 x^2 + \cdots$$
$$+ (n-1)(n)a_n x^{n-2} + (n)(n+1) a_{n+1} x^{n-1} + (n+1)(n+2) a_{n+2} x^n + \cdots.$$

By the uniqueness of power series representations, the equation $y''(x) = -y(x)$ leads to the following equalities for the coefficients $a_1, a_2, a_3, \ldots$:

$$-2a_2 = a_0,$$
$$-(2)(3)a_3 = a_1,$$
$$-(3)(4)a_4 = a_2,$$
$$\vdots$$
$$-(n+1)(n+2) a_{n+2} = a_n,$$
$$\vdots$$

## 9.5. POWER SERIES: PART 1

We obtain $a_2$ and $a_4$ in terms of $a_0$:

$$a_2 = -\frac{1}{2}a_0,$$
$$a_4 = -\frac{1}{(3)(4)}a_2 = \frac{1}{(2)(3)(4)}a_0 = \frac{1}{4!}a_0.$$

This suggests the inductive hypothesis that

$$a_{2k} = \frac{(-1)^k}{(2k)!}a_0, \ k = 1, 2, 3, \ldots$$

(you may wish to compute $a_6$ and $a_8$ if you need more evidence pointing in that direction). We can obtain the expression for $a_{2k+2}$:

$$a_{2k+2} = -\frac{1}{(2k+1)(2k+2)}a_{2k} = -\frac{1}{(2k+1)(2k+2)}\left(\frac{(-1)^k}{(2k)!}a_0\right) = \frac{(-1)^{k+1}}{(2k+2)!}a_0.$$

Similarly,

$$a_3 = -\frac{1}{(2)(3)}a_1 = -\frac{1}{3!}a_1,$$

and

$$a_{2k+1} = \frac{(-1)^k}{(2k+1)!}a_1, \ k = 1, 2, 3, \ldots.$$

The differential equation $y''(x) = -y(x)$ comes with initial conditions that involve $y(x_0)$ and $y'(x_0)$ at some $x_0$. Assume that

$$y(0) = 1 \text{ and } y'(0) = 0.$$

Since

$$y(x) = a_0 + a_1 x + a_2 x^2 + a_3 x^3 + a_4 x^4 + \cdots + a_n x^n + a_{n+1} x^{n+1} + a_{n+2} x^{n+2} + \cdots,$$

we have

$$y(0) = a_0 \text{ and } y'(0) = a_1.$$

Therefore, we must have

$$a_0 = 1 \text{ and } a_1 = 0.$$

Thus,

$$a_{2k} = \frac{(-1)^k}{(2k)!}a_0 = \frac{(-1)^k}{(2k)!}, \ k = 1, 2, 3, \ldots,$$

and

$$a_{2k+1} = \frac{(-1)^k}{(2k+1)!}a_1 = 0, \ k = 1, 2, 3, \ldots.$$

Therefore, we must have

$$y(x) = 1 - \frac{1}{2!}x^2 + \frac{1}{4!}x^4 - \frac{1}{6!}x^6 + \cdots + \frac{(-1)^k}{(2k)!}x^{2k} + \cdots = 1 + \sum_{k=1}^{\infty}\frac{(-1)^k}{(2k)!}x^{2k} = \sum_{k=0}^{\infty}\frac{(-1)^k}{(2k)!}x^{2k}.$$

You will recognize the above series as the Taylor series for cosine in powers of $x$. If we were not familiar with cosine, we would have invented that special function all over again, as the solution of the initial value problem,

$$y'' = -y, \ y(0) = 1, \ y'(0) = 0.$$

Similarly, the power series solution of the initial value problem

$$y''(0) = -y,\ y(0) = 0,\ y'(0) = 1,$$

leads to the power series

$$x - \frac{1}{3!}x^3 + \frac{1}{5!}x^5 - \frac{1}{7!}x^7 + \cdots + \frac{(-1)^k}{(2k+1)!}x^{2k+1} + \cdots$$

(fill in the details). This is the Taylor series for sine in powers of $x$ □

**Remark** It is possible to *define* sine and cosine via power series, and deduce all their properties rigorously. Recall that these functions had been introduced geometrically, and we had to assume that arclength was measurable on the unit circle. That was the pragmatic approach, since we did not wish to wait until our discussion of power series before introducing these important functions. Nevertheless, you may find it of interest that an alternative rigorous approach to trigonometric functions is feasible. ◊

## Problems

In problems 1-8 determine the radius of convergence and the open interval of convergence of the given power series (you need not investigate the convergence of the series at the endpoints of the interval)

1. $$\sum_{n=1}^{\infty} \frac{2^n}{n} x^n$$

2. $$\sum_{n=1}^{\infty} (-1)^{n-1} \frac{n^2}{3^n} x^n$$

3. $$\sum_{n=1}^{\infty} (-1)^{n+1} \frac{4^n}{n^4} x^n$$

4. $$\sum_{n=0}^{\infty} (-1)^n \frac{x^{2n+1}}{(2n+1)!}$$

5. $$\sum_{n=1}^{\infty} \frac{1}{e^n} (x-e)^n$$

6. $$\sum_{n=1}^{\infty} (-1)^{n+1} \frac{\pi^n}{n^2+1} (x+\pi)^n$$

7. $$\sum_{n=1}^{\infty} (-1)^n \frac{1}{n!} x^{2n}$$

8. $$\sum_{n=1}^{\infty} \sqrt{\frac{n-1}{2n+3}} (x-4)^n$$

Given that

$$\frac{1}{1-x} = 1 + x + x^2 + x^3 + \cdots + x^n + \cdots = \sum_{n=0}^{\infty} x^n,$$

$$e^x = 1 + x + \frac{1}{2}x^2 + \frac{1}{3!}x^3 + \cdots + \frac{1}{n!}x^n + \cdots = \sum_{n=0}^{\infty} \frac{1}{n!}x^n,$$

$$\sin(x) = x - \frac{1}{3!}x^3 + \frac{1}{5!}x^5 - \frac{1}{7!}x^7 + \cdots + \frac{(-1)^n}{(2n+1)!}x^{2n+1} + \cdots = \sum_{n=0}^{\infty} \frac{(-1)^n}{(2n+1)!}x^{2n+1},$$

$$\cos(x) = 1 - \frac{1}{2}x^2 + \frac{1}{4!}x^4 - \frac{1}{6!}x^6 + \cdots + \frac{(-1)^n}{(2n)!}x^{2n} + \cdots = \sum_{n=0}^{\infty} \frac{(-1)^n}{(2n)!}x^{2n},$$

$$(1+x)^r = 1 + rx + \frac{(r-1)r}{2!}x^2 + \frac{(r-2)(r-1)r}{3!}x^3 + \cdots + \frac{(r-n+1)\cdots(r-1)r}{n!}x^n + \cdots,$$

## 9.6. POWER SERIES: PART 2

obtain the Maclaurin series of the given function via suitable substitutions and differentiation in problems 9-16 (display the first four nonzero terms and the general expression):

9. $$f(x) = \frac{1}{1+x^2}$$

10. $$f(x) = \frac{d^2}{dx^2}\left(\frac{1}{1-x^2}\right)$$

11. $$f(x) = e^{-x^2}$$

12. $$f(x) = \sinh(x)$$

13.

14. $$f(x) = \cosh(x)$$

15. $$f(x) = \frac{\cos(x) - 1 + \frac{1}{2}x^2}{x^4}$$

16. $$f(x) = \frac{\sin(x) - x + \frac{x^3}{6}}{x^5}$$

$$f(x) = \frac{1}{\sqrt{9+x^2}}$$

In problems 17-20 determine the required limit by making use of the appropariate Taylor series (do not use L'Hôpital's rule).

17. $$\lim_{x \to 0} \frac{\cos(x) - 1}{x^2}$$

18. $$\lim_{x \to 0} \frac{\sin(x) - x}{x^3}$$

19. $$\lim_{x \to 0} \frac{\sin(x) - x + \frac{x^3}{6}}{x^5}$$

20. $$\lim_{x \to 0} \frac{e^x - 1 - x}{x^2}.$$

## 9.6 Power Series: Part 2

In Section 9.5 we determined new Taylor series by substitutions in Taylor series that had been derived earlier and by termwise differentiation. In this section, we will determine new Taylor series by termwise integration, multiplication and division. We will also discuss binomial series.

### Termwise Integration of Power Series

We can integrate a power series termwise:

**Theorem 1** Assume that

$$f(t) = a_0 + a_1(t-c) + a_2(t-c)^2 + \cdots + a_n(t-c)^n + \cdots$$

**for all $t$ in the open interval $J$, and that**

$$F(x) = \int_c^x f(t)\,dt.$$

**Then,**

$$F(x) = \int_c^x \left(a_0 + a_1(t-c) + a_2(t-c)^2 + \cdots + a_n(t-c)^n + \cdots\right) dt$$
$$= a_0(x-c) + a_1\frac{(x-c)^2}{2} + a_2\frac{(x-c)^3}{3} + \cdots + a_n\frac{(x-c)^n}{n} + \cdots$$

**for each $x \in J$.**

The proof of Theorem 1 is left to a course in advanced calculus.

**Example 1** Determine the Maclaurin series for $\ln(1+x)$ by making use of the fact that

$$\frac{d}{dt}\ln(1+t) = \frac{1}{1+t}$$

if $t > -1$.

**Solution**

By the Fundamental Theorem of Calculus,

$$\ln(1+x) = \ln(1+x) - \ln(1) = \int_0^x \frac{d}{dt}\ln(1+t)\,dt = \int_0^x \frac{1}{1+t}\,dt,\ x > -1,$$

As in Example 12 of Section 9.5,

$$\frac{1}{1+t} = 1 - t + t^2 - t^3 + \cdots + (-1)^{n-1} t^{n-1} + \cdots$$

if $|t| < 1$. Therefore, by Theorem 1,

$$\ln(1+x) = \int_0^x \frac{1}{1+t}\,dt$$

$$= \int_0^x \left(1 - t + t^2 - t^3 + \cdots + (-1)^{n-1} t^{n-1} + \cdots\right) dt$$

$$= x - \frac{1}{2}x^2 + \frac{1}{3}x^3 - \frac{1}{4}x^4 + \cdots + (-1)^{n-1}\frac{1}{n}x^n + \cdots$$

$$= \sum_{n=1}^{\infty} (-1)^{n-1} \frac{1}{n} x^n.$$

for $-1 < x < 1$. □

**Example 2** Determine the Maclaurin series for $\arctan(x)$ by making use of the fact that

$$\frac{d}{dt}\arctan(t) = \frac{1}{1+t^2}.$$

**Solution**

By the Fundamental Theorem of Calculus,

$$\arctan(x) = \arctan(x) - \arctan(0) = \int_0^x \frac{d}{dt}\arctan(t)\,dt = \int_0^x \frac{1}{1+t^2}\,dt,\ x \in R.$$

Thanks to our previous experience with geometric series,

$$\frac{1}{1+t^2} = \frac{1}{1-(-t)^2}$$

$$= 1 + (-t^2) + (-t^2)^2 + (-t^2)^3 + \cdots + (-t^2)^n + \cdots$$

$$= 1 - t^2 + t^4 - t^6 + \cdots + (-1)^n t^{2n} + \cdots$$

$$= \sum_{n=0}^{\infty} (-1)^n t^{2n}$$

## 9.6. POWER SERIES: PART 2

if $\left|-t^{2}\right|=t^{2}<1$, i.e., if $-1<t<1$. Therefore, if $-1<x<1$,

$$\arctan(x) = \int_0^x \left(1 - t^2 + t^4 - t^6 + \cdots + (-1)^n t^{2n} + \cdots\right) dt$$

$$= x - \frac{1}{3}x^3 + \frac{1}{5}x^5 - \frac{1}{7}x^7 + \cdots + (-1)^n \frac{x^{2n+1}}{2n+1} + \cdots$$

$$= \sum_{n=0}^{\infty} (-1)^n \frac{x^{2n+1}}{2n+1},$$

thanks to Theorem 1.

Theorem 1 does not provide any information at the endpoints $-1$ and $1$ of the open interval $(-1, 1)$, but it can be shown that the above equality is valid at $-1$ and $1$ as well. □

**Example 3** Recall that the error function **erf** is defined by the expression

$$\operatorname{erf}(x) = \frac{2}{\sqrt{\pi}} \int_0^x e^{-t^2} dt.$$

a) Determine the Maclaurin series for erf by making use of the Taylor series for the natural exponential function. Specify the open interval of convergence of the resulting series with the help of Theorem 1.

b) Let $P_{2n+1}(x)$ be the Maclaurin polynomial of order $2n+1$ for erf. Determine $P_{2n+1}(x)$. Make use of your calculator to compare the graph of $P_{11}$ with the graph of erf. Does the picture indicate that $P_{11}(x)$ approximates $\operatorname{erf}(x)$ well if $x$ is not too far away from the basepoint 0?

c) Calculate $P_{2n+1}(1)$ and $|\operatorname{erf}(1) - P_{2n+1}(1)|$ for $n = 3, 4, 5, 6, 7$. Obtain $\operatorname{erf}(1)$ from your have a computational utility if erf is a built-in function. Otherwise, treat the approximation that you obtain for erf(1) via the numerical integrator of you computational utility as the exact value of $\operatorname{erf}(1)$. Do the numbers indicate that it should be possible to approximate $\operatorname{erf}(1)$ with desired accuracy provided that the order of the Taylor polynomial is sufficiently large?

**Solution**

a) We have

$$e^{-t^2} = 1 - t^2 + \frac{1}{2!}t^4 - \frac{1}{3!}t^6 + \cdots + (-1)^n \frac{1}{n!}t^{2n} + \cdots$$

for each $x \in \mathbb{R}$, as in Example 13 of Section 9.5. Therefore,

$$\operatorname{erf}(x) = \frac{2}{\sqrt{\pi}} \int_0^x \left(1 - t^2 + \frac{1}{2!}t^4 - \frac{1}{3!}t^6 + \cdots + (-1)^n \frac{1}{n!}t^{2n} + \cdots\right) dt$$

$$= \frac{2}{\sqrt{\pi}} \left(x - \frac{x^3}{3} + \frac{1}{2!}\frac{x^5}{5} - \frac{1}{3!}\frac{x^7}{7} + \cdots + (-1)^n \frac{1}{n!} \frac{x^{2n+1}}{(2n+1)} + \cdots\right)$$

for any $x \in \mathbb{R}$, by Theorem 1. Thus, the open interval of convergence of the resulting series is $(-\infty, +\infty)$.

b) By part a)

$$P_{2n+1}(x) = \frac{2}{\sqrt{\pi}} \left(x - \frac{x^3}{3} + \frac{1}{2!}\frac{x^5}{5} - \frac{1}{3!}\frac{x^7}{7} + \cdots + (-1)^n \frac{1}{n!} \frac{x^{2n+1}}{(2n+1)}\right).$$

In particular,

$$P_{11}(x) \frac{2}{\sqrt{\pi}} \left(x - \frac{x^3}{3} + \frac{1}{2!}\frac{x^5}{5} - \frac{1}{3!}\frac{x^7}{7} + \frac{1}{4!}\frac{x^9}{9} - \frac{1}{5!}\frac{x^{11}}{11}\right)$$

Figure 1 compares the graph of $P_{11}$ with the graph of erf (the graph of $P_{11}$ is the dashed curve). The picture indicates that $P_{11}(x)$ approximates $\operatorname{erf}(x)$ very well if $x \in [-1.5, 1.5]$.

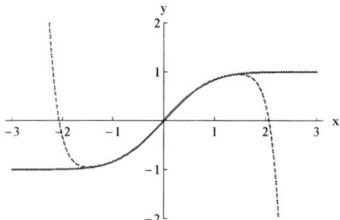

Figure 1

c) Table 1 displays $P_{2n+1}(1)$ and $|\text{erf}(1) - P_{2n+1}(1)|$ for $n = 3, 4, 5, 6, 7$ ($\text{erf}(1) \cong .842\,701$). the magnitude of the error in the approximation of $\text{erf}(1)$ by $P_{2n+1}(1)$ is decreases rapidly as $n$ increases. Thus, the numbers indicate that it should be possible to approximate $\text{erf}(1)$ with desired accuracy provided that the order of the Taylor polynomial is sufficiently large.

| $n$ | $P_{2n+1}(1)$ | $|\text{erf}(1) - P_{2n+1}(1)|$ |
|---|---|---|
| 3 | .838 225 | $4.5 \times 10^{-3}$ |
| 4 | .843 449 | $7.5 \times 10^{-4}$ |
| 5 | .842 594 | $1.1 \times 10^{-4}$ |
| 6 | .842 714 | $1.3 \times 10^{-5}$ |
| 7 | .842 699 | $1.5 \times 10^{-6}$ |

Table 1

If you do not have access to a computer algebra system, you can compare the results of the above approximation with the approximation of the values of erf via the numerical integrator of your calculator. □

## Arithmetic Operations on Taylor Series

We will discuss some facts about arithmetic operations on Taylor series within the framework of Maclaurin series. Each fact has an obvious generalization to Taylor series that are based at a point other than 0.

**Theorem 2 (The Maclaurin Series of Sums and Differences) Assume that**

$$f(x) = a_0 + a_1 x + a_2 x^2 + \cdots + a_n x^n + \cdots$$

and

$$g(x) = b_0 + b_1 x + b_2 x^2 + \cdots + b_n x^n + \cdots$$

**for each $x$ in the interval $J$. Then**

$$f(x) \pm g(x) = (a_0 \pm b_0) + (a_1 \pm b_1) x + (a_2 \pm b_2) x^2 + \cdots + (a_n \pm b_n) x^n + \ldots$$

**for each $x \in J$.**

The proof of Theorem 2 is left as an exercise, since it follows from the general result about the addition of infinite series.

**Example 4** Determine the Maclaurin series for

$$\ln\left(\frac{1+x}{1-x}\right)$$

by making use of the Maclaurin series for $\ln(1+x)$. Specify the open interval of convergence of the resulting series.

## 9.6. POWER SERIES: PART 2

**Solution**

In Example 1 we showed that

$$\ln(1+x) = x - \frac{1}{2}x^2 + \frac{1}{3}x^3 - \frac{1}{4}x^4 + \cdots + (-1)^{n-1}\frac{1}{n}x^n + \cdots$$

if $|x| < 1$. If we replace $x$ by $-x$, we obtain

$$\ln(1-x) = (-x) - \frac{1}{2}(-x)^2 + \frac{1}{3}(-x)^3 - \frac{1}{4}(-x)^4 + \cdots + (-1)^{n-1}\frac{1}{n}(-x)^n + \cdots$$

$$= -x - \frac{1}{2}x^2 - \frac{1}{3}x^3 - \frac{1}{4}x^4 - \cdots - \frac{1}{n}x^n - \cdots,$$

where $|-x| = |x| < 1$. Therefore, if $|x| < 1$,

$$\ln\left(\frac{1+x}{1-x}\right) = \ln(1+x) - \ln(1-x)$$

$$= \left(x - \frac{1}{2}x^2 + \frac{1}{3}x^3 - \frac{1}{4}x^4 + \cdots + (-1)^{n-1}\frac{1}{n}x^n + \cdots\right)$$

$$- \left(-x - \frac{1}{2}x^2 - \frac{1}{3}x^3 - \frac{1}{4}x^4 - \cdots - \frac{1}{n}x^n - \cdots\right)$$

$$= \left(x - \frac{1}{2}x^2 + \frac{1}{3}x^3 - \frac{1}{4}x^4 + \cdots + (-1)^{n-1}\frac{1}{n}x^n + \cdots\right)$$

$$+ \left(x + \frac{1}{2}x^2 + \frac{1}{3}x^3 + \frac{1}{4}x^4 + \cdots + \frac{1}{n}x^n + \cdots\right)$$

$$= 2\left(x + \frac{1}{3}x^3 + \frac{1}{5}x^5 + \cdots + \frac{1}{2n-1}x^{2n-1} + \cdots\right)$$

$$= 2\sum_{n=1}^{\infty} \frac{1}{2n-1}x^{2n-1}.$$

Thus,

$$\frac{1}{2}\ln\left(\frac{1+x}{1-x}\right) = \sum_{n=1}^{\infty} \frac{1}{2n-1}x^{2n-1}, \quad |x| < 1.$$

Recall that

$$\operatorname{arctanh}(x) = \frac{1}{2}\ln\left(\frac{1+x}{1-x}\right), \quad |x| < 1.$$

Thus, we have computed the Taylor series for arctanh in powers of $x$. □

We can determine the Taylor series of products of functions by "multiplying" their Taylor series as if they were polynomials:

**Theorem 3 (The Maclaurin Series of Products)** Assume that

$$f(x) = a_0 + a_1 x + a_2 x^2 + a_3 x^3 + \cdots,$$
$$g(x) = b_0 + b_1 x + b_2 x^2 + b_3 x^3 + \cdots,$$

for each $x$ in the interval $J$. Then

$$f(x)g(x) = a_0 b_0 + (a_0 b_1 + a_1 + b_0)x + (a_0 b_2 + a_1 b_1 + a_0 b_1)x$$
$$+ (a_0 b_3 + a_1 b_2 + a_2 b_1 + a_3 b_0)x^3 + \cdots,$$

$x \in J.$

The proof of Theorem 3 is left to a course in advanced calculus. Theorem 2 and Theorem 3 have obvious generalizations to power series in powers of $(x - c)$ where $c$ is arbitrary.

**Example 5** Let
$$F(x) = e^{-x} \sin(x).$$

Determine the Maclaurin polynomial of order 5 for $F$ by making use of the Maclaurin series for the natural exponential function and sine.

**Solution**

We have
$$e^u = 1 + u + \frac{1}{2}u^2 + \frac{1}{3!}u^3 + \frac{1}{4!}u^4 + \frac{1}{5!}u^5 + \cdots$$

for each $u \in \mathbb{R}$, so that
$$e^{-x} = 1 - x + \frac{1}{2}x^2 - \frac{1}{3!}x^3 + \frac{1}{4!}x^4 - \frac{1}{5!}x^5 + \cdots$$

for each $x \in \mathbb{R}$. We also have
$$\sin(x) = x - \frac{1}{3!}x^3 + \frac{1}{5!}x^5 - \frac{1}{7!}x^7 + \cdots$$

for each $x \in \mathbb{R}$. By Theorem 3,
$$e^{-x}\sin(x) = x - x^2 + \left(\frac{1}{2} - \frac{1}{3!}\right)x^3 + \left(-\frac{1}{3!} + \frac{1}{3!}\right)x^4 + \left(\frac{1}{4!} + \left(\frac{1}{2}\right)\left(-\frac{1}{3!}\right) + \frac{1}{5!}\right)x^5 + \cdots$$
$$= x - x^2 + \frac{1}{3}x^3 - \frac{1}{30}x^5 + \cdots$$

for each $x \in \mathbb{R}$. The Maclaurin polynomial of order 5 for $F$ is
$$x - x^2 + \frac{1}{3}x^3 - \frac{1}{30}x^5.$$

□

We can compute the Maclaurin series for quotients of functions by "dividing" their Maclaurin series:

**Theorem 4 (The Maclaurin Series of Quotients)** Assume that
$$f(x) = a_0 + a_1 x + a_2 x^2 + a_3 x^3 + \cdots,$$
$$g(x) = b_0 + b_1 x + b_2 x^2 + b_3 x^3 + \cdots,$$

**and $g(x) \neq 0$ for each $x$ in the interval $J$. Then,**
$$\frac{f(x)}{g(x)} = d_0 + d_1 x + d_2 x^2 + d_3 x^3 + \cdots, \quad x \in J,$$

**where the coefficients $d_0, d_1, d_2, \ldots$ can be computed as follows:**

**We need to have**
$$f(x) = g(x)\left(d_0 + d_1 x + d_2 x^2 + d_3 x^3 + \cdots\right),$$

## 9.6. POWER SERIES: PART 2

**i.e.,**

$$a_0 + a_1 x + a_2 x^2 + a_3 x^3 + \cdots$$
$$= \left(b_0 + b_1 x + b_2 x^2 + b_3 x^3 + \cdots\right)\left(d_0 + d_1 x + d_2 x^2 + d_3 x^3 + \cdots\right)$$
$$= b_0 d_0 + (b_1 d_0 + b_0 d_1) x + (b_2 d_0 + b_1 d_1 + b_0 d_2) x^2 + \cdots.$$

**By the uniqueness of power series representations, we have to equate the coefficients of the corresponding powers of $x$:**

$$b_0 d_0 = a_0,$$
$$b_0 d_1 + b_1 d_0 = a_1,$$
$$b_0 d_2 + b_1 d_1 + b_2 d_0 = a_2,$$
$$\vdots$$

**Notice that $b_0 = g(0) \neq 0$, so that**

$$d_0 = \frac{a_0}{b_0},$$
$$d_1 = \frac{a_1 - b_1 d_0}{b_0},$$
$$d_2 = \frac{a_2 - b_2 d_0 - b_1 d_1}{b_0},$$
$$\vdots$$

**Therefore, we can compute $d_0, d_1, d_2, \ldots$ successively.**

We will leave the proof of the convergence of the resulting series to $f(x)/g(x)$ for each $x \in J$ to a course in advanced calculus.

**Example 6** Compute the Maclaurin of order 5 for $\tan(x)$ by making use of the Maclaurin series for $\sin(x)$ and $\cos(x)$. Specify the open interval of convergence for the resulting series by making use of Theorem 4.

**Solution**

We have
$$\tan(x) = \frac{\sin(x)}{\cos(x)} = \frac{x - \frac{1}{3!}x^3 + \frac{1}{5!}x^5 - \frac{1}{7!}x^7 + \cdots}{1 - \frac{1}{2}x^2 + \frac{1}{4!}x^4 - \frac{1}{6!}x^6 + \cdots}.$$

Since $\cos(x) \neq 0$ if $x \in (-\pi/2, \pi/2)$, we expect that
$$\tan(x) = d_0 + d_1 x + d_2 x^2 + d_3 x^3 + d_4 x^4 + d_5 x^5 + \cdots, \ x \in (-\pi/2, \pi/2).$$

Note that
$$d_0 = \tan(0) = 0.$$

We must have
$$\frac{x - \frac{1}{3!}x^3 + \frac{1}{5!}x^5 - \frac{1}{7!}x^7 + \cdots}{1 - \frac{1}{2}x^2 + \frac{1}{4!}x^4 - \frac{1}{6!}x^6 + \cdots} = d_1 x + d_2 x^2 + d_3 x^3 + d_4 x^4 + d_5 x^5 + \cdots,$$

so that

$$x - \frac{1}{3!}x^3 + \frac{1}{5!}x^5 - \frac{1}{7!}x^7 + \cdots = \left(1 - \frac{1}{2}x^2 + \frac{1}{4!}x^4 - \frac{1}{6!}x^6 + \cdots\right)\left(d_1 x + d_2 x^2 + d_3 x^3 + d_4 x^4 + d_5 x^5 + \cdots\right)$$

$$= d_1 x + d_2 x^2 + \left(d_3 - \frac{1}{2}d_1\right)x^3 + \left(d_4 - \frac{1}{2}d_2\right)x^4$$

$$+ \left(d_5 - \frac{1}{2}d_3 + \frac{1}{4!}d_1\right)x^5 + \cdots.$$

Therefore,

$$d_1 = 1,$$
$$d_2 = 0,$$
$$d_3 - \frac{1}{2}d_1 = -\frac{1}{3!} \Rightarrow d_3 = -\frac{1}{3!} + \frac{1}{2} = \frac{1}{3},$$
$$d_4 - \frac{1}{2}d_2 = 0 \Rightarrow d_4 = 0,$$
$$d_5 - \frac{1}{2}d_3 + \frac{1}{4!}d_1 = \frac{1}{5!} \Rightarrow d_5 = \frac{1}{5!} + \frac{1}{2}\left(\frac{1}{3}\right) - \frac{1}{4!} = \frac{2}{15}.$$

Thus, the Maclaurin polynomial of order 5 for $\tan(x)$ is

$$x + \frac{1}{3}x^3 + \frac{2}{15}x^5.$$

□

Taylor series shed light on some familiar limits, as in the following example:

**Example 7** We know that
$$\lim_{x \to 0} \frac{\sin(x)}{x} = 1.$$

Set
$$f(x) = \begin{cases} \dfrac{\sin(x)}{x} & \text{if } x \neq 0, \\ 1 & \text{if } x = 0. \end{cases}$$

a) Show that $f$ is infinitely differentiable on the entire number line, and determine the Maclaurin series for $f$.
b) Discuss the absolute error in the approximation of 1 by $\sin(x)/x$ if $|x|$ is small, in the light of the result of part a).

**Solution**

a) We know that
$$\sin(x) = x - \frac{1}{3!}x^3 + \frac{1}{5!}x^5 - \frac{1}{7!}x^7 + \cdots + (-1)^n \frac{1}{(2n+1)!}x^{2n+1} + \cdots$$

for all $x \in R$. By Theorem 4,
$$\frac{\sin(x)}{x} = 1 - \frac{1}{3!}x^2 + \frac{1}{5!}x^4 - \frac{1}{7!}x^6 + \cdots + (-1)^n \frac{1}{(2n+1)!}x^{2n} + \cdots$$

if $x \neq 0$. Let's set
$$g(x) = 1 - \frac{1}{3!}x^2 + \frac{1}{5!}x^4 - \frac{1}{7!}x^6 + \cdots + (-1)^n \frac{1}{(2n+1)!}x^{2n} + \cdots.$$

## 9.6. POWER SERIES: PART 2

The power series converges for each $x \in \mathbb{R}$ (say, by the ratio test). Therefore, $g$ is infinitely differentiable, by Theorem 2 of Section 10.7. We have $f(x) = g(x)$ for each $x \neq 0$. We also have $f(0) = 1$ and $g(0) = 1$, so that $f(0) = g(0)$ as well. Therefore, $f = g$. In particular, $f$ is infinitely differentiable on the entire number line, including $x = 0$.

Figure 2 shows the graphs of $f$ and

$$P_6(x) = 1 - \frac{1}{3!}x^2 + \frac{1}{5!}x^4 - \frac{1}{7!}x^6$$

(the graph of $P_6$ is the dashed curve). The picture indicates that $P_6(x)$ approximates $\sin(x)/x$ very well if $x \in [-2\pi, 2\pi]$.

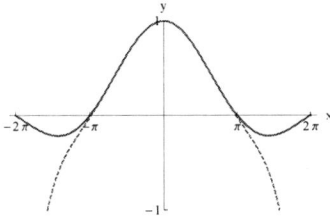

Figure 2

b) By part a),

$$\left| \frac{\sin(x)}{x} - 1 \right| = \left| -\frac{1}{3!}x^2 + \frac{1}{5!}x^4 - \frac{1}{7!}x^6 + \cdots + (-1)^n \frac{1}{(2n+1)!}x^{2n} + \cdots \right|$$

$$= x^2 \left| -\frac{1}{3!} + \frac{1}{5!}x^2 - \frac{1}{7!}x^4 + \cdots + (-1)^n \frac{1}{(2n+1)!}x^{2n-2} + \cdots \right|$$

$$= x^2 h(x),$$

where

$$h(x) = \left| -\frac{1}{3!} + \frac{1}{5!}x^2 - \frac{1}{7!}x^4 + \cdots + (-1)^n \frac{1}{(2n+1)!}x^{2n-2} + \cdots \right| \cong \frac{1}{6}$$

if $|x|$ is small. Therefore,

$$\left| \frac{\sin(x)}{x} - 1 \right| = x^2 h(x) \cong \frac{1}{6}x^2$$

if $|x|$ is small. The above approximate equality shows that $\sin(x)/x$ approximates 1 with an absolute error that is comparable to $x^2/6$ if $|x|$ is small. □

### The Binomial Series

A discussion of power series cannot be complete without the binomial series:

**Theorem 5 (The Binomial Series)** Let $f(x) = (1+x)^r$ where $r$ is an arbitrary exponent. The Maclaurin series for $f$ is

$$1 + rx + \frac{(r-1)r}{2!}x^2 + \frac{(r-2)(r-1)r}{3!}x^3 + \cdots + \frac{(r-n+1)\cdots(r-1)r}{n!}x^n + \cdots.$$

**Proof**

We have

$$f(x) = (1+x)^r,$$
$$f'(x) = r(1+x)^{r-1},$$
$$f''(x) = (r-1)r(1+x)^{r-2},$$
$$f^{(3)}(x) = (r-2)(r-1)r(1+x)^{r-3},$$
$$\vdots$$
$$f^{(n)}(x) = (r-n+1)\cdots(r-1)r(1+x)^{r-n},$$
$$\vdots$$

Therefore,

$$f(0) = 1,$$
$$f'(0) = r,$$
$$f''(0) = (r-1)r.$$
$$f^{(3)}(0) = (r-2)(r-1)r,$$
$$\vdots$$
$$f^{(n)}(0) = (r-n+1)\cdots(r-1)r.$$

Thus, the Taylor series for $f$ is

$$f(0) + f'(0)x + \frac{1}{2!}f''(0)x^2 + \frac{1}{3!}f^{(3)}(0)x^3 + \cdots + \frac{1}{n!}f^{(n)}(0)x^n + \cdots$$
$$= 1 + rx + \frac{(r-1)r}{2!}x^2 + \frac{(r-2)(r-1)r}{3!}x^3 + \cdots + \frac{(r-n+1)\cdots(r-1)r}{n!}x^n + \cdots.$$

■

Note that the binomial series reduces to the ordinary binomial expansion if $r$ is a positive integer.

Let's assume that $r$ is not a nonnegative integer, so that the binomial series is not reduced to a finite sum. We can determine the open interval of convergence of the series by the ratio test:

$$\lim_{n\to\infty} \frac{\left|\frac{(r-n)(r-n+1)\ldots(r-1)r}{(n+1)!}x^{n+1}\right|}{\left|\frac{(r-n+1)\ldots(r-1)r}{n!}x^n\right|} = \lim_{n\to\infty}\left|\frac{r-n}{n+1}\right||x| = |x|.$$

Therefore, the open interval of convergence of the series is $(-1,1)$. It can be shown that

$$f(x) = 1 + rx + \frac{(r-1)r}{2!}x^2 + \frac{(r-2)(r-1)r}{3!}x^3 + \cdots + \frac{(r-n+1)\cdots(r-1)r}{n!}x^n + \cdots$$

for each $x \in (-1,1)$ with some effort, but we will leave the proof to a course in advanced calculus.

## 9.6. POWER SERIES: PART 2

**Example 8** If we set $r = 1/2$ in Theorem 5, we obtain the Maclaurin series for $\sqrt{1+x}$:

$$\sqrt{1+x} = (1+x)^{\frac{1}{2}}$$
$$= 1 + \frac{1}{2}x + \frac{\left(\frac{1}{2}-1\right)\left(\frac{1}{2}\right)}{2}x^2 + \frac{\left(\frac{1}{2}-2\right)\left(\frac{1}{2}-1\right)\left(\frac{1}{2}\right)}{3!}x^3 +$$
$$\cdots + \frac{\left(\frac{1}{2}-n+1\right)\cdots\left(\frac{1}{2}-1\right)\left(\frac{1}{2}\right)}{n!}x^n + \cdots$$
$$= 1 + \frac{1}{2}x - \frac{1}{8}x^2 + \frac{1}{16}x^3 - \frac{5}{128}x^4 + \cdots .$$

□

**Example 9** If we set $r = -1/2$, we obtain the Taylor series for $1/\sqrt{1+x}$:

$$\frac{1}{\sqrt{1+x}} = (1+x)^{-1/2} = 1 - \frac{1}{2}x + \frac{\left(-\frac{1}{2}-1\right)\left(-\frac{1}{2}\right)}{2!}x^2 +$$
$$\cdots + \frac{\left(-\frac{1}{2}-n+1\right)\cdots\left(-\frac{1}{2}-1\right)\left(-\frac{1}{2}\right)}{n!}x^n + \cdots$$
$$= 1 - \frac{1}{2}x + \frac{3}{8}x^2 - \frac{5}{16}x^3 + \frac{35}{128}x^4 - \frac{63}{256}x^5 + \cdots .$$

□

**Example 10**

a) Determine $P_5$, the Maclaurin polynomial of order 5 for arcsine.
b) Compare the graph of $P_5$ with the graph of arcsine. Does the picture indicate that $P_5(x)$ approximates $\arcsin(x)$ well if $x$ is near the basepoint 0?

**Solution**

a) By the Fundamental Theorem of Calculus,

$$\arcsin(x) = \arcsin(x) - \arcsin(0) = \int_0^x \frac{d}{dt}\arcsin(t)\, dt = \int_0^x \frac{1}{\sqrt{1-t^2}}dt.$$

If we substitute $-t^2$ for $x$ in Example 9, we obtain

$$\frac{1}{\sqrt{1-t^2}} = 1 - \frac{1}{2}\left(-t^2\right) + \frac{\left(-\frac{1}{2}-1\right)\left(-\frac{1}{2}\right)}{2!}\left(-t^2\right)^2 +$$
$$\cdots + \frac{\left(-\frac{1}{2}-n+1\right)\cdots\left(-\frac{1}{2}-1\right)\left(-\frac{1}{2}\right)}{n!}\left(-t^2\right)^n + \cdots$$
$$= 1 + \frac{1}{2}t^2 + \frac{3}{8}t^4 + \frac{5}{16}t^6 + \cdots .$$

Therefore,

$$\arcsin(x) = \int_0^x \frac{1}{\sqrt{1-t^2}}dt = \int_0^x \left(1 + \frac{1}{2}t^2 + \frac{3}{8}t^4 + \cdots\right)dt$$
$$= x + \frac{1}{2}\left(\frac{x^3}{3}\right) + \frac{3}{8}\left(\frac{x^5}{5}\right) + \cdots .$$

The Macluarin polynomial of order 5 for arcsine is

$$P_5(x) = x + \frac{x^3}{6} + \frac{3}{40}x^5.$$

b) Figure 3 shows the graphs of arcsine and $P_5$. The picture indicates that $P_5(x)$ approximates $\arcsin(x)$ very well if $x \in [-0.9, 0.9]$. □

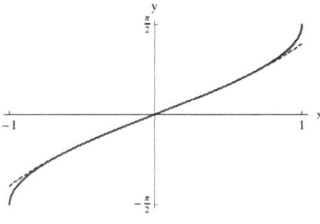

Figure 3

## Problems

Given that

$$\frac{1}{1-x} = 1 + x + x^2 + x^3 + \cdots + x^n + \cdots = \sum_{n=0}^{\infty} x^n,$$

$$e^x = 1 + x + \frac{1}{2}x^2 + \frac{1}{3!}x^3 + \cdots + \frac{1}{n!}x^n + \cdots = \sum_{n=0}^{\infty} \frac{1}{n!}x^n,$$

$$\sin(x) = x - \frac{1}{3!}x^3 + \frac{1}{5!}x^5 - \frac{1}{7!}x^7 + \cdots + \frac{(-1)^n}{(2n+1)!}x^{2n+1} + \cdots = \sum_{n=0}^{\infty} \frac{(-1)^n}{(2n+1)!}x^{2n+1},$$

$$\cos(x) = 1 - \frac{1}{2}x^2 + \frac{1}{4!}x^4 - \frac{1}{6!}x^6 + \cdots + \frac{(-1)^n}{(2n)!}x^{2n} + \cdots = \sum_{n=0}^{\infty} \frac{(-1)^n}{(2n)!}x^{2n},$$

$$(1+x)^r = 1 + rx + \frac{(r-1)r}{2!}x^2 + \frac{(r-2)(r-1)r}{3!}x^3 + \cdots + \frac{(r-n+1)\cdots(r-1)r}{n!}x^n + \cdots,$$

obtain the Maclaurin series of the given function in problems 1-5 via suitable substitutions, arithmetic operations, differentiation or integration. Specify the open interval of the resulting power series. Display the first 4 nonzero terms and the general term.

1. $$F(x) = \int_0^x e^{-t^2/4} dt, \; x \in \mathbb{R}.$$

2. $$F(x) = \int_0^x \frac{1}{1-t^2} dt, \; 1 < x < 1.$$

3. $$\text{Si}(x) = \int_0^x \frac{\sin(t)}{t} dt, \; x \in R$$

4. $$F(x) = \int_0^x \frac{e^t - 1 - t}{t^2} dt, \; x \in \mathbb{R}.$$

5. $$F(x) = \int_0^x t^2 e^{-t^2} dt, \; x \in \mathbb{R}.$$

In problems 6 and 7 determine an antiderivative of the given function as a Maclaurin series. You need to display only the first four nonzero terms of the series. The starting point is the same list of Maclaurin series as in problems 1-5.
Hint: For a given continuous function $f$ if we set

$$F(x) = \int_a^x f(t)\, dt,$$

where $a$ is a fixed point, we have
$$F'(x) = f(x),$$
by the Fundamental Theorem of Calculus:

6.
$$f(x) = \sin(x^2).$$

7.
$$f(x) = \frac{e^x}{1+x^2}.$$

In problems 8 and 9, the starting point is the same list of Maclaurin series as in problems 1-7. Display only the first 4 nonzero terms of the Maclaurin series of the given finction:

8.
$$f(x) = e^{-x/2} \cos(x)$$

9.
$$f(x) = \frac{x^2}{1+x^2}$$

## 9.7 The Integral Test and Comparison Tests

In this section we will discuss **the integral test** for absolute convergence. We will also discuss **comparison tests** which enable us to predict the convergence or divergence of a given series by comparing its terms with the terms of a series that is known to be convergent or divergent.

### The Integral Test

We discussed improper integrals in sections 5.6 and 5.7. The integral test for the convergence of an infinite series establishes a connection between the convergence of an infinite series and an improper integral.

A point of terminology: If $1 \leq x_1 \leq x_2$ implies that $f(x_1) \leq f(x_2)$, we will say that $f$ is decreasing on the interval $[0, +\infty)$, even though the adjective "nonincreasing" is more precise.

**Theorem 1 (The Integral Test) Assume that $f$ is continuous and decreasing on the interval $[1, +\infty)$ and that $f(x) \geq 0$ if $x \geq 1$.**

**a) The series $\sum c_n$ converges absolutely if $|c_n| = f(n)$ for each $n$ and the improper integral**
$$\int_1^\infty f(x)\,dx$$
**converges.**

**b) If $c_n = f(n)$ for each $n$ and the improper integral**
$$\int_1^\infty f(x)\,dx$$
**diverges, the infinite series $\sum c_n$ diverges as well.**

**Proof**

a) Assume that $\int_1^\infty f(x)\,dx$ converges.

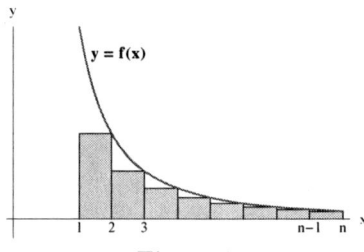

Figure 1

With reference to Figure 1, the area between the graph of $f$ and the interval $[1, n]$ is greater than the sum of the areas of the rectangles, and the length of the base of each rectangle is 1. Therefore,
$$\int_1^n f(x)\,dx \geq f(2) + f(3) + \cdots + f(n) = |c_2| + |c_3| + \cdots + |c_n|.$$
Thus,
$$|c_1| + |c_2| + |c_3| + \cdots + |c_n| \leq |c_1| + \int_1^n f(x)\,dx.$$
Since the improper integral
$$\int_1^\infty f(x)\,dx$$
converges, and $f(x) \geq 0$ for each $x \geq 1$, we have
$$\int_1^n f(x)\,dx \leq \int_1^\infty f(x)\,dx.$$
Therefore,
$$|c_1| + |c_2| + |c_3| + \cdots + |c_n| \leq |c_n| + \int_1^n f(x)\,dx \leq c_1 + \int_1^\infty f(x)\,dx$$
for each $n$. Thus, the sequence of partial sums for the infinite series $\sum |c_n|$ is bounded from above. By Theorem 2 of Section 9.4, the series $\sum |c_n|$ converges.

b) Now let's assume that the improper integral $\int_1^\infty f(x)\,dx$ diverges. Since $f(x) \geq 0$ for $x \geq 1$, this means that
$$\lim_{n \to \infty} \int_1^n f(x)\,dx = +\infty,$$
as we discussed in Section 6.7.

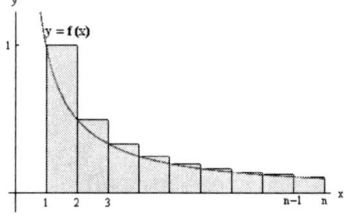

Figure 2

## 9.7. THE INTEGRAL TEST AND COMPARISON TESTS

With reference to Figure 2, the sum of the areas of the rectangles is greater than the area of the region between the graph of $f$ and the interval $[1, n+1]$. Therefore,

$$c_1 + c_2 + c_3 + \cdots + c_n = f(1) + f(2) + \cdots + f(n) \geq \int_1^{n+1} f(x)\,dx.$$

Since

$$\lim_{n \to \infty} \int_1^{n+1} f(x)\,dx = +\infty,$$

we have

$$\lim_{n \to \infty} (c_1 + c_2 + c_3 + \cdots + c_n) = +\infty$$

as well. Thus, the series $\sum c_n$ diverges. ∎

**Remark** As we noted earlier, the convergence or divergence of an infinite series is not affected by adding, removing or changing a finite number of terms. Therefore, the integral test remains valid if we assume that $f$ is continuous, nonnegative and decreasing on an interval of the form $[N, +\infty)$, where $N$ is some positive integer, and $|c_n| = f(n)$ for $n = N, N+1, N+2, \ldots$ for part a), and $c_n = f(n)$, $n = N, N+1, N+2, \ldots$ for part b). ◊

**Example 1** Consider the infinite series

$$\sum_{n=1}^{\infty} (-1)^{n-1} \frac{1}{n^2+1}.$$

Apply the integral test to determine whether the series converges absolutely.

**Solution**
Set

$$f(x) = \frac{1}{x^2+1}.$$

Figure 3 shows the graph of $f$.

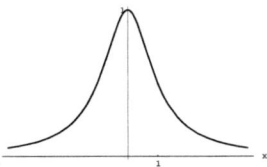

Figure 3

The function $f$ is continuous, positive-valued and decreasing on the interval $[1, +\infty)$. Therefore, the integral test is applicable. We have

$$\int_1^A f(x)\,dx = \int_1^A \frac{1}{x^2+1} = \arctan(x)\Big|_1^A = \arctan(A) - \arctan(1) = \arctan(A) - \frac{\pi}{4}.$$

Therefore,

$$\lim_{A \to \infty} \int_1^A f(x)\,dx = \lim_{A \to \infty} \left(\arctan(A) - \frac{\pi}{4}\right) = \frac{\pi}{2} - \frac{\pi}{4} = \frac{\pi}{4}.$$

Thus, the improper integral
$$\int_1^\infty f(x)\,dx$$
converges (and has the value $\pi/4$). Therefore, the series
$$\sum_{n=1}^\infty \left|(-1)^{n-1}\frac{1}{n^2+1}\right| = \sum_{n=1}^\infty \frac{1}{n^2+1} = \sum_{n=1}^\infty f(n)$$
converges as well. □

**Example 2** Consider the series
$$\sum_{n=2}^\infty \frac{1}{n\ln(n)} = \frac{1}{2\ln(2)} + \frac{1}{3\ln(3)} + \frac{1}{4\ln(4)} + \cdots.$$
Apply the integral test to determine whether the series converges absolutely or diverges.

**Solution**
Let's set
$$f(x) = \frac{1}{x\ln(x)}.$$
Figure 4 shows the graph of $f$.

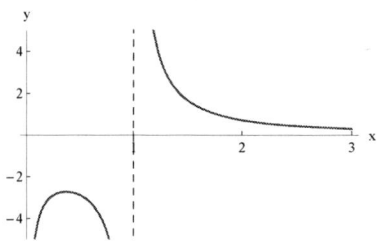

Figure 4

The function $f$ is continuous, nonnegative and decreasing on the interval $[2,+\infty)$. Thus, we can apply the integral test to the given series (the starting value of $n$ can be different from 1 (Remark ??)). The relevant improper integral is
$$\int_2^\infty f(x)\,dx = \int_2^\infty \frac{1}{x\ln(x)}\,dx.$$
If $u = \ln(x)$, we have $du = (1/x)\,dx$, so that
$$\int \frac{1}{\ln(x)}\frac{1}{x}\,dx = \int \frac{1}{u}\,du = \ln(u) = \ln(\ln(x)).$$
Thus,
$$\int_2^A \frac{1}{x\ln(x)}\,dx = \ln(\ln(A)) - \ln(\ln(2)).$$
Therefore,
$$\lim_{A\to\infty} \int_2^A \frac{1}{x\ln(x)}\,dx = \lim_{A\to\infty}(\ln(\ln(A)) - \ln(\ln(2))) = +\infty,$$

## 9.7. THE INTEGRAL TEST AND COMPARISON TESTS

so that the improper integral
$$\int_2^\infty \frac{1}{x \ln(x)} dx$$
diverges. By the integral test, the given series also diverges. □

A series of the form $\sum_{n=1}^\infty 1/n^p$ is referred to as a **p-series**. The harmonic series is the special case $p = 1$.

**Proposition 1** A p-series converges if $p > 1$ and diverges if $p \leq 1$.

**Proof**

Let's first eliminate the cases where $p \leq 0$. In such a case the series diverges, since
$$\lim_{n \to \infty} \frac{1}{n^p} \neq 0.$$

If $p > 0$, set
$$f(x) = \frac{1}{x^p}.$$
Then, $f$ is continuous, positive and decreasing on the interval $[1, +\infty)$, so that the integral test is applicable to the series
$$\sum_{n=1}^\infty \frac{1}{n^p} = \sum_{n=1}^\infty f(n).$$
We know that the improper integral
$$\int_1^\infty f(x) \, dx = \int_1^\infty \frac{1}{x^p} dx$$
converges if $p > 1$, and diverges if $p \leq 1$ (Proposition 1 of Section 6.6). Therefore, the integral test implies that the infinite series
$$\sum_{n=1}^\infty \frac{1}{n^p}$$
converges if $p > 1$, diverges if $0 < p \leq 1$. ■

### Error Estimates Related to the Integral Test

The technique that was used in establishing the integral test leads to a useful estimate for the error in the approximation of the sum of an infinite series by a partial sum, provided that the conditions of the integral test are met:

**Proposition 2** Assume that $f$ is continuous and decreasing on the interval $[0, +\infty)$, $f(x) \geq 0$ for each $x \geq 1$, and the improper integral
$$\int_1^\infty f(x) \, dx$$
**converges. If $S_n = c_1 + c_2 + \cdots + c_n$ is the $n$th partial sum and $S = c_1 + c_2 + \cdots + c_n + \cdots$ is the sum of the infinite series $\sum c_n = \sum f(n)$, we have**
$$S_n \leq S \leq S_n + \int_n^\infty f(x) \, dx.$$

**Proof**

With reference to Figure 5, the area between the graph of $f$ and the interval $[n, n+k]$ is greater than or equal to the sum of the areas of the rectangles. Therefore,

$$c_{n+1} + c_{n+2} + \cdots + c_{n+k} \leq \int_n^{n+k} f(x)\,dx \leq \int_n^\infty f(x)\,dx.$$

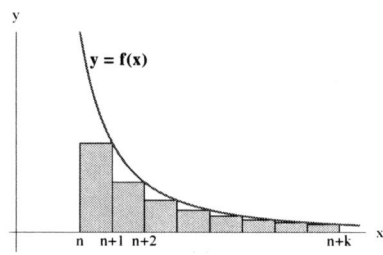

Figure 5

Since the above inequality is valid for any $k$, and the upper bound is independent of $k$, we have

$$\lim_{k \to \infty} (c_{n+1} + c_{n+2} + \cdots + c_{n+k}) \leq \int_n^\infty f(x)\,dx,$$

i.e.,

$$c_{n+1} + c_{n+2} + \cdots + c_{n+k} + \ldots \leq \int_n^\infty f(x)\,dx.$$

But,

$$S - S_n = (c_1 + c_2 + \cdots + c_n + c_{n+1} + c_{n+2} + \cdots + c_{n+k} + \ldots) - (c_1 + c_2 + \cdots + c_n)$$
$$= c_{n+1} + c_{n+2} + \cdots + c_{n+k} + \ldots \geq 0$$

Therefore,

$$0 \leq S - S_n \leq \int_n^\infty f(x)\,dx,$$

so that

$$S_n \leq S \leq S_n + \int_n^\infty f(x)\,dx,$$

as claimed. ∎

The inequality of Proposition 2 ais useful, since it enable us to approximate the sum of the series with desired accuracy, even though we may not be able to calculate the sum exactly.

**Example 3** Let $S$ denote the sum of the infinite series

$$\sum_{n=1}^\infty \frac{1}{n^2}$$

and let $S_n$ be its $n$th partial sum. Approximate $S$ by $S_n$ with absolute error less than $10^{-3}$.

## Solution

a) We apply Proposition 2 with $f(x) = 1/x^2$. Thus,

$$S_n \leq S \leq S_n + \int_n^\infty \frac{1}{x^2}dx = \frac{1}{n}$$

Therefore, we can choose $n$ to be 1001. We have

$$S_{1001} = \sum_{k=1}^{1001} \frac{1}{k^2} \cong 1.6439$$

Therefore,

$$1.6439 \leq S \leq 1.6439 + 10^{-3} \Rightarrow 1.6439 \leq S \leq 1.6449$$

In particular

$$\sum_{n=1}^\infty \frac{1}{n^2} \cong 1.6449.$$

It is known that

$$\sum_{n=1}^\infty \frac{1}{n^2} = \frac{\pi^2}{6} \cong 1.64493,$$

so that.

$$\sum_{n=1}^\infty \frac{1}{n^2} - 1.6449 \cong 3.4067 \times 10^{-5} < 10^{-3}.$$

□

## Comparison Tests

The tests that we have discussed until now are quite adequate for the discussion of the absolute convergence or divergence of many series that you will encounter in this course. Nevertheless, we may encounter some cases where it is not easy to apply any of the previous tests. It may be convenient to compare a given series with a series that is known to be convergent or divergent.

**Theorem 2 (The Comparison Test)**

**1. If there exists a positive integer $N$ such that $|c_n| \leq d_n$ for $n = N, N+1, N+2, N+3, \ldots$, and the series $\sum_{n=1}^\infty d_n$ converges, then the series $\sum_{n=1}^\infty c_n$ converges absolutely.**
**2. If there exists a positive integer $N$ such that $c_n \geq d_n \geq 0$ for $n = N, N+1, N+2, N+3, \ldots$, and the series $\sum_{n=1}^\infty d_n$ diverges, then the series $\sum_{n=1}^\infty c_n$ diverges as well.**

**Proof**

1. If we assume the first set of conditions, we have

$$|c_N| + |c_{N+1}| + |c_{N+2}| + \cdots |c_{N+k}| \leq d_N + d_{N+1} + d_{N+2} + \cdots + d_{N+k}$$
$$\leq d_N + d_{N+1} + d_{N+2} + \cdots + d_{N+k} + d_{N+k+1} + \cdots$$
$$\leq \sum_{n=1}^\infty d_n$$

since the series $\sum d_n$ converges. Therefore,

$$|c_1| + |c_2| + \cdots |c_{N-1}| + |c_N| + |c_{N+1}| + |c_{N+2}| + \cdots |c_{N+k}| \leq |c_1| + |c_2| + \cdots |c_{N-1}| + \sum_{n=1}^\infty d_n$$

for each $k$. Thus, the number

$$|c_1| + |c_2| + \cdots |c_{N-1}| + \sum_{n=1}^{\infty} d_n$$

is an upper bounded for the sequence of partial sums corresponding to the infinite series $\sum |c_n|$. By Theorem 2 of Section 10.4, the series $\sum |c_n|$ converges, i.e., $\sum c_n$ converges absolutely.

2. Now we assume the second set of conditions. Then,

$$c_N + c_{N+1} + c_{N+2} + \cdots + c_{N+k} \geq d_N + d_{N+1} + d_{N+2} + \cdots + d_{N+k}$$

for $k = 0, 1, 2, \ldots$. The sequence $\{d_N + d_{N+1} + d_{N+2} + \cdots + d_{N+k}\}_{k=0}^{\infty}$ does not have an upper bound: Otherwise, the infinite series $\sum_{k=0}^{\infty} d_{N+k}$, hence the infinite series $\sum_{n=1}^{\infty} d_n$ converges, and that is not the case. By the above inequality, the sequence $\{c_N + c_{N+1} + c_{N+2} + \cdots + c_{N+k}\}_{k=0}^{\infty}$ does not have an upper bound either. Therefore, the infinite series $\sum_{k=0}^{\infty} c_{N+k}$, hence the infinite series $\sum_{n=1}^{\infty} c_n$ diverges. ■

**Example 4** Show that the series

$$\sum_{n=1}^{\infty} \frac{\cos(n)}{n^2}$$

converges absolutely.

**Solution**

We have

$$\left| \frac{\cos(n)}{n^2} \right| = \frac{|\cos(n)|}{n^2} \leq \frac{1}{n^2}, \; n = 1, 2, 3, \ldots,$$

since $|\cos(x)| \leq 1$ for any real number $x$. We know that the series

$$\sum_{n=1}^{\infty} \frac{1}{n^2}$$

converges. By the comparison test, the given series converges absolutely. □

**Example 5** Determine whether the series

$$\sum_{n=2}^{\infty} \frac{\ln(n)}{\sqrt{n^2 - 1}}$$

converges or diverges.

**Solution**

Since $\ln(n) > 1$ for $n \geq 3 > e$, we have

$$\frac{\ln(n)}{\sqrt{n^2 - 1}} > \frac{1}{\sqrt{n^2 - 1}}, \; n = 3, 4, 5, \ldots.$$

We also have

$$\sqrt{n^2 - 1} < \sqrt{n^2} = n \Rightarrow \frac{1}{\sqrt{n^2 - 1}} > \frac{1}{n}.$$

Therefore,

$$\frac{\ln(n)}{\sqrt{n^2 - 1}} > \frac{1}{n} \text{ if } n \geq 3.$$

The harmonic series $\sum_{n=1}^{\infty} 1/n$ diverges. Therefore, the divergence clause of the comparison test is applicable, and we conclude that the given series diverges. $\square$

Sometimes it is more convenient to apply the limit-comparison test, rather than the comparison test:

**Theorem 3 (The Limit-Comparison Test)**

**1. Assume that**
$$\lim_{n \to \infty} \frac{|c_n|}{d_n}$$
**exists (as a finite limit)**, $d_n > 0$, $n = 1, 2, 3, \ldots$, **and that** $\sum_{n=1}^{\infty} d_n$ **converges. Then, the series** $\sum_{n=1}^{\infty} c_n$ **converges absolutely.**

**2. If** $d_n > 0$, $n = 1, 2, 3, \ldots$, $\sum_{n=1}^{\infty} d_n$ **diverges, and**
$$\lim_{n \to \infty} \frac{c_n}{d_n} = L \text{ exists and } L > 0, \text{ or } \lim_{n \to \infty} \frac{c_n}{d_n} = +\infty,$$
**then the series** $\sum_{n=1}^{\infty} c_n$ **diverges.**

**Proof**

1. If we set
$$L = \lim_{n \to \infty} \frac{|c_n|}{d_n},$$
there exists an integer $N$ such that
$$\frac{|c_n|}{d_n} \leq L + 1$$
for all $n \geq N$. Therefore,
$$|c_n| \leq (L+1)d_n, \ n \geq N.$$
Since $\sum_{n=N}^{\infty} d_n$ converges, so does $\sum_{n=N}^{\infty} (L+1)d_n$. Therefore, $\sum_{n=N}^{\infty} |c_n|$ converges, by the comparison test. This implies that $\sum_{n=1}^{\infty} |c_n|$ converges, i.e., $\sum_{n=1}^{\infty} c_n$ converges absolutely.

2. Now let us assume the conditions of the divergence clause of the limit-comparison test. If
$$\lim_{n \to \infty} \frac{c_n}{d_n} = L > 0,$$
there exists an integer $N$ such that
$$\frac{c_n}{d_n} \geq \frac{L}{2}$$
for all $n \geq N$. Therefore,
$$c_n \geq \left(\frac{L}{2}\right) d_n, \ n = N, N+1, N+2, \ldots.$$

Therefore,
$$c_N + c_{N+1} + \cdots + c_{N+k} \geq \frac{L}{2}(d_N + d_{N+1} + \cdots + d_{N+k}),$$
for $k = 0, 1, 2, \ldots$. Since the series $\sum_{n=1}^{\infty} d_n$ diverges and $d_n > 0$ for each $n$, we have
$$\lim_{n \to \infty} (d_N + d_{N+1} + \cdots + d_{N+k}) = +\infty.$$

By the above inequality, we also have
$$\lim_{k \to \infty} (c_N + c_{N+1} + \cdots + c_{N+k}) = +\infty,$$

since $L > 0$. Therefore, $\sum_{n=1}^{\infty} c_n$ diverges also.
If we assume that
$$\lim_{n \to \infty} \frac{c_n}{d_n} = +\infty,$$
there exists a positive integer $N$ such that
$$\frac{c_n}{d_n} \geq 1 \text{ for each } n \geq N.$$
Therefore, $c_n \geq d_n$ if $n \geq N$, and $\sum d_n$ diverges, so that $\sum c_n$ diverges, by the comparison test. ∎

**Remark 1** In the divergence clause of the limit-comparison test, $\lim_{n \to \infty} c_n/d_n$ it is essential that the limit in question is positive (or $+\infty$). For example, we have
$$\lim_{n \to \infty} \frac{1/n^2}{1/n} = \lim_{n \to \infty} \frac{1}{n} = 0,$$
and the series $\sum 1/n$ diverges, but the series $\sum 1/n^2$ converges. ◊

**Example 6** Determine whether the series
$$\sum_{n=2}^{\infty} \frac{\ln(n)}{n^{3/2}}$$
converges or diverges.

**Solution**

Since $\ln(n)$ grows more slowly than any power of $n$, we will compare the given series with the series
$$\sum_{n=2}^{\infty} \frac{n^{1/4}}{n^{6/4}} = \sum_{n=2}^{\infty} \frac{1}{n^{5/4}}.$$
The series $\sum 1/n^{5/4}$ converges, since it is a $p$-series with $p = 5/4 > 1$. We have
$$\lim_{n \to \infty} \frac{\frac{\ln(n)}{n^{3/2}}}{\frac{1}{n^{5/4}}} = \lim_{n \to \infty} \frac{\ln(n)}{n^{1/4}} = 0.$$
Therefore, the given series converges. □

**Example 7** Determine whether the series
$$\sum_{n=3}^{\infty} \frac{1}{\sqrt{n^2 - 4}}$$
converges or diverges.

**Solution**

Since
$$\frac{1}{\sqrt{n^2 - 4}} \cong \frac{1}{\sqrt{n^2}} = \frac{1}{n}$$

## 9.7. THE INTEGRAL TEST AND COMPARISON TESTS

if $n$ is large, the harmonic series is a good candidate for the series which will be chosen for comparison. We have

$$\lim_{n\to\infty} \frac{\frac{1}{\sqrt{n^2-4}}}{\frac{1}{n}} = \lim_{n\to\infty} \frac{n}{\sqrt{n^2-4}} = \lim_{n\to\infty} \frac{n}{n\sqrt{1-\frac{4}{n^2}}} = \lim_{n\to\infty} \frac{1}{\sqrt{1-\frac{4}{n^2}}} = 1 > 0.$$

The divergence clause of the limit-comparison test implies that the given series diverges, since $\sum_{n=3}^{\infty} 1/n$ diverges. $\square$

## Problems

In problems 1 - 4 apply the **integral test** to determine whether the given series converges absolutely or whether it diverges. **You need to verify that the conditions for the applicability of the integral test are met:**

1. $$\sum_{n=1}^{\infty} \frac{1}{n^2+1}$$

2. $$\sum_{n=2}^{\infty} \frac{n}{n^2-1}$$

3. $$\sum_{n=2}^{\infty} \frac{1}{n\ln^3(n)}$$

4. $$\sum_{n=1}^{\infty} (-1)^{n-1} \frac{n}{e^n}$$

Assume that $c_n = f(n)$ for each $n$ and that $f(x)$ satisfies the conditions of the integral test for $x \geq 1$. Of $S_n$ is the $n$th partial sum of the series $\sum_{k=1}^{\infty} c_k$ and $S$ is its sum, we have shown that

$$S_n \leq S \leq S_n + \int_n^{\infty} f(x)\,dx.$$

In problems 5 and 6 make use of the above inequality to approximate the given series with absolute error less than $10^{-3}$. You need to check that the conditions are met for the validity of the estimates. Check your response if you have access to a CAS that can compute the sums.

5 [C] $$\sum_{n=1}^{\infty} \frac{1}{n^2+4}$$

6 [C] $$\sum_{n=1}^{\infty} \frac{n}{e^n}$$

In problems 7 - 10 use **the comparison test** to establish the absolute convergence or the divergence of the series.

7 $$\sum_{n=1}^{\infty} \frac{1}{n^2+\sqrt{n}}.$$

8. $$\sum_{n=1}^{\infty} e^{-\frac{1}{4}n} \sin(n^2).$$

9. $$\sum_{n=1}^{\infty} \frac{\cos(4n)}{n^{\frac{3}{2}}}.$$

10. $$\sum_{n=1}^{\infty} \frac{\ln(n+1)}{\sqrt{n}}.$$

In problems 11 - 14 use **the limit-comparison test** to show that the given series converges absolutely or that the series diverges.

11.
$$\sum_{n=2}^{\infty} \frac{1}{n - \sqrt{n}}.$$

12.
$$\sum_{n=2}^{\infty} (-1)^n \frac{1}{\sqrt{n^3 - 2n}}.$$

13.
$$\sum_{n=2}^{\infty} \frac{1}{\sqrt{n^2 - 1}}.$$

14.
$$\sum_{n=2}^{\infty} (-1)^n \frac{\ln(n)}{n^{\frac{3}{2}}}.$$

## 9.8 Conditional Convergence

In this section we will discuss a theorem on "**alternating series**" that will enable us to show that certain series converge conditionally. The discussion of the theorem will have the added benefit of error estimates in the approximation of the sums of alternating series by partial sums, regardless of absolute or conditional convergence. The theorem on alternating series will be the last addition to our toolbox of convergence tests for infinite series. In the second half of this section we will summarize possible strategies for the implementation of these tests.

### Alternating Series

**Definition 1** A series in the form

$$\sum_{n=1}^{\infty} (-1)^{n-1} a_n = a_1 - a_2 + a_3 - a_4 + \cdots \text{ or } \sum_{n=1}^{\infty} (-1)^n a_n = -a_1 + a_2 - a_3 + \cdots$$

where $a_n \geq 0$ for each $n$ is called an **alternating series**.

Since

$$\sum_{n=1}^{\infty} (-1)^n a_n = -\sum_{n=1}^{\infty} (-1)^{n-1} a_n,$$

and the multiplication of a series by $-1$ does not alter the convergence or divergence of a series, we will state the general results of this section in terms of series in the form $\sum_{n=1}^{\infty} (-1)^{n-1} a_n$. The following theorem predicts the convergence of an alternating series under certain conditions:

**Theorem 1 (The Theorem on Alternating Series)** Assume that $a_n \geq 0$ and that the sequence $\{a_n\}_{n=1}^{\infty}$ is decreasing, i.e., $a_n \geq a_{n+1}$ for each $n$. If $\lim_{n \to \infty} a_n = 0$, the alternating series

$$\sum_{n=1}^{\infty} (-1)^{n-1} a_n = a_1 - a_2 + a_3 - a_4 + \cdots$$

**converges. If $S$ is the sum of the series and**

$$S_n = \sum_{k=1}^{n} (-1)^{k-1} a_k$$

**is the $n$th partial sum of the series, we have**

$$S_{2k} \leq S \leq S_{2k+1},$$
$$S_{2k+2} \leq S \leq S_{2k+1}$$

**for $k = 1, 2, 3, \ldots$, and**

$$|S - S_n| \leq a_{n+1}$$

**for $n = 1, 2, 3, \ldots$**

## 9.8. CONDITIONAL CONVERGENCE

Thus, the magnitude of the error in the approximation of the sum of the series by the $n$th partial sum is at most equal to the magnitude of the first term that is left out.

Note that the theorem on alternating series predicts the convergence of the series $\sum_{n=1}^{\infty} (-1)^{n-1} a_n$ if the necessary condition for the convergence of the series, $\lim_{n \to \infty} a_n = 0$, is met, provided that the magnitude of the $n$th term decreases towards 0 *monotonically*. If $a_n$ does not approach 0 as $n \to \infty$, the series diverges anyway.

**The Proof of Theorem 1**

Consider the partial sums $S_{2k}$, $k = 1, 2, 3, \ldots$, so that each $S_{2k}$ corresponds to the addition of an even number of the terms of the series $\sum_{n=1}^{\infty} (-1)^{n-1} a_n$. We claim that the sequence $S_{2k}$, $k = 1, 2, 3, \ldots$ is increasing:

$$S_2 \leq S_4 \leq S_6 \leq \cdots \leq S_{2k} \leq S_{2k+2} \leq \cdots.$$

Indeed,

$$S_{2k+2} = a_1 - a_2 + a_3 - a_4 + \cdots + a_{2k-1} - a_{2k} + a_{2k+1} - a_{2k+2}$$
$$= S_{2k} + (a_{2k+1} - a_{2k+2}) \geq S_{2k},$$

since $a_{2k+1} \geq a_{2k+2}$.

On the other hand, the sequence $S_{2k+1}$, $k = 0, 1, 2, \ldots$, corresponding to the addition of odd numbers of terms, is decreasing:

$$S_1 \geq S_3 \geq S_5 \geq \cdots \geq S_{2k+1} \geq S_{2k+3} \geq \cdots.$$

Indeed,

$$S_{2k+3} = a_1 - a_2 + \cdots + a_{2k+1} - a_{2k+2} + a_{2k+3}$$
$$= S_{2k+1} - (a_{2k+2} - a_{2k+3}) \leq S_{2k+1},$$

since $a_{2k+2} \geq a_{2k+3}$.

The following is also true: Any partial sum which corresponds to the addition of an even number of terms does not exceed any partial sum which is obtained by adding an odd number of terms:

$$S_{2k} \leq S_{2m+1},$$

where $k$ is an arbitrary positive integer, and $m$ is an arbitrary nonnegative integer. Let us first assume that $k \leq m$. Then,

$$S_{2m+1} = a_1 - a_2 + a_3 - a_4 + \cdots - a_{2m} + a_{2m+1} = S_{2m} + a_{2m+1} \geq S_{2m} \geq S_{2k}.$$

If $k \geq m + 1$, we have $2k - 1 \geq 2m + 1$, so that

$$S_{2k} = S_{2k-1} - a_{2k} \leq S_{2k-1} \leq S_{2m+1}.$$

Thus, the sequence of partial sum of the series are lined up on the number line as shown in Figure 1.

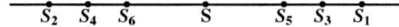

Figure 1

Since the sequence $\{S_{2k}\}_{k=1}^{\infty}$ is an increasing sequence and $S_{2k} \leq S_1 = a_1$ for each $k$, $L_1 = \lim_{k \to \infty} S_{2k}$ exists, by the Monotone Convergence Principle (Theorem 1 of Section 9.4), and $S_{2k} \leq L_1$ for each $k$. The Monotone Convergence Principle has a counterpart for decreasing sequences: If $c_n \geq c_{n+1}$ and there exists a lower bound $m$ so that $c_n \geq m$ for each $m$, the sequence$\{c_n\}_{n=1}^{\infty}$ has a limit and $c_n \geq \lim_{k \to \infty} c_k$ for each $n$ (you can deduce this from the Monotone Convergence Principle for increasing sequences by considering the increasing sequence $\{-c_n\}_{n=1}^{\infty}$). Since $\{S_{2k+1}\}_{k=0}^{\infty}$ is a decreasing sequence and $S_2 \leq S_{2m+1}$ for each $m$, $L_2 = \lim_{m \to \infty} S_{2m+1}$ exists and $L_2 \leq S_{2m+1}$ for each $m$. Since $S_{2k} \leq S_{2m+1}$ for each $k$ and $m$, we have
$$L_1 = \lim_{k \to \infty} S_{2k} \leq S_{2m+1}$$
for each $m$. Therefore,
$$L_1 \leq \lim_{m \to \infty} S_{2m+1} = L_2.$$

We claim that $L_1 = L_2$. Indeed,
$$S_{2k} \leq L_1 \leq L_2 \leq S_{2k+1}$$
for each $k$. Therefore,
$$0 \leq L_2 - L_1 \leq S_{2k+1} - S_{2k} = a_{2k+1}$$
for each $k$, Since $\lim_{k \to \infty} a_{2k+1} = 0$, $L_2 - L_1$ is a nonnegative real number that is arbitrarily small. This is the case if and only if that number is 0. Thus, $L_2 - L_1 = 0$, i.e., $L_2 = L_1 = S$. Therefore, the sequence of partial sums that corresponds to the infinite series $\sum (-1)^{n-1} a_n$ has to converge to $S$, i.e. the series $\sum (-1)^{n-1} a_n$ converges and has the sum $S$. We have
$$S_{2k} \leq L_1 = S = L_2 \leq S_{2k+1},$$
so that
$$S_{2k} \leq S \leq S_{2k+1}$$
for each $k$, as claimed. Similarly,
$$S_{2k+2} \leq S \leq S_{2k+1}$$
for each $k$.

Now Let's establish the estimate of the error in the approximation of the sum $S$ by a partial sum. Since
$$S_{2k} \leq S \leq S_{2k+1},$$
we have
$$0 \leq S - S_{2k} \leq S_{2k+1} - S_{2k} = a_{2k+1}.$$
Since
$$S_{2k+2} \leq S \leq S_{2k+1}$$
also, we have
$$0 \leq S_{2k+1} - S \leq S_{2k+1} - S_{2k+2} = a_{2k+2}.$$
Therefore,
$$|S - S_n| = \left| S - \sum_{k=1}^{n} (-1)^{k-1} a_k \right| \leq a_{n+1}$$
for $n = 1, 2, 3, \ldots$. ∎

## 9.8. CONDITIONAL CONVERGENCE

**Example 1** Consider the series,

$$\sum_{n=1}^{\infty} (-1)^{n-1} \frac{1}{\sqrt{n}} = 1 - \frac{1}{\sqrt{2}} + \frac{1}{\sqrt{3}} - \frac{1}{\sqrt{4}} + \cdots.$$

a) Does the series converge absolutely or conditionally?
b) If the series converges, determine $n$ so that $|S_n - S| \leq 10^{-2}$, where $S_n$ is the $n$th partial sum and $S$ is the sum of the series, and an interval of length less than $10^{-2}$ that contains $S$.

**Solution**

a) The series does not converge absolutely, since

$$\sum_{n=1}^{\infty} \left|(-1)^{n-1} \frac{1}{\sqrt{n}}\right| = \sum_{n=1}^{\infty} \frac{1}{\sqrt{n}} = \sum_{n=1}^{\infty} \frac{1}{n^{1/2}}$$

is a $p$-series with $p = \frac{1}{2} < 1$. On the other hand, the theorem on the convergence of an alternating series is applicable, since

$$\frac{1}{\sqrt{n+1}} < \frac{1}{\sqrt{n}} \quad \text{and} \quad \lim_{n \to \infty} \frac{1}{\sqrt{n}} = 0.$$

Therefore, the series converges. The series converges conditionally since it does not converge absolutely.

b) The theorem on alternating series provides an error estimate:

$$\left| S - \sum_{k=1}^{n} (-1)^{k-1} \frac{1}{\sqrt{k}} \right| \leq \frac{1}{\sqrt{n+1}}.$$

Therefore, in order that $|S - S_n| \leq 10^{-2}$, it is sufficient to have

$$\frac{1}{\sqrt{n+1}} \leq 10^{-2} \Leftrightarrow \sqrt{n+1} \geq 10^2 \Leftrightarrow n+1 \geq 10^4.$$

Thus, it is sufficient to have $n = 10^4$.
By Theorem 1,

$$S_{10^4} \leq S \leq S_{10^4+1},$$

so that $S$ is contained in the interval $[S_{10^4}, S_{10^4+1}]$. The length of the interval $[S_{10^4}, S_{10^4+1}]$ is less than $10^{-2}$. Indeed,

$$0 \leq S_{10^4+1} - S_{10^4} = \frac{1}{\sqrt{10^4+1}} < 10^{-2},$$

We have
$$S_{10^4} \cong 0.599\,899 \quad \text{and} \quad S_{10^4+1} \cong 0.609\,898$$

Therefore,
$$0.599899 \leq S \leq 0.609898$$

(The actual value of $S$ is $0.604899$, rounded to 6 significant digits). $\square$

You should not get the impression that a series is conditionally convergent whenever the theorem on alternating series is applicable to the series:

**Example 2** Consider the series

$$\sum_{n=1}^{\infty} (-1)^{n-1} \frac{n}{2^n}.$$

a) Show that the series converges absolutely.

b) Apply the theorem on alternating series in order to determine $n$ so that $|S - S_n| < 10^{-3}$, where $S$ is the sum of the series and $S_n$ is its $n$th partial sum, and an interval of length less than $10^{-3}$ that contains $S$.

**Solution**

a) We will apply the ratio test:

$$\lim_{n \to \infty} \frac{\left|(-1)^n \frac{n+1}{2^{n+1}}\right|}{\left|(-1)^{n-1} \frac{n}{2^n}\right|} = \lim_{n \to \infty} \left(\left(\frac{2^n}{2^{n+1}}\right)\left(\frac{n+1}{n}\right)\right) = \lim_{n \to \infty} \left(\frac{1}{2}\left(\frac{n+1}{n}\right)\right)$$

$$= \frac{1}{2} \lim_{n \to \infty} \frac{n+1}{n} = \frac{1}{2}(1) = \frac{1}{2} < 1.$$

Therefore, the series converges absolutely.

b) We can also apply the theorem on alternating series to the given series. Indeed,

$$(-1)^{n-1} \frac{n}{2^n} = (-1)^{n-1} f(n),$$

where

$$f(x) = \frac{x}{2^x}.$$

The function $f$ is decreasing function on $[2, +\infty)$ (confirm with the help of the derivative test for monotonicity), and

$$\lim_{x \to +\infty} f(x) = 0$$

(with or without L'Hôpital's rule). By Theorem 1,

$$|S - S_n| \leq f(n+1).$$

We have

$$f(13) \cong 1.558691 \times 10^{-3} \text{ and } f(14) \cong 8.54492 \times 10^{-4} < 10^{-3}.$$

Therefore, it is sufficient to approximate $S$ by $S_{13}$. We have

$$|S - S_{13}| \leq f(14) < 10^{-3}.$$

By Theorem 1,

$$S_{14} \leq S \leq S_{13},$$

and

$$0 \leq S_{13} - S_{14} = f(14) < 10^{-3}.$$

Thus, the interval $[S_{14}, S_{13}]$ contains $S$ and has length less than $10^{-3}$. We have

$$S_{13} \cong 0.222778 \text{ and } S_{14} \cong 0.221924.$$

Therefore,

$$0.221924 \leq S \leq 0.222778$$

(The actual value of $S$ is $2/9 \cong 0.222\ldots$). □

## A Strategy to Test Infinite Series for Convergence or Divergence

Let's map a strategy to test infinite series for convergence or divergence based on all the tests that we discussed. Let $\sum a_n$ be a given series:

- **If** $\lim_{n \to \infty} a_n \neq 0$**, the series** $\sum a_n$ **diverges**, since a necessary condition for the convergence of an infinite series is that the $n$th term should converge to 0 as $n$ tends to infinity. There is nothing more to be done.

- If $\lim_{n \to \infty} a_n = 0$, we can test the series $\sum a_n$ for **absolute convergence**. If we conclude that the series converges absolutely, we are done, since absolute convergence implies convergence. Usually, **the ratio test** is the easiest test to apply, provided that it is conclusive. In some cases, **the root test** may be more convenient. If these tests are not conclusive, we may try **the integral test**. We may also try **the comparison test** or **the limit comparison test**, if we spot a "comparison series" without too much difficulty. Usually, the limit comparison test is easier to apply than the "basic" comparison test.

- If we conclude that the given series does not converge absolutely, we may still investigate **conditional convergence**. We have only **the theorem on alternating series** that we can count on for help in order to determine conditional convergence. You may discuss other tests for conditional convergence in a course on advanced calculus.

Let's illustrate the implementation of the suggested strategy by a few examples.

**Example 3** Determine whether the series

$$\sum_{n=1}^{\infty} (-1)^{n-1} \frac{10^n}{(n!)^2}$$

converges absolutely, converges conditionally or diverges.

**Solution**

Whenever we see the factorial sign in the expression for the terms of a series, it is a good idea to try the ratio test:

$$\lim_{n \to \infty} \frac{\left|(-1)^n \frac{10^{n+1}}{((n+1)!)^2}\right|}{\left|(-1)^{n-1} \frac{10^n}{(n!)^2}\right|} = \lim_{n \to \infty} \left(\left(\frac{10^{n+1}}{10^n}\right)\left(\frac{n!}{(n+1)!}\right)^2\right)$$

$$= \lim_{n \to \infty} \left(10 \left(\frac{1}{n+1}\right)^2\right)$$

$$= 10 \left(\lim_{n \to \infty} \frac{1}{n+1}\right)^2 = 10 \,(0) = 0 < 1.$$

Therefore, the series converges absolutely. $\square$

**Example 4** Determine whether the series

$$\sum_{n=2}^{\infty} (-1) \frac{1}{n \ln^2(n)}$$

converges absolutely, converges conditionally or diverges.

**Solution**

The ratio test and the root test are inconclusive, since the required limit in either case is 1 (confirm). Let's try the integral test for absolute convergence. Thus, we set

$$f(x) = \frac{1}{x \ln^2(x)},$$

so that

$$\left|(-1)\frac{1}{n \ln^2(n)}\right| = \frac{1}{n \ln^2(n)} = f(n).$$

The function $f$ is continuous, positive-valued and decreasing on $[2, +\infty)$. Therefore, the integral test is applicable. If we set $u = \ln(x)$,

$$\int \frac{1}{(\ln(x))^2} \frac{1}{x} dx = \int \frac{1}{u^2} \frac{du}{dx} dx = \int \frac{1}{u^2} du = \int u^{-2} du = -u^{-1} = -\frac{1}{u} = -\frac{1}{\ln(x)}.$$

Therefore,

$$\int_2^b f(x) \, dx = \int_2^b \frac{1}{x \ln^2(x)} dx = -\frac{1}{\ln(x)}\bigg|_2^b = -\frac{1}{\ln(b)} + \frac{1}{\ln(2)}.$$

Thus,

$$\lim_{b \to \infty} \int_2^b f(x) \, dx = \lim_{b \to \infty} \left(-\frac{1}{\ln(b)} + \frac{1}{\ln(2)}\right) = \frac{1}{\ln(2)}.$$

Therefore, the improper integral

$$\int_2^b f(x) \, dx = \int_2^\infty \frac{1}{x \ln^2(x)} dx$$

converges. Therefore. the series

$$\sum_{n=2}^\infty (-1) \frac{1}{n \ln^2(n)}$$

converges absolutely.
Note that the theorem on alternating series is applicable to the given series, but does not lead to the fact that the series converges absolutely. □

**Example 5** Determine whether the series

$$\sum_{n=1}^\infty (-1)^{n-1} \frac{1}{\sqrt{n^2 + 4n + 1}}$$

converges absolutely, converges conditionally or diverges.

**Solution**

The ratio test and the root test are inconclusive, since the required limits are equal to 1 (confirm). The integral test will lead to the conclusion that the series does not converge absolutely after some hard work involving antidifferentiation. We will choose to apply the limit-comparison test. Since

$$\frac{1}{\sqrt{n^2 + 4n + 1}} \cong \frac{1}{\sqrt{n^2\left(1 + \frac{4}{n} + \frac{1}{n^2}\right)}} \cong \frac{1}{\sqrt{n^2}} = \frac{1}{n}$$

## 9.8. CONDITIONAL CONVERGENCE

for large $n$, the harmonic series appears to be a good choice as a "comparison series". Let's evaluate the limit that is required by the limit-comparison test:

$$\lim_{n\to\infty} \frac{\frac{1}{\sqrt{n^2+4n+1}}}{\frac{1}{n}} = \lim_{n\to\infty} \frac{n}{\sqrt{n^2+4n+1}}$$

$$= \lim_{n\to\infty} \frac{n}{\sqrt{n^2\left(1+\frac{4}{n}+\frac{1}{n^2}\right)}}$$

$$= \lim_{n\to\infty} \frac{n}{n\sqrt{1+\frac{4}{n}+\frac{1}{n^2}}} = \lim_{n\to\infty} \frac{1}{\sqrt{1+\frac{4}{n}+\frac{1}{n^2}}} = 1 \neq 0.$$

Since the harmonic series diverges, so does the series

$$\sum_{n=1}^{\infty} \frac{1}{\sqrt{n^2+4n+1}}$$

Therefore, the series

$$\sum_{n=1}^{\infty} (-1)^{n-1} \frac{1}{\sqrt{n^2+4n+1}}$$

does not converge absolutely.

The theorem on alternating series is applicable to the given alternating series. Clearly, the sequence

$$\left\{ \frac{1}{\sqrt{n^2+4n+1}} \right\}_{n=1}^{\infty}$$

is decreasing, and we have

$$\lim_{n\to\infty} \frac{1}{\sqrt{n^2+4n+1}} = 0.$$

Therefore, the series converges. Since the series does not converge absolutely, the series converges conditionally. $\square$

## Problems

In problems 1-4 use the theorem on **alternating series** to show that the given series converges. **You need to verify that the conditions of the theorem are met**:

1.
$$\sum_{n=1}^{\infty} (-1)^{n-1} \frac{1}{2n-1}$$

1.
$$\sum_{n=1}^{\infty} (-1)^{n-1} \frac{1}{n}$$

3.
$$\sum_{n=1}^{\infty} (-1)^{n-1} \frac{1}{(n+1)\ln(n+1)}$$

4.
$$\sum_{n=1}^{\infty} (-1)^{n-1} \frac{1}{n 2^n}$$

In problems 5-14, determine whether the given series diverges, or whether it converges **absolutely or conditionally**. Use any means at your disposal.

5. $$\sum_{n=1}^{\infty}(-1)^{n-1}\frac{1}{\sqrt{n}}.$$

6. $$\sum_{n=0}^{\infty}(-1)^{n}\frac{10^{n}}{n!}$$

7. $$\sum_{k=0}^{\infty}(-1)^{k-1}\frac{3^{k}}{k}.$$

8. $$\sum_{k=0}^{\infty}(-1)^{k}\frac{1}{3^{2k+1}(2k+1)}.$$

9. $$\sum_{n=2}^{\infty}(-1)^{n-1}\frac{1}{n\ln(n)}.$$

10. $$\sum_{n=1}^{\infty}(-1)^{n-1}\frac{1}{n4^{n}}.$$

11. $$\sum_{k=0}^{\infty}(-1)^{k}\frac{\pi^{2k+1}}{6^{2k+1}(2k+1)!}.$$

12. $$\sum_{n=2}^{\infty}(-1)^{n}\frac{1}{\ln^{2}(n)}.$$

13. $$\sum_{n=1}^{\infty}e^{-n^{2}}\cos(10n)$$

14. $$\sum_{n=1}^{\infty}\frac{\sin(n^{2})}{n^{3/2}}$$

In problems 15-18, determine **the interval of convergence** of the given power series. **You need to discuss whether the series converges absolutely or conditionally at the endpoints of the open interval of convergence.**

15. $$\sum_{n=1}^{\infty}\frac{1}{\sqrt{n}}x^{n}$$

16. $$\sum_{n=1}^{\infty}\frac{1}{n^{3}}(x-1)^{n}$$

17. $$\sum_{n=1}^{\infty}n(x-2)^{n}$$

18. $$\sum_{n=2}^{\infty}(-1)^{n}\frac{1}{n\ln(n)}(x-3)^{n}$$

## 9.9 Fourier Series

### Fourier Series of $2\pi$-periodic Functions

The basic building blocks of power series are powers of $(x-c)$ where $c$ is the basepoint. In this section we will discuss **Fourier series** for which the basic building blocks are trigonometric functions of the form $\sin(nx)$ or $\cos(nx)$, where $n$ is a nonnegative integer.

We will consider the approximation of functions by **trigonometric polynomials** of the form

$$a_0+\sum_{k=1}^{n}(a_k\cos(kx)+b_k\sin(kx))=a_0+(a_1\cos(x)+b_1\sin(x))+\cdots+(a_n\cos(nx)+b_n\sin(nx)).$$

We will refer to $n$ as **the order** of the trigonometric polynomial. Recall that a trigonometric polynomial is periodic with period $2\pi$ since $\cos(kx)$ and $\sin(kx)$ are periodic with period $2\pi$ if $k$ is an integer. The term $a_0$ is the **constant term**, $a_k$ is **the coefficient of** $\sin(kx)$ and $b_k$ is **the coefficient of** $\cos(kx)$.

**Example 1** Let

$$F(x)=\frac{4}{\pi}\left(\sin(x)+\frac{1}{3}\sin(3x)+\frac{1}{5}\sin(5x)\right).$$

## 9.9. FOURIER SERIES

Then, $F(x)$ is a trigonometric polynomial of order 5. The constant term, the coefficients of $\sin(2x)$ and $\sin(4x)$, and the coefficients of $\cos(kx)$ ($k = 1, 2, \ldots, 5$) are 0. The coefficient of $\sin(5x)$ is $4/(5\pi)$. $\square$

The following expressions for the constant term and the coefficients of $\sin(kx)$ and $\cos(kx)$ are crucial in the approximation of arbitrary $2\pi$-periodic functions by trigonometric polynomials:

**Proposition 1** Let
$$F(x) = a_0 + \sum_{k=1}^{n} (a_k \cos(kx) + b_k \sin(kx))$$
be a trigonometric polynomial is of order $n$. Then,
$$a_0 = \frac{1}{2\pi} \int_{-\pi}^{\pi} F(x)\, dx,$$
$$a_k = \frac{1}{\pi} \int_{-\pi}^{\pi} F(x) \cos(kx)\, dx,\ k = 1, 2, \ldots, n,$$
$$b_k = \frac{1}{\pi} \int_{-\pi}^{\pi} F(x) \sin(kx)\, dx,\ k = 1, 2, \ldots, n.$$

In particular, $a_0$ is the average value of $F$ on the interval $[-\pi, \pi]$. Note that the integrals can be evaluated on any interval of length $2\pi$ since the functions have period $2\pi$.

**Proof**

The proof of Proposition 1 is based on the following properties of $\sin(kx)$ and $\cos(kx)$: Assume that $k$ and $l$ are nonnegative integers. Then,
$$\int_{-\pi}^{\pi} \cos(kx) \cos(lx)\, dx = 0 \text{ if } k \neq l,$$
$$\int_{-\pi}^{\pi} \sin(kx) \sin(lx)\, dx = 0 \text{ if } k \neq l,$$
$$\int_{-\pi}^{\pi} \cos(kx) \sin(lx)\, dx = 0,$$
$$\int_{-\pi}^{\pi} \cos^2(kx)\, dx = \pi\ (k \neq 0),$$
$$\int_{-\pi}^{\pi} \sin^2(kx)\, dx = \pi\ (k \neq 0)$$

(Proposition 1 of Section 6.3).
We have
$$\int_{-\pi}^{\pi} F(x)\, dx = \int_{-\pi}^{\pi} \left( a_0 + \sum_{k=1}^{n} (a_k \cos(kx) + b_k \sin(kx)) \right) dx$$
$$= a_0 \int_{-\pi}^{\pi} 1\, dx + \sum_{k=1}^{n} a_k \int_{-\pi}^{\pi} \cos(kx)\, dx + \sum_{k=1}^{n} b_k \int_{-\pi}^{\pi} \sin(kx)\, dx = 2\pi a_0,$$
since
$$\int_{-\pi}^{\pi} \cos(kx)\, dx = \int_{-\pi}^{\pi} \sin(kx)\, dx = 0,\ k = 1, 2, 3, \ldots$$
Thus,
$$a_0 = \frac{1}{2\pi} \int_{-\pi}^{\pi} F(x)\, dx.$$

Let $k = 1, 2, \ldots, n$. We have

$$\int_{-\pi}^{\pi} F(x) \cos(kx) \, dx = \int_{-\pi}^{\pi} \left( a_0 + \sum_{l=1}^{n} (a_l \cos(lx) + b_l \sin(lx)) \right) \cos(kx) dx$$

$$= a_0 \int_{-\pi}^{\pi} \cos(kx) dx + \sum_{l=1}^{n} a_l \int_{-\pi}^{\pi} \cos(lx) \cos(kx) dx + \sum_{l=1}^{n} b_l \int_{-\pi}^{\pi} \sin(lx) \cos(kx)$$

$$= a_k \int_{-\pi}^{\pi} \cos^2(kx) \, dx = \pi a_k,$$

since

$$\int_{-\pi}^{\pi} \cos(lx) \cos(kx) dx = 0 \text{ if } l \neq k, \int_{-\pi}^{\pi} \cos^2(kx) \, dx = \pi \text{ and } \int_{-\pi}^{\pi} \sin(lx) \cos(kx).$$

Thus,

$$a_k = \frac{1}{\pi} \int_{-\pi}^{\pi} F(x) \cos(kx) \, dx.$$

Similarly,

$$b_k = \frac{1}{\pi} \int_{-\pi}^{\pi} F(x) \sin(kx) \, dx, \ k = 1, 2, \ldots, n.$$

■

Assume that $f$ is an integrable $2\pi$-periodic function (for example, $f$ is continuous or piecewise continuous). If $f$ were a trigonometric polynomial of order $n$, we would have

$$f(x) = a_0 + \sum_{k=1}^{n} (a_k \cos(kx) + b_k \sin(kx)),$$

where

$$a_0 = \frac{1}{2\pi} \int_{-\pi}^{\pi} f(x) \, dx,$$

$$a_k = \frac{1}{\pi} \int_{-\pi}^{\pi} f(x) \cos(kx) \, dx, \ k = 1, 2, \ldots, n,$$

$$b_k = \frac{1}{\pi} \int_{-\pi}^{\pi} f(x) \sin(kx) \, dx, \ k = 1, 2, \ldots, n,$$

by Proposition 1. Therefore, it seems to be reasonable to approximate $f$ by the trigonometric polynomial

$$F_n(x) = a_0 + \sum_{k=1}^{n} (a_k \cos(kx) + b_k \sin(kx)),$$

where the constant term $a_0$ and the coefficients of $\sin(kx)$ and $\cos(kx)$ are given by the above expressions. It can be shown that such a trigonometric polynomial is **the best approximation** to $f$ among all trigonometric polynomials of order $n$, in the sense that

$$\int_{-\pi}^{\pi} (f(x) - F_n(x))^2 \, dx \leq \int_{-\pi}^{\pi} (f(x) - G(x))^2 \, dx,$$

where $G$ is an arbitrary trigonometric polynomial of order $n$. We leave the proof of this fact to a post-calculus course.

## 9.9. FOURIER SERIES

We may hope that $F_n(x)$ approximates $f(x)$ with desired accuracy if $n$ is sufficiently large and if $f$ does not behave too erratically near $x$. Thus, we would like to have

$$\lim_{n \to \infty} F_n(x) = \lim_{n \to \infty} \left( a_0 + \sum_{k=1}^{n} (a_k \cos(kx) + b_k \sin(kx)) \right)$$
$$= a_0 + \sum_{k=1}^{\infty} (a_k \cos(kx) + b_k \sin(kx)) = f(x).$$

**Definition 1** Let $f$ be an integrable $2\pi$-periodic function. The infinite series

$$a_0 + \sum_{k=1}^{\infty} (a_k \cos(kx) + b_k \sin(kx)),$$

where

$$a_0 = \frac{1}{2\pi} \int_{-\pi}^{\pi} f(x)\, dx,$$
$$a_k = \frac{1}{\pi} \int_{-\pi}^{\pi} f(x) \cos(kx)\, dx, \ k = 1, 2, 3, \ldots,$$
$$b_k = \frac{1}{\pi} \int_{-\pi}^{\pi} f(x) \sin(kx)\, dx, \ k = 1, 2, 3, \ldots,$$

is **the Fourier series for the function** $f$. The numbers $a_0$ and $a_k, b_k$, $k = 1, 2, 3, \ldots$ that are determine by the above expressions are called **the Fourier coefficients of** $f$.

**Fourier** was a French mathematician who introduced such series. He was led to these series by mathematical models that he had constructed in connection with heat conduction. It turned out Fourier series are indispensable tools in many areas of applied Mathematics such as wave propagation and signal processing.

**Example 2** Let's define the function $f$ that is periodic with period $2\pi$ as follows:

$$f(x) = \begin{cases} -1 & \text{if } -\pi < x < 0, \\ 1 & \text{if } 0 < x < \pi, \end{cases}$$

and

$$f(x) = f(x + 2k\pi)$$

if $k$ is an integer such that $x + 2k\pi \in (-\pi, \pi)$.

For example, if $x \in (\pi, 3\pi)$,

$$f(x) = f(x - 2\pi) = \begin{cases} -1 & \text{if } \pi < x < 2\pi, \\ 1 & \text{if } 2\pi < x < 3\pi, \end{cases}$$

if $x \in (-3\pi, -\pi)$,

$$f(x) = f(x + 2\pi) = \begin{cases} -1 & \text{if } -3\pi < x < -2\pi, \\ 1 & \text{if } -2\pi < x < -\pi, \end{cases}$$

Such a function is referred to as **the $2\pi$-periodic extension** of the function $F$, where

$$F(x) = \begin{cases} -1 & \text{if } -\pi < x < 0, \\ 1 & \text{if } 0 < x < \pi, \end{cases}$$

Both $F$ and its $2\pi$-periodic extension $f$ are not defined if $x$ is an integer multiple of $\pi$. Figure 1 shows the graph of $f$ on the interval $(-3\pi, 3\pi)$.

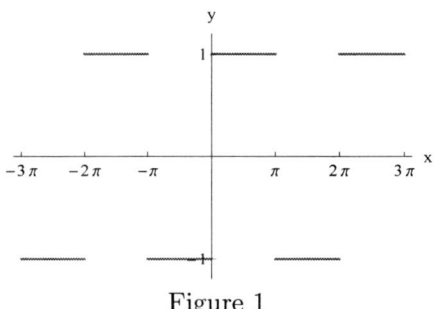

Figure 1

a) Determine the Fourier series

$$a_0 + \sum_{k=1}^{\infty} (a_k \cos(kx) + b_k \sin(kx))$$

for $f$.

b) Let

$$F_n(x) = a_0 + \sum_{k=1}^{n} (a_k \cos(kx) + b_k \sin(kx)).$$

Compare the graphs of $F_5$ and $F_{11}$ with the graph of $f$ on the interval $[-3\pi, 3\pi]$. What does the picture indicate with respect to the sum of the Fourier series versus the values of $f$?

**Solution**

a) We have

$$a_0 = \frac{1}{2\pi} \int_{-\pi}^{\pi} f(x)dx = \frac{1}{2\pi} \int_{-\pi}^{0} f(x)dx + \frac{1}{2\pi} \int_{0}^{\pi} f(x)dx = \frac{1}{2\pi}(-\pi) + \frac{1}{2\pi}(\pi) = 0.$$

Indeed, since $f$ is an odd function and $\cos(kx)$ defines an even function, $f(x)\cos(kx)$ defines an odd function for any integer $k$, so that

$$a_k = \frac{1}{\pi} \int_{-\pi}^{\pi} f(x) \cos(kx) dx = 0, \ k = 1, 2, 3, \ldots.$$

As for the coefficient of $\sin(kx)$, $f(x)\sin(kx)$ defines an even function, and we have

$$b_k = \frac{1}{\pi} \int_{-\pi}^{\pi} f(x) \sin(kx) dx = \frac{2}{\pi} \int_{0}^{\pi} f(x) \sin(kx) dx$$

$$= \frac{2}{\pi} \int_{0}^{\pi} \sin(kx) dx = \frac{2}{\pi} \left( -\frac{1}{k} \cos(kx) \Big|_0^{\pi} \right) = \frac{2}{\pi} \left( -\frac{1}{k} \cos(k\pi) + \frac{1}{k} \right).$$

Since $\cos(kx) = 1$ if $k$ is even, and $\cos(kx) = -1$ if $k$ is odd, we have

$$b_k = \begin{cases} 0 & \text{if } k \text{ is even,} \\ \dfrac{4}{\pi k} & \text{if } k \text{ is odd.} \end{cases}$$

## 9.9. FOURIER SERIES

Therefore, the Fourier series for $f$ is

$$\frac{4}{\pi}\left(\sin(x) + \frac{1}{3}\sin(3x) + \frac{1}{5}\sin(5x) + \cdots + \frac{1}{(2j+1)}\sin\left((2j+1)x\right) + \cdots\right)$$
$$= \frac{4}{\pi}\sum_{k=0}^{\infty}\frac{1}{(2k+1)}\sin\left((2k+1)x\right).$$

b) Let's set

$$F_{2n+1}(x) = \frac{4}{\pi}\sum_{k=0}^{n}\frac{1}{(2k+1)}\sin\left((2k+1)x\right).$$

Thus, the sum of the Fourier series at $x$ is $f(x)$ iff $\lim_{n\to\infty} F_{2n+1}(x) = f(x)$. Figure 2 shows the graphs of $f$ and $F_{11}$ on the interval $[-3\pi, 3\pi]$. The picture indicates that it should be possible to approximate $f(x)$ by $F_{2n+1}(x)$ if $x \notin \{n\pi : n = 0, \pm 1, \pm 2, \ldots\}$ and $n$ is large enough.

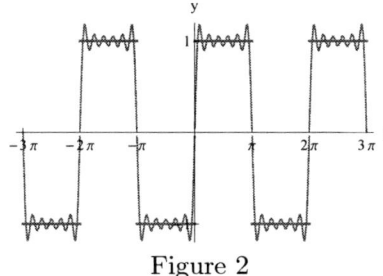

Figure 2

Thus, we should have

$$\frac{4}{\pi}\sum_{k=0}^{\infty}\frac{1}{(2k+1)}\sin\left((2k+1)x\right) = f(x),$$

if $x \notin \{n\pi : n = 0, \pm 1, \pm 2, \ldots\}$. □

We will quote a theorem about the convergence of the Fourier series for certain types of functions.

**Definition 2 A function $f$ is sectionally smooth if $f$ and $f'$ are piecewise continuous on an arbitrary interval $[a, b]$.**

Thus, $f$ and $f'$ are continuous on any interval $[a, b]$, with the possible exception of a finite number of points. If $c$ is a point of discontinuity for $f$ or $f'$, the one-sided limits

$$\lim_{x\to c-} f(x),\ \lim_{x\to c+} f(x),\ \lim_{x\to c-} f'(x), \text{and}\ \lim_{x\to c+} f'(x)$$

exist.

**Example 3** The function $f$ of Example 2 is sectionally smooth. Indeed, the only discontinuities of $f$ are at $n\pi$, where $n = 0, \pm 1, \pm 2, \ldots$. If $c$ has such a point, the one-sided limits of $f$ at $c$ are $+1$ or $-1$. Since $f'(x) = 0$ if $x \notin \{n\pi : n = 0, \pm 1, \pm 2, \ldots\}$, we have $\lim_{x\to c-} f'(x) = \lim_{x\to c+} f'(x) = 0$ if $c \in \{n\pi : n = 0, \pm 1, \pm 2, \ldots\}$. Thus, $f'$ is piecewise continuous. In fact, the only discontinuities of $f'$ are removable discontinuities. □

**Theorem 1 Assume that $f$ is periodic with period $2\pi$ and that it is sectionally smooth. Let**

$$a_0 + \sum_{k=1}^{\infty}(a_k\cos(kx) + b_k\sin(kx))$$

be the Fourier series for $f$. Then,

$$a_0 + \sum_{k=1}^{\infty} (a_k \cos(kx) + b_k \sin(kx)) = f(x)$$

if $f$ is continuous at $x$. If $c$ is a point of discontinuity for $f$, the Fourier series converges to the mean of the one-sided limits of $f$ at $c$. Thus,

$$a_0 + \sum_{k=1}^{\infty} (a_k \cos(kc) + b_k \sin(kc)) = \frac{1}{2} \left( \lim_{x \to c-} f(x) + \lim_{x \to c+} f(x) \right).$$

**Example 4** If $f$ is the function of Example 2 and Example 3, the Fourier series for $f$ is

$$\frac{4}{\pi} \sum_{k=0}^{\infty} \frac{1}{(2k+1)} \sin\left((2k+1)x\right).$$

Since $f$ is sectionally smooth, and the only discontinuities of $f$ are at $n\pi$, $n = 0, \pm 1, \pm 2, \ldots$, we have

$$\frac{4}{\pi} \sum_{k=0}^{\infty} \frac{1}{(2k+1)} \sin\left((2k+1)x\right) = f(x)$$

if $x \notin \{n\pi : n = 0, \pm 1, \pm 2, \ldots\}$, by Theorem 1. We have

$$\left. \frac{4}{\pi} \sum_{k=0}^{\infty} \frac{1}{(2k+1)} \sin\left((2k+1)x\right) \right|_{x=n\pi} = 0.$$

Since 0 is the mean of the one-sided limits of $f$ at a point of the form $n\pi$, $n = 0, \pm 1, \pm 2, \ldots$, the prediction of Theorem 1 is confirmed at the points where $f$ is discontinuous. $\square$

**Example 5** Let

$$F(x) = |x|, \quad -\pi \leq x \leq \pi,$$

and let $f$ be the $2\pi$-periodic extension of $F$. Thus,

$$f(x) = |x + 2k\pi|$$

if $k$ is an integer such that $x + 2k\pi \in [-\pi, \pi]$.

a) Sketch the graph of $f$ on the interval $[-2\pi, 2\pi]$.
b) Determine the Fourier series for $f$.
c) Show that the sum of the Fourier series at $x$ is $f(x)$ for each $x \in \mathbb{R}$.

**Solution**

a) Figure 3 shows the graph of $f$ on $[-3\pi, 3\pi]$.

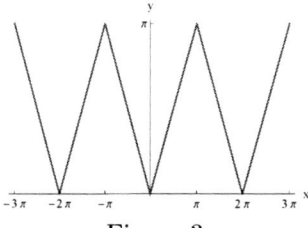

Figure 3

## 9.9. FOURIER SERIES

b) Let
$$a_0 + \sum_{k=1}^{\infty}(a_k\cos(kx) + b_k\sin(kx))$$
be the Fourier series for $f$. Since $f(x)\sin(kx)$ defines an odd function for each $k$,
$$b_k = \frac{1}{\pi}\int_{-\pi}^{\pi} f(x)\sin(kx)dx = 0, \; k = 1, 2, 3, \ldots.$$

We have
$$a_0 = \frac{1}{2\pi}\int_{-\pi}^{\pi} f(x)dx = \frac{1}{2\pi}\int_{-\pi}^{\pi} |x|\,dx = \frac{2}{2\pi}\int_{0}^{\pi} x\,dx = \frac{2}{2\pi}\left(\frac{\pi^2}{2}\right) = \frac{\pi}{2}.$$

Since $f(x)\cos(kx)$ defined an even function for each $k$,
$$a_k = \frac{1}{\pi}\int_{-\pi}^{\pi} f(x)\cos(kx)dx = \frac{2}{\pi}\int_{0}^{\pi} x\cos(kx)dx$$
$$= \frac{2}{\pi}\left(\frac{\cos(k\pi)-1}{k^2}\right) = \begin{cases} 0 & \text{if } k \text{ is even,} \\ -\frac{4}{\pi k^2} & \text{if } k \text{ is odd.} \end{cases}$$

Therefore, the Fourier series for $f$ is
$$\frac{\pi}{2} - \frac{4}{\pi}\left(\cos(x) + \frac{1}{3^2}\cos(3x) + \frac{1}{5^2}\cos(5x) + \cdots\right) = \frac{\pi}{2} - \frac{4}{\pi}\sum_{k=0}^{\infty}\frac{1}{(2k+1)^2}\cos((2k+1)x).$$

c) The function $f$ is continuous on $\mathbb{R}$. We have
$$f'(x) = \begin{cases} -1 & \text{if } -\pi < x < 0, \\ 1 & \text{if } 0 < x < \pi. \end{cases}$$

The function $f'$ is a $2\pi$-periodic extension of its restriction to $(-\pi, \pi)$. Therefore, its only discontinuities are at integer multiples of $\pi$. At such a point, the one-sided limits of $f'$ are 1 or $-1$. Thus, $f'$ is piecewise continuous. Thus, $f$ is sectionally smooth, so that Theorem 1 is applicable. Since $f$ is continuous on the entire number line,
$$f(x) = \frac{\pi}{2} - \frac{4}{\pi}\left(\cos(x) + \frac{1}{3^2}\cos(3x) + \frac{1}{5^2}\cos(5x) + \cdots\right)$$
for each $x \in \mathbb{R}$.
Set
$$F_{2n+1}(x) = \frac{\pi}{2} - \frac{4}{\pi}\sum_{k=0}^{n}\frac{1}{(2k+1)^2}\cos((2k+1)x).$$

Figure 4 shows the graphs of $f$ and $F_3$ on the interval $[-2\pi, 2\pi]$. The graph of $F_3$ is hardly distinguishable from the graph of $f$, except near the points $-2\pi, 0$ and $2\pi$ at which $f'$ is discontinuous.
□

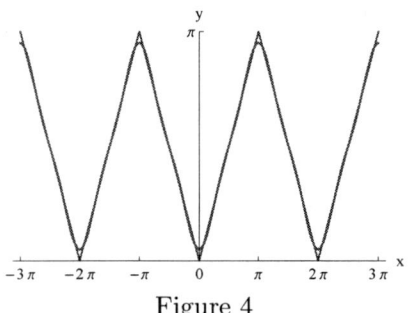
Figure 4

**Remark** The convergence of the Fourier series of Example 5 can be predicted with the help of the machinery that has been developed in the previous sections of this chapter. Indeed,

$$\left| \frac{1}{(2k+1)^2} \cos\left((2k+1)x\right) \right| = \frac{|\cos\left((2k+1)x\right)|}{(2k+1)^2} \leq \frac{1}{(2k+1)^2}$$

since $|\cos(\theta)| \leq 1$ for any $\theta \in \mathbb{R}$. The series

$$\sum_{k=0}^{\infty} \frac{1}{(2k+1)^2}$$

converges (you can apply the integral test, or compare the series with $\sum 1/(4k^2)$). By the comparison test, the Fourier series converges absolutely at each $x \in \mathbb{R}$. ◊

## Fourier Series when the Period is Different from $2\pi$

The idea of a Fourier series can be extended to periodic functions which have period other than $2\pi$. Assume that $L > 0$ and $f$ is periodic with period $2L$. In this case, the basic functions are

$$1,\ \sin\left(\frac{\pi k x}{L}\right),\ \cos\left(\frac{\pi k x}{L}\right),\ k = 1, 2, 3, \ldots.$$

These functions are periodic with period $2L$ (check). We consider the approximation of a function $f$ which has period $2L$ by trigonometric polynomials of the form

$$F_n(x) = a_0 + \sum_{k=1}^{n} \left( a_k \cos\left(\frac{\pi k x}{L}\right) + b_k \sin\left(\frac{\pi k x}{L}\right) \right).$$

As in the special case $L = \pi$, we have

$$a_0 = \frac{1}{2L} \int_{-L}^{L} F_n(x)\, dx,$$

$$a_k = \frac{1}{L} \int_{-L}^{L} F_n(x) \cos\left(\frac{\pi k x}{L}\right) dx,\ k = 1, 2, \ldots, n,$$

$$b_k = \frac{1}{L} \int_{-L}^{L} F_n(x) \sin\left(\frac{\pi k x}{L}\right) dx,\ k = 1, 2, \ldots, n.$$

Thus, we define the Fourier series for the $2L$-periodic function $f$ to be

$$a_0 + \sum_{k=1}^{\infty} \left( a_k \cos\left(\frac{\pi k x}{L}\right) + b_k \sin\left(\frac{\pi k x}{L}\right) \right),$$

## 9.9. FOURIER SERIES

where

$$a_0 = \frac{1}{2L} \int_{-L}^{L} f(x)\, dx,$$

$$a_k = \frac{1}{L} \int_{-L}^{L} f(x) \cos\left(\frac{\pi k x}{L}\right) dx, \ k = 1, 2, 3, \ldots,$$

$$b_k = \frac{1}{L} \int_{-L}^{L} f(x) \sin\left(\frac{\pi k x}{L}\right) dx, \ k = 1, 2, 3, \ldots.$$

As in the special case $L = \pi$, if $f$ is sectionally smooth, the sum of the Fourier series for $f$ at $x$ is $f(x)$ if $f$ is continuous at $x$. At a point of discontinuity $c$, the sum of the Fourier series is the mean of the one-sided limits of $f$ at $c$:

$$a_0 + \sum_{k=1}^{\infty}\left(a_k \cos\left(\frac{\pi k c}{L}\right) + b_k \sin\left(\frac{\pi k c}{L}\right)\right) = \frac{1}{2}\left(\lim_{x \to c-} f(x) + \lim_{x \to c+} f(x)\right).$$

**Example 6** Let $F(x) = x$ if $-1 < x < 1$, and let $f$ be the function which extends $F$ as a periodic function with period 2. Thus, $f(x) = F(x + 2k) = x + 2k$ if $k$ is an integer such that $x + 2k \in (-1, 1)$. The function $f$ is sectionally smooth and has jump discontinuities at points of the form $\pm(2n+1)$, $n = 0, 1, 2, 3, \ldots$. Figure 5 shows the graph of $f$ on the interval $(-3, 3)$.

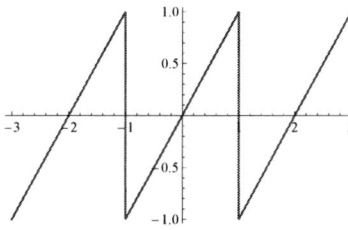

Figure 5

The Fourier series for $f$ is in the form

$$a_0 + \sum_{k=1}^{\infty}\left(a_k \cos(\pi k x) + b_k \sin(\pi k x)\right).$$

We have

$$a_0 = \frac{1}{2} \int_{-1}^{1} f(x)\, dx = \frac{1}{2} \int_{-1}^{1} x\, dx = 0,$$

and

$$a_k = \frac{1}{1} \int_{-1}^{1} f(x) \cos(\pi k x) = \int_{-1}^{1} x \cos(\pi k x)\, dx = 0, \ k = 1, 2, 3, \ldots$$

($x \cos(\pi k x)$ is an odd function). As for the coefficient of $\sin(\pi k x)$,

$$b_k = \frac{1}{1} \int_{-1}^{1} x \sin(\pi k x)\, dx = 2 \int_{0}^{1} x \sin(\pi k x)\, dx$$

$$= 2 \left(-\frac{\cos(k\pi)}{\pi k}\right) = \begin{cases} \dfrac{2}{\pi k} & \text{if } k \text{ is odd,} \\ -\dfrac{2}{\pi k} & \text{if } k \text{ is even.} \end{cases}$$

Therefore, the Fourier series of $f$ is

$$\frac{2}{\pi}\left(\sin(\pi x) - \frac{1}{2}\sin(2\pi x) + \frac{1}{3}\sin(3\pi x) - \frac{1}{4}\sin(4\pi x) + \cdots\right) = \frac{2}{\pi}\sum_{k=1}^{\infty}(-1)^{k+1}\frac{1}{k}\sin(k\pi x).$$

Let's set

$$F_n(x) = \frac{2}{\pi}\sum_{k=1}^{n}(-1)^{k+1}\frac{1}{k}\sin(k\pi x).$$

Figure 6 shows the graph of $f$ (and spurious line segments) and the graph of $F_{10}$ (the dashed curve) Note that the graph of $F_{10}$ is very close to the graph of $f$ away from the discontinuities of $f$. Indeed, since $f$ is sectionally smooth and $f$ is continuous at any $x \notin \{\pm(2n+1), n = 0, 1, 2, \ldots\}$, the general theory predicts that

$$\frac{2}{\pi}\sum_{k=1}^{\infty}(-1)^{k+1}\frac{1}{k}\sin(k\pi x) = f(x), \ x \notin \{\pm(2n+1), n = 0, 1, 2, \ldots\}.$$

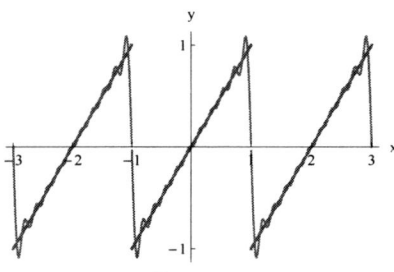

Figure 6

The general theory also predicts that the Fourier series converges to the mean of the right and left limits of $f$ at a point of discontinuity. That value is 0. Indeed, if we set $x = \pm(2n+1)$, $n = 0, 1, 2, \ldots$, in the series

$$\frac{2}{\pi}\sum_{k=1}^{\infty}(-1)^{k+1}\frac{1}{k}\sin(k\pi x),$$

we have

$$\frac{2}{\pi}\sum_{k=1}^{\infty}(-1)^{k+1}\frac{1}{k}\sin(\pm k\pi(2n+1)) = 0 + 0 + \cdots = 0.$$

If you look closely at Figure 6, you will see that there is considerable difference between $F_{10}(x)$ and $f(x)$ if $x$ is close to a discontinuity of $f$. This **"Gibbs phenomenon"** is observable even if we consider the approximation of $f$ by $F_n$ where $n$ is very large (try some large values for $n$).
□

## Problems

In problems 1 and 2,
a) Let $f$ be periodic with period $2\pi$ with the given values on $[-\pi, \pi]$. Determine the Fourier series

$$a_0 + \sum_{k=1}^{\infty}(a_k\cos(kx) + b_k\sin(kx))$$

of $f$.

## 9.9. FOURIER SERIES

b) Show that the series converges if $x = \pi/2$ and $x = \pi$.

c) [C] Set
$$F_n(x) = a_0 + \sum_{k=1}^{n} (a_k \cos(kx) + b_k \sin(kx))$$

Calculate $f(x)$ and $F_n(x)$ at $x = \pi/2$ and $x = \pi$ for $n = 4, 8, 16, 32, 64$. Do the numbers indicate that $\lim_{n\to\infty} F_n(x) = f(x)$? If not, do the numbers give an indication of $\lim_{n\to\infty} F_n(x)$?

d) [C] Plot the graphs of $f$ and $F_8$ on the interval $[-3\pi, 3\pi]$ with the help of your graphing utility. What does the picture indicate with respect to the approximation of $f$ by $F_8$?

1.
$$f(x) = \pi^2 - x^2 \text{ if } -\pi \leq x \leq \pi.$$

2.
$$f(x) = x \text{ if } -\pi \leq x \leq \pi.$$

3. Let $f$ be periodic with period 2 and
$$f(x) = \begin{cases} -1 & \text{if } -1 < x < 0, \\ 1 & \text{if } 0 < x < 1, \end{cases}$$

a) Determine the Fourier series
$$a_0 + \sum_{k=1}^{\infty} (a_k \cos(\pi k x) + b_k \sin(\pi k x))$$

of $f$,

b) Show that the series converges if $x = 1/2$ and $x = 1$.

c) [C] Calculate $f(x)$ and $F_n(x)$ at $x = 1/2$ and $x = 1$ for $n = 4, 8, 16, 32, 64$. Do the numbers indicate that $\lim_{n\to\infty} F_n(x) = f(x)$? If not, do the numbers give an indication of $\lim_{n\to\infty} F_n(x)$?

d) [C] Set
$$F_n(x) = a_0 + \sum_{k=1}^{n} (a_k \cos(\pi k x) + b_k \sin(\pi k x))$$

Plot the graphs of $f$ and $F_8$ on the interval $[-3, 3]$ with the help of your graphing utility. What does the picture indicate with respect to the approximation of $f$ by $F_8$?

4. Let $f$ be periodic with period 4 and
$$f(x) = |x| \text{ if } -2 \leq x \leq 2.$$

a) Determine the Fourier series
$$a_0 + \sum_{k=1}^{\infty} \left( a_k \cos\left(\frac{\pi k x}{2}\right) + b_k \sin\left(\frac{\pi k x}{2}\right) \right)$$

of $f$,

b) Show that the series converges if $x = 1$ and $x = 2$.

c) [C] Calculate $f(x)$ and $F_n(x)$ at $x = 1$ and $x = 2$ for $n = 4, 8, 16, 32, 64$. Do the numbers indicate that $\lim_{n\to\infty} F_n(x) = f(x)$? If not, do the numbers give an indication of $\lim_{n\to\infty} F_n(x)$?

d) [C] Set
$$F_n(x) = a_0 + \sum_{k=1}^{n} \left( a_k \cos\left(\frac{\pi k x}{2}\right) + b_k \sin\left(\frac{\pi k x}{2}\right) \right)$$

Plot the graphs of $f$ and $F_8$ on the interval $[-6, 6]$ with the help of your graphing utility. What does the picture indicate with respect to the approximation of $f$ by $F_8$?

# Chapter 10

# Parametrized Curves and Polar Coordinates

In this chapter we introduce **parametrized curves** and **polar coordinates**. A more complete discussion needs the concept of a vector that will be taken up in volume III.

## 10.1 Parametrized Curves

Until now we have dealt with functions of a single variable whose values are real numbers. In this section we will discuss functions that assign points in the plane to real numbers. A set formed by such points is referred to as a parametrized curve.

### Terminology and Examples

We will use the notation $\mathbb{R}^2$ to refer to the Cartesian coordinate plane.

**Definition 1** Assume that $\sigma$ is a function from an interval $J$ into the plane $\mathbb{R}^2$. If $\sigma(t) = (x(t), y(t))$, then $x(t)$ and $y(t)$ are the **component functions** of $\sigma$. The letter $t$ is **the parameter**. The image of $\sigma$, i.e., the set of points $C = \{(x(t), y(t)) : t \in J\}$ is said to be **the curve that is parametrized by the function** $\sigma$. We say that the function $\sigma$ is a **parametric representation of the curve** $C$.

We will use the notation $\sigma : J \to \mathbb{R}^2$ to indicate that $\sigma$ is a function from $J$ into $\mathbb{R}^2$

**Example 1** Let
$$\sigma(t) = (\cos(t), \sin(t)), \ t \in [0, 2\pi].$$

The coordinate functions of the function $\sigma$ are $x(t) = \cos(t)$ and $y(t) = \sin(t)$, where $t \in [0, 2\pi]$. Since
$$x^2(t) + y^2(t) = 1$$

for each $t$, the image of $\sigma$ is the unit circle. The unit circle is parametrized by the function $\sigma$. The function $\sigma$ is a parametrization of the unit circle.

You can imagine that a particle moves along the unit circle in such a way that its position at time $t$ is the point $(\cos(t), \sin(t))$, as indicated by arrows in Figure 1. □

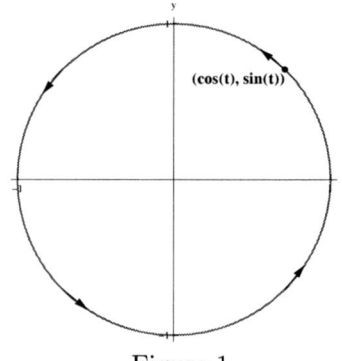

Figure 1

As in Example 1, it is useful to view a parametrized curve $C$ that is parametrized by the function $\sigma(t) = (x(t), y(t))$ in terms of the motion of a particle that is at the point $(x(t), y(t))$ at time $t$. **We must distinguish between the function $\sigma$ and the curve that is parametrized by $\sigma$**, as illustrated by the following example:

**Example 2** Let
$$\tilde{\sigma}(t) = (\cos(2t), \sin(2t)), \ t \in [0, 2\pi].$$
Since
$$\cos^2(2t) + \sin^2(2t) = 1$$
for each $t$, the function $\tilde{\sigma}$ parametrizes the unit circle, just like the function
$$\sigma(t) = (\cos(t), \sin(t)), \ t \in [0, 2\pi],$$
of Example 1. Nevertheless, the functions $\sigma$ and $\tilde{\sigma}$ are distinct functions. Note that the point $\tilde{\sigma}(t)$ traverses the unit circle twice as $t$ varies from 0 to $2\pi$, whereas, the point $\sigma(t)$ traverses the unit circle only once as $t$ varies from 0 to $2\pi$. You can imagine that a particle whose position is $\tilde{\sigma}(t)$ at time $t$ moves with twice the angular speed of the particle whose position is $\sigma(t)$ at time $t$. □

**Example 3** Let $\sigma(t) = (\cosh(t), \sinh(t)), t \in \mathbb{R}$. Identify the curve that is parametrized by $\sigma$.

**Solution**

We have
$$\cosh^2(t) - \sinh^2(t) = 1 \text{ for each } t \in \mathbb{R},$$
so that the point $\sigma(t)$ is on the hyperbola $x^2 - y^2 = 1$ for each $t \in \mathbb{R}$. Since $\cosh(t) \geq 1$ for each $t$ the curve that is parametrized by $\sigma$ is the branch of the hyperbola that lies in the right half-plane, as in indicated in Figure 2. You can imagine that a particle moves in the direction indicated by the arrows as $t$ takes on increasing values. □

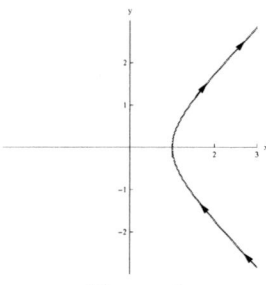

Figure 2

## 10.1. PARAMETRIZED CURVES

The graph of a real-valued function $f$ of a single variable on an interval $J$ can be viewed as a parametrized curve. Indeed, if we set $\sigma(x) = (x, f(x))$, $x \in J$, the image of $\sigma$ is the graph of $f$. on $J$. Thus, we have chosen $x$ to be the parameter, and we have parametrized the graph of $f$ by the function $\sigma$.

**Example 4** Let $f(x) = x^2$, $x \in \mathbb{R}$. Describe the graph of $f$ as a parametrized curve.

**Solution**

We will set $\sigma(x) = (x, x^2)$, $x \in \mathbb{R}$. The function $\sigma$ parametrizes the graph of $f$ that is the parabola illustrated in Figure 3. □

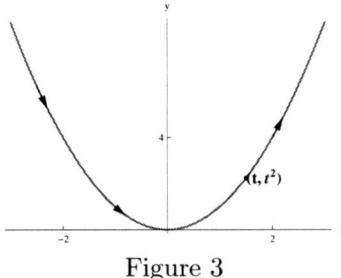

Figure 3

A parametrized curve that is not the graph of a function need not pass the vertical line, as in Example 1 and Example 3. Here is an example of a parametrized curve that intersects itself several times:

**Example 5** Let $\sigma(t) = (\cos(3t), \sin(2t))$, $0 \leq t \leq 2\pi$. Figure 4 shows the curve that is parametrized by $\sigma$. As in the previous examples, you can imagine that a particle moves along the curve as $t$ increases from 0 to $2\pi$, as indicated by the arrows. □

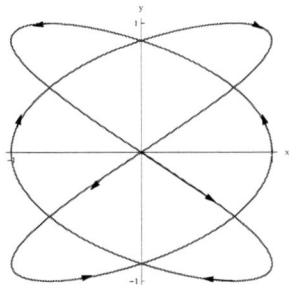

Figure 4: A Lissajous curve

Example 5 belongs to the family of curves that are referred to as **Lissajous curves**. These are curves that are parametrized by coordinate functions of the form

$$\sigma(t) = (a_1 \cos(mt + \theta_1), a_2 \sin(nt + \theta_2)),$$

where $a_1$, $a_2$, $\theta_1$ and $\theta_2$ are arbitrary real numbers, $m$ and $n$ are integers.

**Example 6** Imagine that a circle of radius $r$ rolls along a line, and that $P(\theta)$ is a point on the circle, as in Figure 5 ($P(0) = 0$).

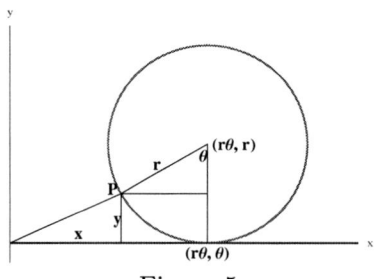

Figure 5

We have
$$x(\theta) = r\theta - r\sin(\theta) = r(\theta - \sin(\theta))$$
and
$$y(\theta) = r - r\cos(\theta) = r(1 - \cos(\theta)).$$

The parameter $\theta$ can be any real number. The curve that is parametrized by the function $\boldsymbol{\sigma}$ : $\mathbb{R} \to \mathbb{R}^2$, where

$$\boldsymbol{\sigma}(\theta) = (x(\theta), y(\theta)) = (r(\theta - \sin(\theta)), r(1 - \cos(\theta))), \; -\infty < \theta < +\infty,$$

is referred to as a **cycloid**. Figure 6 shows the part of the cycloid that is parametrized by $\boldsymbol{\sigma}(\theta) = (2\theta - 2\sin(\theta), 2 - 2\cos(\theta))$ $(r = 2)$, where $\theta \in [0, 8\pi]$. □

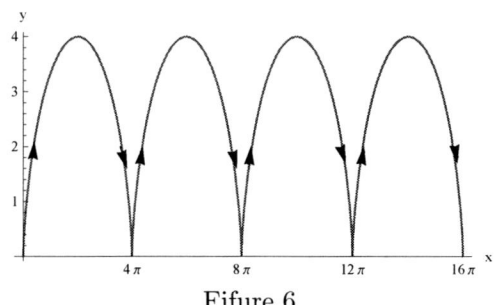

Fifure 6

### Tangents to Parametrized Curves

Sometimes we can eliminate the parameter $t$ from the equations $x = x(t)$ and $y = y(t)$ and describe the curve that is parametrized by $\sigma(t) = (x(t), y(t))$ as the graph of an equation of the form $F(x, y) = C$, where $C$ is a constant, as in Example 1 and Example 3. If $y(x)$ is defined implicitly by the equation $F(x, y) = C$, we can compute $dy/dx$ by implicit differentiation, even if $y(x)$ is not available explicitly. We can determine $dy/dx$ from the expressions $x(t)$ and $y(t)$, even if a relationship of the form $F(x, y) = C$ is not available:

**Proposition 1** *Assume that $\sigma(t) = (f(t), g(t))$, the coordinate functions $f$ and $g$ have continuous derivatives at $t_0$ and $f'(t_0) \neq 0$. Then, there exists an open interval $J$ which contains $t_0$, an open interval $I$ which contains $x_0 = f(t_0)$, and a function $G$ such that*

$$x = f(t) \text{ and } y = g(t) \Leftrightarrow y = G(x),$$

## 10.1. PARAMETRIZED CURVES

where $t \in J$ and $x \in I$. We have

$$\frac{dy}{dx} = \frac{\frac{dy}{dt}}{\frac{dx}{dt}}.$$

**Proof**

Since $f'(t_0) \neq 0$ and $f'$ is continuous at $t_0$, there exists an open interval $J$ containing $t_0$ such that $f'(t) > 0$ for each $t \in J$ or $f'(t) < 0$ for each $t \in J$. In either case, the restriction of $f$ to $J$ has an inverse that will be denoted by $f^{-1}$. Thus, we have

$$t = f^{-1}(x) \Leftrightarrow x = x(t)$$

where $x \in I$ and $t \in J$. Therefore, $y = g(t) = g\left(f^{-1}(x)\right)$ for each $x \in I$. We set $y = G(x) = g\left(f^{-1}(x)\right)$, where $x \in J$. By the relationship between the derivative of a function and its inverse,

$$\frac{dt}{dx} = \frac{1}{\frac{dx}{dt}}.$$

By the chain rule,

$$\frac{dy}{dx} = \frac{dy}{dt}\frac{dt}{dx} = \frac{dy}{dt}\left(\frac{1}{\frac{dx}{dt}}\right) = \frac{\frac{dy}{dt}}{\frac{dx}{dt}}$$

for each $x \in I$. ∎

**Example 7** Let $x(t) = \cos(t)$ and $y(t) = \sin(t)$, where $t \in [0, 2\pi]$, as in Example 1. We have

$$\frac{dy}{dx} = \frac{\frac{dy}{dt}}{\frac{dx}{dt}} = \frac{\cos(t)}{-\sin(t)} = -\frac{\cos(t)}{\sin(t)}$$

if

$$\frac{dx}{dt} = -\sin(t) \neq 0,$$

i.e., if $t \neq 0$, $t \neq \pi$ and $t \neq 2\pi$.

We can obtain the same result via implicit differentiation. Since

$$x^2 + y^2 = 1$$

for each $(x, y)$ on the curve that is parametrized by $\cos(t)$ and $\sin(t)$, we have

$$2x + 2y\frac{dy}{dx} = 0 \Rightarrow \frac{dy}{dx} = -\frac{x}{y} = -\frac{\cos(t)}{\sin(t)}$$

if $\sin(t) \neq 0$. The above expression is not valid if $\sin(t) = 0$. Indeed, the relationship $x^2 + y^2 = 1$ does not define $y$ as a function of $x$ in any open interval that contains $(\pm 1, 0)$. □

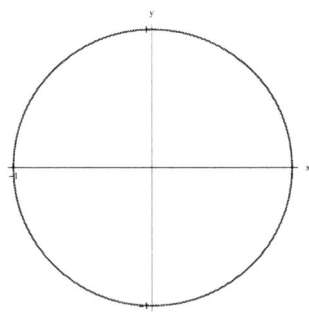

Figure 7

**Example 8** Consider the cycloid that is parametrized by the coordinate functions

$$x(\theta) = 2\theta - 2\sin(\theta), \ y(\theta) = 2 - 2\cos(\theta),$$

as in Example 6.

a) Determine $dy/dx$
b) Determine the tangent line to the cycloid at $(x(\pi/2), y(\pi/2))$.
c) Show that the cycloid has a cusp at $(4\pi, 0)$.
**Solution**

a) We have

$$\frac{dx}{d\theta} = 2 - 2\cos(\theta) \text{ and } \frac{dy}{d\theta} = 2\sin(\theta).$$

Therefore,

$$\frac{dy}{dx} = \frac{\frac{dy}{d\theta}}{\frac{dx}{d\theta}} = \frac{2\sin(\theta)}{2 - 2\cos(\theta)} = \frac{\sin(\theta)}{1 - \cos(\theta)}$$

if $\cos(\theta) \ne 1$, i.e., if $\theta$ is not of the form $2n\pi$, where $n$ is an integer. The corresponding points on the cycloid are $(4n\pi, 0)$.

b) We have

$$x\left(\frac{\pi}{2}\right) = 2\left(\frac{\pi}{2}\right) - 2\sin\left(\frac{\pi}{2}\right) = \pi - 2$$

and

$$y\left(\frac{\pi}{2}\right) = 2 - 2\cos\left(\frac{\pi}{2}\right) = 2.$$

Therefore,

$$\left.\frac{dy}{dx}\right|_{x=\pi-2} = \left.\frac{\sin(\theta)}{1 - \cos(\theta)}\right|_{\theta=\pi/2} = 1.$$

Thus, the tangent line to the cycloid at $(x(\pi/2), y(\pi/2))$ is the graph of the equation

$$y = 2 + (x - (\pi - 2)) = x - \pi + 4$$

Figure 8 shows the tangent line.

## 10.1. PARAMETRIZED CURVES

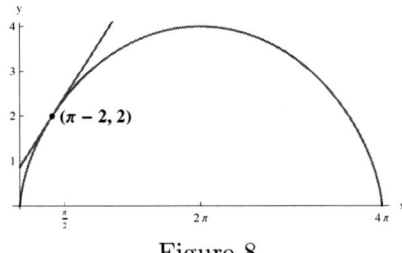

Figure 8

c) If $\theta = 2\pi$, the corresponding point on the cycloid is $(4\pi, 0)$. We can express $y$ as a function of $x$ just to the right of $4\pi$ and just to the left of $2\pi$, and

$$\frac{dy}{dx} = \frac{\sin(\theta)}{1 - \cos(\theta)}$$

on either side. We have

$$\frac{dx}{d\theta} = 2 - 2\cos(\theta),$$

so that $x'(\theta) > 0$ if $\theta$ is slightly to the left or to the right of $2\pi$. Therefore, $x(\theta)$ is an increasing function in some open interval containing $2\pi$. Thus,

$$\lim_{x \to 4\pi+} \frac{dy}{dx} = \lim_{\theta \to 2\pi+} \frac{\sin(\theta)}{1 - \cos(\theta)} = \lim_{\theta \to 2\pi+} \frac{\cos(\theta)}{\sin(\theta)} = +\infty$$

(with the help of L'Hôpital's rule at the intermediate step). Similarly,

$$\lim_{x \to 4\pi-} \frac{dy}{dx} = -\infty.$$

Therefore the cycloid has a cusp at $(4\pi, 0)$. Figure 6 reflects this fact. □

**Example 9** Let

$$x(t) = \cos(t) \text{ and } y(t) = \sin(2t), \ t \in [0, 2\pi].$$

Figure 9 shows the curve that is parametrized by $\sigma(t) = (x(t), y(t))$. The picture indicates that the curve intersects itself at the origin, in the sense that the origin corresponds to two distinct values of the parameter in the interval $[0, 2\pi]$. Indeed,

$$\cos(t) = 0 \text{ and } \sin(2t) = 0$$

if $t = \pi/2$ or $t = 3\pi/2$.

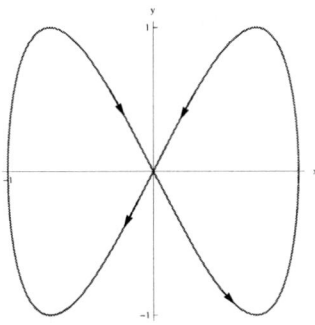

Figure 9

Figure 10 shows the part of the curve that corresponds to the values of $t$ in the interval $[0, \pi]$:

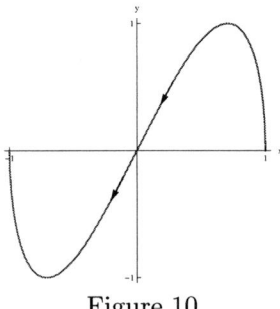

Figure 10

Figure 11 shows the part of the curve that corresponds to the values of $t$ in the interval $[\pi, 2\pi]$:

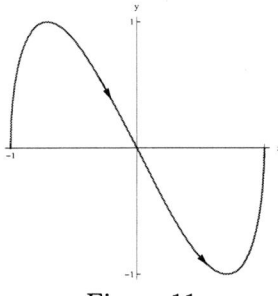

Figure 11

Determine the tangent lines to the curves shown in Figure 10 and Figure 11 at the origin.

**Solution**

We have
$$\frac{dy}{dx} = \frac{\frac{dy}{dt}}{\frac{dx}{dt}} = \frac{\frac{d}{dt}\sin(2t)}{\frac{d}{dt}\cos(t)} = \frac{2\cos(2t)}{-\sin(t)}.$$

The slope of the tangent line to the curve shown in Figure 10 at the origin corresponds to $t = \pi/2$:
$$\frac{dy}{dx} = \frac{\frac{dy}{dt}}{\frac{dx}{dt}} = \frac{2\cos(2t)}{-\sin(t)} = \frac{2\cos(\pi)}{-\sin(\pi/2)} = 2.$$

The equation of the tangent line is
$$y = 2x.$$

The slope of the tangent line to the curve shown in Figure 11 at the origin corresponds to $t = 3\pi/2$:
$$\frac{dy}{dx} = \frac{\frac{dy}{dt}}{\frac{dx}{dt}} = \frac{2\cos(2t)}{-\sin(t)} = \frac{2\cos(3\pi)}{-\sin(3\pi 2)} = \frac{-2}{1} = -2.$$

The equation of the tangent line is
$$y = -2x.$$

□

## Problems

In problems 1 and 2,
a) [C] Plot the curve $C$ that is parametrized by the function $\sigma$ with the help of your graphing utility.
b) Identify $C$ as part of a familiar shape.

**1.**
$$\sigma(t) = (4 + 2\cos(t), 3 + 2\sin(t)), \ 0 \leq t \leq 2\pi.$$

Hint: $\cos^2(t) + \sin^2(t) = 1$.

**2.**
$$\sigma(t) = (\sinh(t), \cosh(t)), \ -2 \leq t \leq 2$$

Hint: $\cos^2(t) - \sinh^2(t) = 1$.

[C] In problems 3 and 4, plot the curve $C$ that is parametrized by the function $\sigma$ with the help of your graphing utility.

**3.**
$$\sigma(t) = (\sin(4t), \cos(3t)), \ 0 \leq t \leq 2\pi.$$

**4..**
$$\sigma(t) = (3t - 3\cos(t), 3 - 3\sin(t)), \ -5\pi/2 \leq t \leq 7\pi/2.$$

In problems 5 and 6,
a) [C] Show that $\sigma_1$ and $\sigma_2$ parametrize the same curve by plotting the curve that is parametrized by the functions $\sigma_1$ and $\sigma_2$ with the help of your graphing utility,
b) Explain why $\sigma_1 \neq \sigma_2$.

**5.**
$$\sigma_1(t) = (\sin(3t), \cos(2t)), \ 0 \leq t \leq 4\pi,$$
$$\sigma_2(t) = (\sin(6t), \cos(4t)), \ 0 \leq t \leq 2\pi.$$

**6.**
$$\sigma_1(t) = (2\sin(2t), 4\cos(2t)), \ 0 \leq t \leq 2\pi,$$
$$\sigma_2(t) = (2\sin(4t), 4\cos(4t)), \ 0 \leq t \leq \pi.$$

In problems 7 and 8, parametrize the part of the graph of $y = f(x)$ that corresponds to the given interval $J$.

**7.**
$$y = \cos(x), \ J = [-2\pi, 2\pi].$$

**8.**
$$y = e^{-x^2}, \ J = [-1, 1].$$

In problems 9 and 10, let $C$ be the curve that is parametrized by the given function $\sigma$
a) Show that there exists an open interval $J$ that contains $t_0$ such that the part of $C$ corresponding to $t \in J$ is the graph of a function $y(x)$.

b) Determine the line that is tangent to $C$ at $\sigma(t_0) = (x_0, y_0)$ (the point-slope with basepoint $x_0$ will do).

c) [C] Plot $C$ and the line that you determined in part b) with the help of your graphing utility. Does the picture suggest that you have determined the required tangent line?

9.
$$\sigma(t) = (\sin(3t), \cos(2t)), \ t_0 = \pi/4.$$

10.
$$\sigma(t) = (4\sin(t), 2\cos(t)), \ t_0 = \pi/3.$$

In problems 11 and 12, let $C$ be the curve that is parametrized by the given function $\sigma(t) = (x(t), y(t))$.

a) [C] Plot the graph of $\sigma$ with the help of your graphing utility. Does the picture indicate that there does not exist an open interval $J$ that contains $t_0$ such that the part of $C$ corresponding to $t \in J$ is the graph of a function of $x$ that is differentiable at $x(t_0)$?.

b) Show that $x'(t_0) = 0$ so that the expression for the calculation of $y'(x)$ is not valid at $x = x(t_0)$.

c) Show that $C$ has a vertical tangent or cusp at $\sigma(t_0)$. Does the picture of part a) support your response?

11.
$$\sigma(t) = (t - \sin(t), 1 - \cos(t)), \ 0 \le t \le 4\pi, \ t_0 = 2\pi.$$

12.
$$\sigma(t) = (4\cos(t), 2\sin(t)), \ -\pi/2 \le t \le \pi/2, \ t_0 = 0.$$

## 10.2 Polar Coordinates

### The Description of Polar Coordinates

Until now we have been using the Cartesian coordinate system in the plane. Now we will introduce the polar coordinate system which is more convenient in certain situations such as the description of the orbits of the planets.

As usual, let's denote the horizontal and vertical axes in the Cartesian coordinate system as the $x$-axis and the $y$-axis, respectively. Let $O = (0,0)$ be the origin and assume that $P = (x,y)$ is a point other than the origin. Let $r$ be the distance of $P$ from the origin, so that

$$r = \sqrt{x^2 + y^2}.$$

We will refer to $r$ as **the radial coordinate of** $P$. Let $\theta$ be an angle (in radians) that is determined by the line segment $OP$ and the positive direction of the $x$-axis. Thus,

$$\cos(\theta) = \frac{x}{r} \text{ and } \sin(\theta) = \frac{y}{r},$$

as in Figure 1. We will refer to $\theta$ as **a polar angle of** $P$. We can assign infinitely many polar angles to the same point. If $\theta$ is a polar angle for $P$, any angle of the form $\theta + 2n\pi$, where $n = 0, \pm 1, \pm 2, \ldots$, is a polar angle for $P$. The ordered pair $(r, \theta)$ is **a polar coordinate pair for** $P$. The origin $O = (0,0)$ has a special status in the polar coordinate system. The radial distance $r$ that corresponds to $O$ is certainly 0. We will declare that any angle $\theta$ can serve as a polar angle for the origin.

## 10.2. POLAR COORDINATES

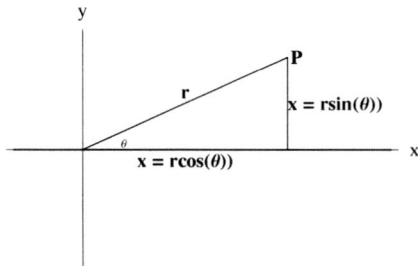

Figure 1

**Example 1** Determine the polar coordinates $r$ and $\theta$ of the point $P$ with the given Cartesian coordinates:

a) $P_1 = (2,0)$ b) $P_2 = (-2,0)$ c) $P_3 = (1, \sqrt{3})$ d) $P_4 = (-1, \sqrt{3})$

**Solution**

a) $r = 2$, $\theta = \pm 2n\pi$, $n = 0, 1, 2, 3, \ldots$.
b) $r = 2$, $\theta = \pi \pm 2n\pi$, $n = 0, 1, 2, 3, \ldots$.
c) $r = \sqrt{1+3} = \sqrt{4} = 2$. The angle $\theta$ is a polar angle for $P_3$ if and only if

$$\cos(\theta) = \frac{1}{2} \text{ and } \sin(\theta) = \frac{\sqrt{3}}{2}.$$

The angle $\pi/6$ satisfies these requirements. Therefore, $\theta$ is a polar angle of $P$ if and only if

$$\theta = \frac{\pi}{6} \pm 2n\pi$$

where $n = 0, 1, 2, 3, \ldots$.
d) As in part c), $r = 2$. We need to have

$$\cos(\theta) = -\frac{1}{2} \text{ and } \sin(\theta) = \frac{\sqrt{3}}{2}.$$

The angle

$$\pi - \frac{\pi}{6} = \frac{5\pi}{6}$$

satisfies these requirements. Therefore, $\theta$ is a polar angle of $P_4$ if and only if

$$\theta = \frac{5\pi}{6} \pm 2n\pi$$

where $n = 0, 1, 2, 3, \ldots$. $\square$

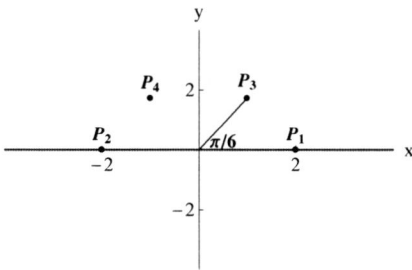

Figure 2

In some cases we can determine the polar angles for a given point by inspection, as in Example 1. In the general case, we can provide a recipe for the calculation of a polar angle that is in the interval $(-\pi, \pi]$:

**Proposition 1** Let $P = (x, y) \neq (0, 0)$. If $r = \sqrt{x^2 + y^2}$, and

$$\theta = \begin{cases} \arccos\left(\dfrac{x}{r}\right) & \text{if } y \geq 0, \\ -\arccos\left(\dfrac{x}{r}\right) & \text{if } y < 0, \end{cases}$$

then $\theta$ is a polar angle for $P$ and $-\pi < \theta \leq \pi$.

Thus, the recipe determines a polar angle between $0$ and $\pi$ if the point is in the upper half-plane ($y > 0$) and angle between $-\pi$ and $0$ if the point is in the lower half-plane ($y < 0$). The recipe prescribes the polar angle $\pi$ for the point $(x, 0)$, if $x < 0$.

**The proof of Proposition 1**

Let $y \geq 0$. If

$$\theta = \arccos\left(\frac{x}{r}\right),$$

we have

$$\cos(\theta) = \frac{x}{r} \text{ and } 0 \leq \theta \leq \pi,$$

by the definition of arccosine. We need to check that

$$\sin(\theta) = \frac{y}{r}.$$

We have

$$\sin^2(\theta) = 1 - \cos^2(\theta) \Rightarrow \sin(\theta) = \pm\sqrt{1 - \cos^2(\theta)}.$$

Since $0 \leq \theta \leq \pi$, $\sin(\theta) \geq 0$. Therefore,

$$\sin(\theta) = \sqrt{1 - \cos^2(\theta)} = \sqrt{1 - \frac{x^2}{r^2}} = \frac{\sqrt{r^2 - x^2}}{r} = \frac{|y|}{r}.$$

Since $y \geq 0$, $|y| = y$. Thus,

$$\sin(\theta) = \frac{|y|}{r} = \frac{y}{r}.$$

Now assume that $y < 0$. In this case we set

$$\theta = -\arccos\left(\frac{x}{r}\right) \Leftrightarrow -\theta = \arccos\left(\frac{x}{r}\right)$$

so that

$$\cos(-\theta) = \frac{x}{r} \text{ and } 0 \leq -\theta \leq \pi.$$

Thus, $-\pi \leq \theta \leq 0$ and

$$\cos(\theta) = \frac{x}{r},$$

since cosine is an even function. We have

$$\sin(\theta) = \pm\sqrt{1 - \cos^2(\theta)}.$$

## 10.2. POLAR COORDINATES

Since $-\pi \leq \theta \leq 0$, $\sin(\theta) \leq 0$. Therefore,
$$\sin(\theta) = -\sqrt{1 - \cos^2(\theta)} = -\frac{|y|}{r}.$$

Since $y < 0$, $|y| = -y$. Thus,
$$\sin(\theta) = -\frac{|y|}{r} = \frac{y}{r}.$$

∎

**Example 2** Determine the polar angle $\theta$ for the given point such that $-\pi < \theta \leq \pi$.

a) $P_1 = (2,1)$ b) $P_2 = (-2,1)$ c) $P_3 = (-2,-1)$ d) $P_4 = (2,-1)$

**Solution**

In all cases $r = \sqrt{2^2 + 1} = \sqrt{5}$.

a) The point lies in the upper half-plane. Therefore,
$$\theta = \arccos\left(\frac{2}{\sqrt{5}}\right) \cong 0.463648$$

b) The point lies in the upper half-plane. Therefore,
$$\theta = \arccos\left(-\frac{2}{\sqrt{5}}\right) \cong 2.67795$$

c) The point lies in the lower half-plane. Therefore,
$$\theta = -\arccos\left(-\frac{2}{\sqrt{5}}\right) \cong -2.67795$$

d) The point lies in the lower half-plane. Therefore,
$$\theta = -\arccos\left(\frac{2}{\sqrt{5}}\right) \cong -0.463648$$

□

The calculation of the Cartesian coordinates of a point whose polar coordinates are given is straightforward: If $(r, \theta)$ is a polar coordinate pair for $P$, we have $x = r\cos(\theta)$ and $y = r\sin(\theta)$.

**Example 3**

a) Assume that $(3, \pi/3)$ is a polar coordinate pair for $P$. Determine the Cartesian coordinates of $P$.

b) Assume that $(3, 1)$ is a polar coordinate pair for $P$. Approximate the Cartesian coordinates of $P$.

**Solution**

a)
$$x = 3\cos\left(\frac{\pi}{3}\right) = 3\left(\frac{1}{2}\right) = \frac{3}{2},$$
$$y = 3\sin\left(\frac{\pi}{3}\right) = 3\left(\frac{\sqrt{3}}{2}\right) = \frac{3\sqrt{3}}{2}.$$

b)
$$x = 3\cos(1) \cong 3(0.540\,302) \cong 1.620\,91,$$
$$y = 3\sin(1) \cong 3(0.841\,471) \cong 2.524\,41$$

□

## Graphs of Equations in Polar Coordinates

We will examine the graphs of equations of the form $F(r, \theta) = 0$ in the $xy$-plane, where $F(r, \theta)$ denote an expression in terms of polar coordinates so that $x = r \cos(\theta)$ and $y = r \sin(\theta)$. In sketching such graphs we will adhere to the following convention: If $r < 0$ and $F(r, \theta) = 0$, the point that corresponds to the polar coordinate pair $(-r, \theta + \pi)$ is on the curve. This amounts to stepping back a distance $-r$ along the ray determined by $\theta$ if $r < 0$.

Let's look at some examples.

**Example 4** Describe the graph of the equation $r = 2$. Express the equation in Cartesian coordinates.

**Solution**

Since $r$ represents the distance from the origin, the graph of the equation $r = 2$ is the circle of radius 2 centered at the origin.

$$r = 2 \Rightarrow \sqrt{x^2 + y^2} = 2 \Rightarrow x^2 + y^2 = 4.$$

□

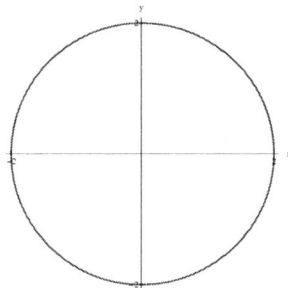

Figure 3

**Example 5** Describe the graph of the equation $\theta = \pi/4$. Express the curve as the graph of an equation in Cartesian coordinates.

**Solution**

If $r \geq 0$, any point with polar coordinates $(r, \pi/4)$ is on the curve. This part of the curve is **the ray** that emanates from the origin at the angle $\pi/4$ with the positive direction of the $x$-axis. If $r < 0$, we associate with the polar coordinate pair $(r, \pi/4)$ the point that has polar coordinates $-r$ and $\pi/4 + \pi$. Therefore, the graph of $\theta = \pi/4$ is the entire line that passes through the origin and has slope 1, as illustrated by Figure 4. □

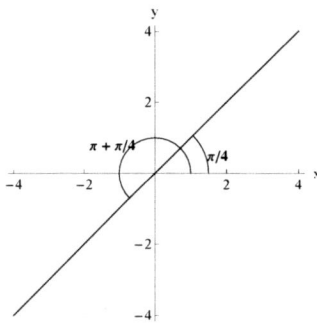

Figure 4

## 10.2. POLAR COORDINATES

Let's consider the graph of an equation of the form $r = f(\theta)$, where $f$ is a given function and $\theta \in [a, b]$. Since $x = r\cos(\theta)$ and $y = r\sin(\theta)$, we can express $x$ and $y$ as functions of $\theta$:

$$x(\theta) = r(\theta)\cos(\theta) = f(\theta)\cos(\theta) \text{ and } y(\theta) = r(\theta)\sin(\theta) = f(\theta)\sin(\theta),$$

where $\theta \in [a, b]$. Thus **the graph of the equation $r = f(\theta)$ is the curve that is parametrized by**

$$\sigma(\theta) = (f(\theta)\cos(\theta), f(\theta)\sin(\theta)).$$

We will refer to this curve as **the polar curve determined by the equation $r = f(\theta)$** ($\theta \in [a, b]$).

**Example 6** Let $f(\theta) = 3\sin(\theta)$, where $\theta \in [0, 2\pi]$.

a) Express the path determined by the equation $r = f(\theta)$.
b) Sketch the graph of $r = f(\theta)$ in the Cartesian $\theta r$-plane ($0 \leq \theta \leq 2\pi$).
c) Sketch the polar curve determined by the equation $r = f(\theta)$.
d) Express the graph of part b) as the graph of an equation in the Cartesian coordinates $x$ and $y$. Identify the curve as a familiar conic section.

**Solution**

a)
$$x(\theta) = f(\theta)\cos(\theta) = 3\sin(\theta)\cos(\theta),$$

and
$$y(\theta) = f(\theta)\sin(\theta) = 3\sin^2(\theta),$$

where $\theta \in [0, 2\pi]$.

b) Figure 5 shows the graph of $r = f(\theta) = 3\sin(\theta)$ ($\theta \in [0, 2\pi]$) in the Cartesian $\theta r$-plane.

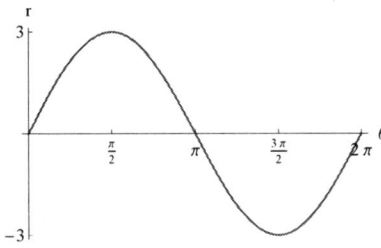

Figure 5

c) As $\theta$ increases from 0 to $\pi/2$, $3\sin(\theta)$ increases from 0 to 3, and as $\theta$ increases from $\pi/2$ to $\pi$, $3\sin(\theta)$ decreases from 3 to 0. Therefore, the point that corresponds to the polar coordinate pair $(3\sin(\theta), \theta)$ traces the curve that is shown in Figure 6.

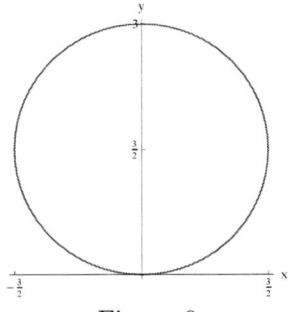

Figure 6

As $\theta$ increases from $\pi$ to $3\pi/2$, $3\sin(\theta)$ decreases from 0 to $-3$, and as $\theta$ increases from $3\pi/2$ to $2\pi$, $3\sin(\theta)$ increases from $-3$ to 0. By the convention of stepping back a distance of $-r$ from the origin along the ray determined by $\theta$ if $r < 0$, the point that corresponds to the polar coordinate pair $(3\sin(\theta), \theta)$ retraces the curve shown in Figure 6.

d) Since $r = \sqrt{x^2 + y^2}$ and

$$\sin(\theta) = \frac{y}{r} = \frac{y}{\sqrt{x^2 + y^2}},$$

$$r = 3\sin(\theta) \Rightarrow \sqrt{x^2 + y^2} = \frac{3y}{\sqrt{x^2 + y^2}} \Rightarrow x^2 + y^2 = 3y.$$

The relationship between $x$ and $y$ is a second-degree equation without any term that involves the product $xy$. As in Section A1.5 of Appendix 1, we can identify the graph of such an equation by completing the square:

$$x^2 + y^2 = 3y \Rightarrow x^2 + y^2 - 3y = 0$$

$$\Rightarrow x^2 + \left(y - \frac{3}{2}\right)^2 - \frac{9}{4} = 0$$

$$\Rightarrow x^2 + \left(y - \frac{3}{2}\right)^2 = \frac{9}{4}$$

The final form of the equation tells us that the curve is a circle that is centered at $(0, 3/2)$ and has radius $3/2$.

Incidentally, if you plot the curve with the help of a graphing utility, the curve may appear to be an ellipse that is not a circle, since the scales on the horizontal and vertical axes are usually different. $\square$

**Remark (Caution)** As in Example 6, the graph of $r = f(\theta)$ ($\theta \in [a, b]$) in the Cartesian $\theta r$-plane and the polar curve determined by the equation $r = f(\theta)$ are distinct entities. The former is the set of points $(\theta, f(\theta))$ in the Cartesian $\theta r$-plane ($\theta \in [a, b]$), whereas the latter is the set of points $(f(\theta)\cos(\theta), f(\theta)\sin(\theta))$ in the Cartesian $xy$-plane ($\theta \in [a, b]$). Usually a graphing utility provides a specific menu to plot the polar curve corresponding to the equation $r = f(\theta)$. $\Diamond$

**Example 7**

a) Sketch the graph of $r = \sin(2\theta)$ in the Cartesian $\theta r$-plane ($0 \leq \theta \leq 2\pi$).
b) Sketch the polar curve determined by the equation $r = \sin(2\theta)$.

**Solution**

a) Figure 7 shows the graph of $r = \sin(2\theta)$ in the Cartesian $\theta r$-plane.

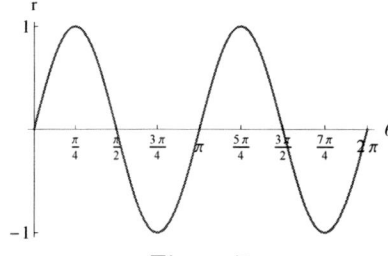

Figure 7

b) As in Example 6, the graph of part a) enables us to trace the polar graph of the equation $r = \sin(2\theta)$ in the $xy$-plane. As $\theta$ increases from 0 to $\pi/4$, $\sin(2\theta)$ increases from 0 to 1, and as $\theta$ increases from $\pi/4$ to $\pi/2$, $\sin(2\theta)$ decreases from 1 to 0. The point that corresponds to the polar coordinate pair $(\sin(2\theta), \theta)$ traces the curve shown in Figure 8.

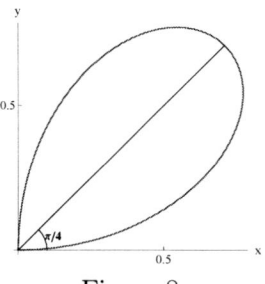

Figure 8

As $\theta$ increases from $\pi/2$ to $3\pi/4$, $\sin(2\theta)$ decreases from 0 to $-1$, and as $\theta$ increases from $3\pi/4$ to $\pi$, $\sin(2\theta)$ increases from $-1$ to 0. By the convention of stepping back a distance of $-r$ from the origin along the ray determined by $\theta$ if $r < 0$, the point that corresponds to the polar coordinate pair $(\sin(2\theta), \theta)$ traces the curve shown in Figure 9.

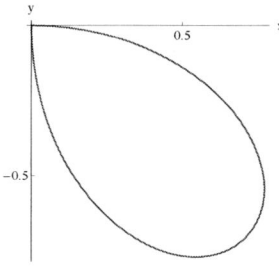

Figure 9

Similarly, as $\theta$ increases from $\pi$ to $3\pi/2$, the point that corresponds to the polar coordinate pair $(\sin(2\theta), \theta)$ traces the curve shown in Figure 10.

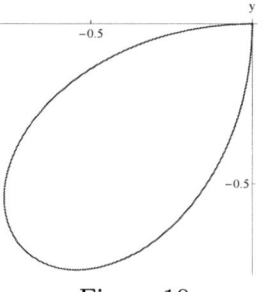

Figure 10

As $\theta$ increases from $3\pi/2$ to $2\pi$, the point that corresponds to the polar coordinate pair $(\sin(2\theta), \theta)$ traces the curve shown in Figure 11.

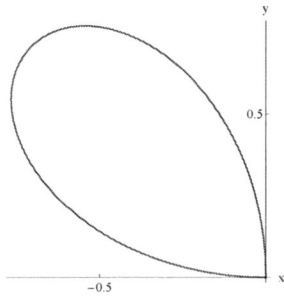

Figure 11

Figure 12 shows the complete polar graph of the equation $r = \sin(2\theta)$. □

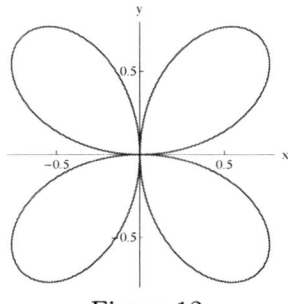

Figure 12

Polar curves that are determined by equations of the form $r = 1 + c\cos(\theta)$ or $r = 1 + c\sin(\theta)$, where $c$ is a constant are referred to as **limaçons**. The next two examples illustrate such curves.

**Example 8**

a) Sketch the graph of $r = 1 + \cos(\theta)$ in the Cartesian $\theta r$-plane ($0 \leq \theta \leq 2\pi$).
b) Sketch the polar curve determined by the equation $r = 1 + \cos(\theta)$.

**Solution**

a) Figure 13 shows the graph of $r = 1 + \cos(\theta)$, where $\theta \in [0, 2\pi]$, in the Cartesian $\theta r$-plane.

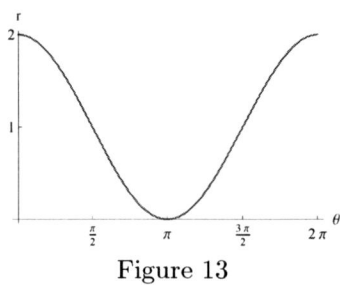

Figure 13

b) As $\theta$ increases from 0 to $\pi$, $1 + \cos(\theta)$ decreases from 2 to 0. As $\theta$ increases from $\pi$ to $2\pi$, $1 + \cos(\theta)$ increases from 0 to 2. Note that

$$1 + \cos(\theta)|_{\pi/2} = 1 + \cos(\theta)|_{3\pi/2} = 1.$$

## 10.2. POLAR COORDINATES

Figure 14 shows the polar graph determined by the equation $r = 1 + \cos(\theta)$. Such a shape may be referred to as **a cardioid**. □

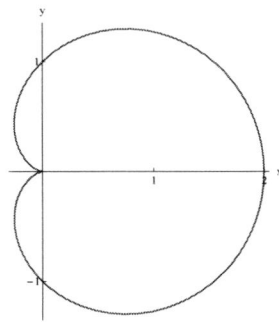

Figure 14

### Example 9

a) Sketch the graph of $r = 1 - 2\sin(\theta)$ in the Cartesian $\theta r$-plane ($0 \leq \theta \leq 2\pi$).
b) Sketch the polar curve corresponding to the equation $1 - 2\sin(\theta)$.

**Solution**

a) Figure 15 shows he graph of $r = 1 - 2\sin(\theta)$ in the Cartesian $\theta r$-plane ($0 \leq \theta \leq 2\pi$).

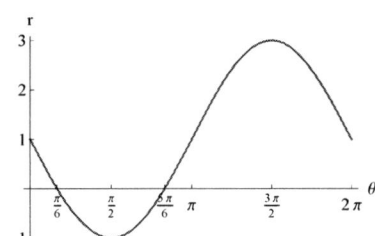

Figure 15

b) Note that
$$1 - 2\sin(\theta) = 0 \Leftrightarrow \sin(\theta) = \frac{1}{2},$$
so that the corresponding values in the interval $[0, 2\pi]$ are
$$\frac{\pi}{6} \text{ and } \pi - \frac{\pi}{6} = \frac{5\pi}{6}.$$
As $\theta$ increases from 0 to $\pi/6$, $1 - 2\sin(\theta)$ decreases from 1 to 0. The points corresponding to the polar coordinate pair $(1 - 2\sin(\theta), \theta)$ trace the curve shown in Figure 16.

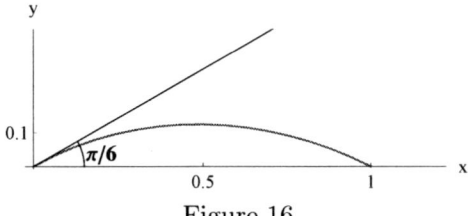

Figure 16

As $\theta$ increases from $\pi/6$ to $\pi/2$, $1 - 2\sin(\theta)$ decreases from 0 to $-1$, and as $\theta$ increases from $\pi/2$ to $5\pi/6$, $1 - 2\sin(\theta)$ increases from $-1$ to 0. Since $1 - 2\sin(\theta) < 0$ for $\theta \in (\pi/6, 5\pi/6)$, the point that corresponds to the polar coordinate pair $(1 - 2\sin(\theta), \theta)$ is obtained by stepping back a distance of $-(1 - 2\sin(\theta))$ from the origin along the ray determined by $\theta \in (\pi/6, 5\pi/6)$. Such points trace the curve shown in Figure 17.

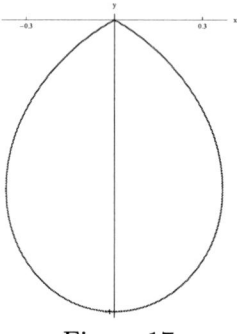

Figure 17

As $\theta$ increases from $5\pi/6$ to $3\pi/2$, $1 - 2\sin(\theta)$ increases from 0 to 3, and as $\theta$ increases from $3\pi/2$ to $2\pi$, $1 - 2\sin(\theta)$ decreases from 3 to 1. The point corresponding to the polar coordinate pair $(1 - 2\sin(\theta), \theta)$ traces the curve shown in Figure 18.

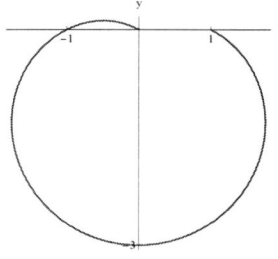

Figure 18

Figure 19 shows the complete polar curve determined by the equation $r = 1 - 2\sin(\theta)$. □

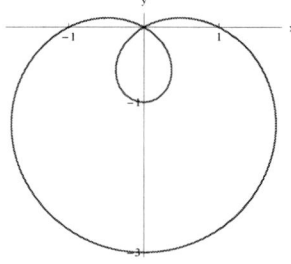

Figure 19

## Problems

In problems 1-6 determine the Cartesian coordinates of the point $P$ with the given polar coordinate pair $(r, \theta)$.

1. $(2, \pi/6)$,
2. $(4, \pi/4)$
3. $(3, 2\pi/3)$
4. $(-2, \pi/3)$
5. $(2, -\pi/6)$
6. $(-4, 3\pi/4)$

In problems 7-12 sketch the point $P$ with the given Cartesian coordinates:and determine polar coordinates $r$ and $\theta$ for $P$ such that $-\pi < \theta \leq \pi$.

7. $(4, 0)$
8. $(-3, 0)$
9. $(0, -2)$
10. $(\sqrt{2}, -\sqrt{2})$
11. $(-2\sqrt{3}, 2)$
12. $(2, -2\sqrt{3})$

In problems 13-22,
a) Sketch the graph of the given equation in polar coordinates $r$ and $\theta$ in the $\theta r$-Cartesian coordinate plane. Restrict $\theta$ so that $0 \leq \theta \leq 2\pi$. If the equation is in the form $r = f(\theta)$ indicate the points at which $f(\theta) = 0$, and the local extrema of $f$.
b) Sketch the graph of the given equation in the $xy$-plane (the polar graph of the equation).

13. $r = 4$
14. $\theta = -\pi/4$
15. $r = \theta$
16. $r = \theta^2$
17. $r = 4\cos(\theta)$
18. $r = \cos(2\theta)$
19. $r = \sin(3\theta)$
20. $r = 1 + \sin(\theta)$
21. $r = 2 - \sin(\theta)$
22. $r = 3 - 6\cos(\theta)$

## 10.3 Tangents and Area in Polar Coordinates

In this section we will discuss the calculation of tangents to polar curves and the calculation of the area of a region whose boundary can be described by polar curve.

### Tangents to Polar Curves

As we remarked earlier, the polar curve determined by the equation $r = r(\theta)$ is the same as the curve that is parametrized by the coordinate functions

$$x(\theta) = r(\theta) \cos(\theta) \text{ and } y(\theta) = r(\theta) \sin(\theta).$$

As we discussed in Section 10.1, if there exists an open interval $J$ containing $\theta_0$ such that

$$\frac{dx}{d\theta} > 0 \text{ for each } \theta \in J \text{ or } \frac{dx}{d\theta} < 0 \text{ for each } \theta \in J,$$

then $\theta$ can be expressed as a function $\theta(x)$ near $x(\theta_0)$. Thus, $y$ can be expressed as a function of $x$ if $\theta$ is restricted to the interval $J$: $y(x) = y(\theta(x)) \sin(\theta(x))$. By the chain rule and the product rule,

$$\frac{dy}{dx} = \frac{dy}{d\theta}\frac{d\theta}{dx} = \frac{\frac{dy}{d\theta}}{\frac{dx}{d\theta}} = \frac{\frac{dr}{d\theta}\sin(\theta) + r(\theta)\cos(\theta)}{\frac{dr}{d\theta}\cos(\theta) - r(\theta)\sin(\theta)}$$

In practice, if $x'(\theta_0) \neq 0$ and $r'(\theta)$ is continuous in a neighborhood of $\theta_0$, the above calculations are valid. A special case is worth mentioning: If $r(\theta_0) = 0$, the point that corresponds to $\theta_0$ is the origin. In this case, the above expression reads

$$\left.\frac{dy}{dx}\right|_{x=0} = \frac{\sin(\theta_0)}{\cos(\theta_0)} = \tan(\theta_0),$$

provided that $r'(\theta_0) \neq 0$ and $r'(\theta)$ is continuous. Therefore, the slope of the tangent line to the polar graph at the origin is $\tan(\theta_0)$.

**Example 1** Let $r(\theta) = 1 - 2\sin(\theta)$. The picture shows the corresponding polar graph.

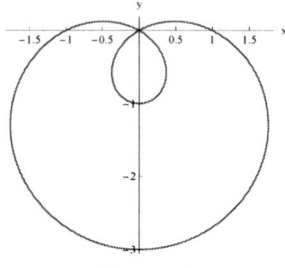

Figure 1

Determine the tangent lines to the curve that correspond to $\theta = \pi/6$, $5\pi/6$ and $0$.

**Solution**

a) We have

$$r(\pi/6) = 1 - 2\sin\left(\frac{\pi}{6}\right) = 1 - 2\left(\frac{1}{2}\right) = 0$$

and

$$r(5\pi/6) = r\left(\pi - \frac{\pi}{6}\right) = 1 - 2\sin\left(\frac{\pi}{6}\right) = 0.$$

Note that

$$r'(\theta) = \frac{d}{d\theta}(1 - 2\sin(\theta)) = -2\cos(\theta) \neq 0$$

at $\theta = \pi/6$ and $5\pi/6$. Therefore, the slope of the tangent line to the curve that correspond to $\pi/6$ and $5\pi/6$ are

$$\tan(\pi/6) = \frac{1}{\sqrt{3}} \text{ and } \tan\left(\frac{5\pi}{6}\right) = -\frac{1}{\sqrt{3}},$$

respectively. Since both tangent lines pass through the origin, their equations are

$$y = \frac{1}{\sqrt{3}}x \text{ and } y = -\frac{1}{\sqrt{3}}x,$$

respectively, Figures 2 and 3 show these tangent lines.

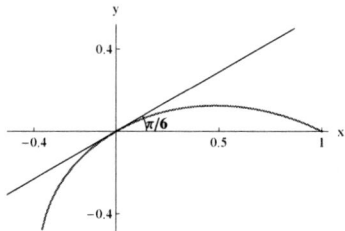

Figure 2

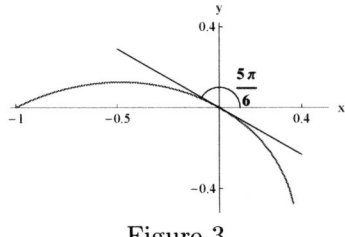

Figure 3

As for the tangent line that corresponds to $\theta = 0$, we have

$$r(0) = 1, \quad \left.\frac{dr}{d\theta}\right|_{\theta=0} = -2\cos(\theta)|_{\theta=0} = -2,$$

so that

$$\frac{dy}{dx} = \frac{\frac{dr}{d\theta}\sin(\theta) + r(\theta)\cos(\theta)}{\frac{dr}{d\theta}\cos(\theta) - r(\theta)\sin(\theta)} = \frac{(-2)(\sin(0)) + (1)\cos(0)}{(-2)(\cos(0)) - (1)(\sin(0))} = -\frac{1}{2}$$

We also have

$$y(0) = r(0)\sin(0) = 0 \text{ and } x(0) = r(0)\cos(0) = 1.$$

Therefore the tangent line is the graph of

$$y = -\frac{1}{2}(x-1).$$

Figure 4 shows that line.

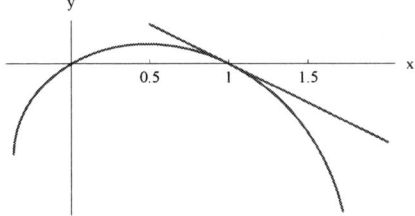

Figure 4

## Area in Polar Coordinates

Assume that the polar curve $C$ determined by the equation $r = r(\theta)$ and that $C$ is traversed exactly once as $\theta$ varies from $\alpha$ to $\beta$. Consider the region $G$ bounded by the rays $\theta = \alpha$, $\theta = \beta$ and $C$.

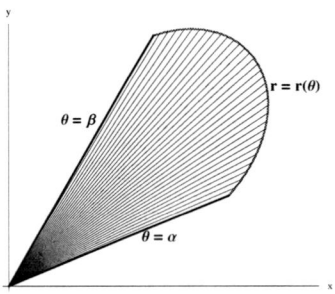

Figure 5

We would like to derive a formula for the area of $G$. Let's subdivide the interval $[\alpha, \beta]$ into the subintervals $[\theta_{k-1}, \theta_k]$, $k = 1, 2, \ldots, n$. With reference to Figure 6, the area of the part of $G$ that corresponds to $[\theta_{k-1}, \theta_k]$ can be approximated by the area of a circular sector of radius $r(\theta_k)$. As in Example 2 of Section 6.4, the area of such a circular sector is

$$\frac{1}{2} r^2(\theta_k) \Delta \theta_k.$$

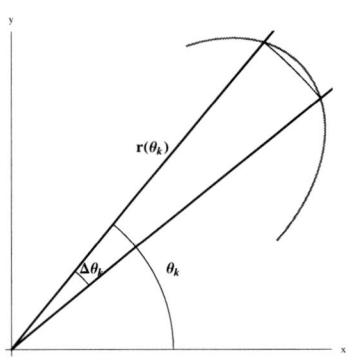

Figure 6

Therefore, the sum

$$\sum_{k=1}^{n} \frac{1}{2} r^2(\theta_k) \Delta \theta_k$$

is an approximation to the area of $G$. Such a sum is a Riemann sum that approximates the integral

$$\int_{\alpha}^{\beta} \frac{1}{2} r^2(\theta) \, d\theta$$

as accurately as desired if $\max_{k=1\ldots,n} \Delta \theta_k$ is small enough. **This is the formula that we will use to calculate the area of $G$.**

**Example 2** Calculate the area inside one loop of the polar curve determined by $r = \sin(2\theta)$.

**Solution**

# 10.3. TANGENTS AND AREA IN POLAR COORDINATES

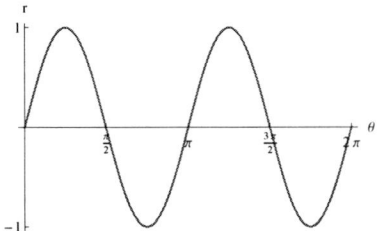

Figure 7

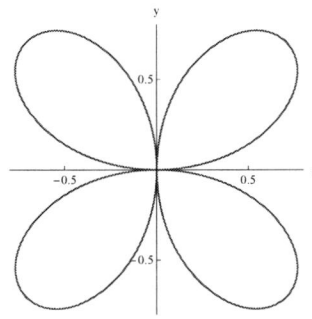

Figure 8

With reference to Figure 7 and Figure 8, one loop is traced if $\theta$ increases from 0 to $\pi/2$. Therefore the area inside one loop is

$$\int_0^{\pi/2} \frac{1}{2} \sin^2(2\theta) \, d\theta.$$

If we set $u = 2\theta$ then $du = 2d\theta$ so that

$$\int_0^{\pi/2} \frac{1}{2} \sin^2(2\theta) \, d\theta = \frac{1}{2} \int_0^{\pi} \sin^2(u) \left(\frac{1}{2}\right) du = \frac{1}{4} \int_0^{\pi} \sin^2(u) \, du.$$

As we discussed in Section 6.3,

$$\int \sin^2(u) \, du = \frac{1}{2} u - \frac{1}{2} \cos(u) \sin(u).$$

Therefore,

$$\frac{1}{4} \int_0^{\pi} \sin^2(u) \, du \neq \left. \frac{1}{8} u - \frac{1}{8} \cos(u) \sin(u) \right|_{u=0}^{u=\pi} = \frac{\pi}{8}.$$

□

## Problems

In problems 1-4,
a) Determine the tangent line to the polar graph at the points that correspond to the given value of $\theta$,
b) [C] Plot the polar graph and the tangent lines with the help of your graphing utility.

**1.**
$$r(\theta) = 2\sin(\theta), \ \theta = \pi/3.$$

**2.**
$$r(\theta) = \cos(2\theta), \ \theta = \pi/6.$$

**3.**
$$r(\theta) = 4 - \cos(\theta), \ \theta = \frac{3\pi}{4}.$$

**4.**
$$r(\theta) = 2 - 3\sin(\theta), \ \theta = \frac{\pi}{6}.$$

In problems 5 and 6,
a) [C] Plot the polar graph with the help of your graphing utility and show the indicated region G.
b) Compute the area of G.

**5.**
$$r(\theta) = 2 - \sin(\theta)$$
and G is inside the graph, between the rays $\theta = \pi/6$ and $\theta = \pi/3$.

**6.**
$$r(\theta) = \cos(4\theta)$$
and G is a region that is contained in one loop of the graph.

In problems 7 and 8,
a) [C] Plot the polar graph and show the indicated region G with the help of your graphing utility,
b) [CAS] Compute the area of G with the help of your computer algebra system

**7.**
$$r(\theta) = 1 - 2\cos(\theta)$$
and G is the region inside the inner loop of the graph.

**8.**
$$r(\theta) = 3 - 2\sin(\theta)$$
and G is inside the graph, between the rays $\theta = \pi/4$ and $\theta = \pi/2$.

## 10.4 Arc Length of Parametrized Curves

In this section we will discuss the calculation of the length of a parametrized curve, and as a special case, the length of a polar curve.

### Arc Length of Parametrized Curves

In Section 7.2 we discussed the calculation of the length of the graph of a function. Now we will consider the more general case of a parametrized curve.

Consider the curve $C$ that is parametrized by $\sigma(t) = (x(t), y(t))$ so that $\sigma(t)$ traces $C$ once as $t$ increases from $a$ to $b$. Assume that $dx/dt$ and $dy/dt$ are continuous on $[a, b]$ (you can interpret the derivatives at the endpoints as one-sided derivatives). Let's subdivide the interval $[a, b]$ into the subintervals $[t_{k-1}, t_k]$, $k = 1, 2, 3, \ldots, n$, by means of the partition $\{t_0, t_1, t_2, \ldots, t_n\}$ of $[a, b]$. As usual, let $\Delta t_k = t_k - t_{k-1}$. We will approximate the length of $C$ by the sum of the lengths of the line segments joining $(x(t_{k-1}), y(t_{k-1}))$ to $(x(t_k), y(t_k))$, $k = 1, 2, \ldots, n$. The length of the line segment that corresponds to $k$ is

$$\sqrt{(x(t_k) - x(t_{k-1}))^2 + (y(t_k) - y(t_{k-1}))^2}.$$

## 10.4. ARC LENGTH OF PARAMETRIZED CURVES

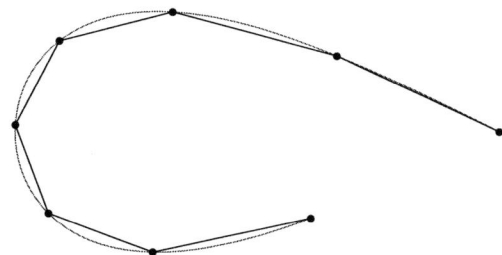

Figure 1: Arc length is approximated by the sum of the lengths of line segments

By the Mean Value Theorem there exists points $t_k^*$ and $\hat{t}_k$ in $[t_{k-1}, t_k]$ such that

$$x(t_k) - x(t_{k-1}) = \frac{dx}{dt}(t_k^*)(t_k - t_{k-1}) = \frac{dx}{dt}(t_k^*) \Delta t_k$$

and

$$y(t_k) - y(t_{k-1}) = \frac{dy}{dt}(\hat{t}_k)(t_k - t_{k-1}) = \frac{dy}{dt}(\hat{t}_k) \Delta t_k.$$

Therefore,

$$\sqrt{(x(t_k) - x(t_{k-1}))^2 + (y(t_k) - y(t_{k-1}))^2} = \sqrt{\left(\frac{dx}{dt}(t_k^*) \Delta t_k\right)^2 + \left(\frac{dy}{dt}(\hat{t}_k) \Delta t_k\right)^2}$$

$$= \sqrt{\left(\frac{dx}{dt}(t_k^*)\right)^2 + \left(\frac{dy}{dt}(\hat{t}_k)\right)^2} \Delta t_k.$$

Thus, the total distance traveled by the object over the time interval $[a,b]$ is approximately

$$\sum_{k=1}^{n} \sqrt{\left(\frac{dx}{dt}(t_k^*)\right)^2 + \left(\frac{dy}{dt}(\hat{t}_k)\right)^2} \Delta t_k.$$

The above sum is almost a Riemann sum that approximates

$$\int_a^b \sqrt{\left(\frac{dx}{dt}\right)^2 + \left(\frac{dy}{dt}\right)^2} dt$$

(we say "almost", since $\hat{t}_k$ is not necessarily the same as $t_k^*$). It can be shown that such a sum approximates the above integral with desired accuracy provided that $\max_k \Delta t_k$ is small enough.

**Definition 1** Assume that the curve $C$ is traced once by $\sigma(t) = (x(t), y(t))$ as $t$ increases from $a$ to $b$, and that $x(t)$ and $y(t)$ are continuously differentiable. The **length (or arc length)** of $C$ is

$$\int_a^b \sqrt{\left(\frac{dx}{dt}\right)^2 + \left(\frac{dy}{dt}\right)^2} dt.$$

**Example 1** Let $\sigma(t) = (\cos(t), \sin(t))$, where $t \in [0, 2\pi]$. Since $(\cos(t), \sin(t))$ traces the unit circle once as $t$ increases from 0 to $2\pi$, the length of the unit circle should be the $2\pi$.

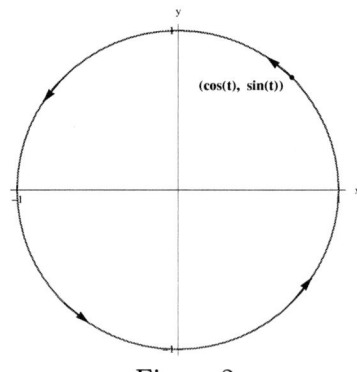

Figure 2

Indeed,

$$\sqrt{\left(\frac{dx}{dt}\right)^2 + \left(\frac{dy}{dt}\right)^2} = \sqrt{\left(\frac{d}{dt}\cos(t)\right)^2 + \left(\frac{d}{dt}\sin(t)\right)^2} = \sqrt{(-\sin(t))^2 + \cos^2(t)}$$
$$= \sqrt{\sin^2(t) + \cos^2(t)} = 1.$$

Therefore,

$$\int_0^{2\pi} \sqrt{\left(\frac{dx}{dt}\right)^2 + \left(\frac{dy}{dt}\right)^2}\, dt = \int_0^{2\pi} 1\, dt = 2\pi.$$

☐

**Remark** With reference to the above example, if $t$ varies from 0 to $4\pi$, then $\sigma(t)$ traces the unit circle twice and

$$\int_0^{4\pi} \sqrt{\left(\frac{dx}{dt}\right)^2 + \left(\frac{dy}{dt}\right)^2}\, dt = \int_0^{4\pi} 1\, dt = 4\pi,$$

as expected. In the general case, we need to make sure that $\sigma(t)$ traces a curve exactly once as the parameter $t$ varies from $a$ to $b$ so that the integral

$$\int_a^b \sqrt{\left(\frac{dx}{dt}\right)^2 + \left(\frac{dy}{dt}\right)^2}\, dt$$

yields the length of the curve.

If a curve $C$ is the graph of a function $f$ on the interval $[a, b]$, we can set $x = t$ and $y(x) = f(x)$. Therefore,

$$\int_a^b \sqrt{\left(\frac{dx}{dx}\right)^2 + \left(\frac{dy}{dx}\right)^2}\, dx = \int_a^b \sqrt{1 + \left(\frac{df}{dx}\right)^2}\, dt.$$

We have used the above formula to compute the length of the graph of a function in Section 7.2. ◊

**Example 2** Let $\sigma(t) = (2\cos(t), \sin(t))$, $t \in [0, 2\pi]$. Show that the curve that is parametrized by $\sigma$ is an ellipse. Make use of your computational utility to approximate the arc length of the ellipse.

## 10.4. ARC LENGTH OF PARAMETRIZED CURVES

**Solution**

Figure 3 shows the curve that is parametrized by $\sigma$. The picture supports the claim that the curve is an ellipse.

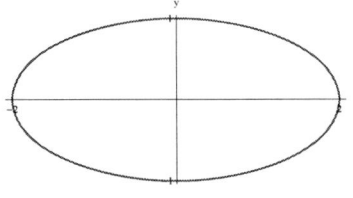

Figure 3

Indeed, if we set $x(t) = 2\cos(t)$ and $y(t) = \sin(t)$, then

$$\frac{x^2(t)}{4} + y^2(t) = \frac{1}{4}\left(4\cos^2(t)\right) + \sin^2(t) = \cos^2(t) + \sin^2(t) = 1.$$

Therefore, $\sigma(t) = (x(t), y(t))$ is on the ellipse

$$\frac{x^2}{4} + y^2 = 1.$$

The major axis is along the $x$-axis. The ellipse is traversed by $\sigma(t)$ exactly once as $t$ increases from 0 to $2\pi$. Therefore, its arc length of the ellipse is

$$\int_0^{2\pi} \sqrt{\left(\frac{dx}{dt}\right)^2 + \left(\frac{dy}{dt}\right)^2}\, dt = \int_0^{2\pi} \sqrt{\left(\frac{d}{dt}(2\cos(t))\right)^2 + \left(\frac{d}{dt}\sin(t)\right)^2}\, dt$$
$$= \int_0^{2\pi} \sqrt{2\sin^2(t) + \cos^2(t)}\, dt.$$

The integral cannot be expressed in terms of the special functions that we are in our portfolio. In any case,

$$\int_0^{2\pi} \sqrt{2\sin^2(t) + \cos^2(t)}\, dt \cong 7.640\,40,$$

with the help of the numerical integrator of a computational utility. $\square$

### Arc Length in Polar Coordinates

Let $r$ and $\theta$ be polar coordinates, and assume that $r = r(\theta)$, where $\theta \in [a, b]$. As we discussed in Section 10.2, if we set

$$\sigma(\theta) = (r(\theta)\cos(\theta), r(\theta)\sin(\theta)),$$

**the polar curve determined by the equation** $r = f(\theta)$ ($\theta \in [a, b]$).is the curve that is parametrized by $\sigma$.

**Proposition 1** Let $r$ and $\theta$ be polar coordinate. Assume that the curve $C$ is the polar curve determined by the equation $r = f(\theta)$ and that $C$ is traced exactly once as $\theta$ varies from $a$ to $b$. The length of $C$ is

$$\int_a^b \sqrt{r^2(\theta) + \left(\frac{dr}{d\theta}\right)^2}\, d\theta.$$

**Proof**

As we discussed in Section 10.2, the curve $C$ is parametrized by

$$\sigma(\theta) = (x(\theta), y(\theta)) = (r(\theta)\cos(\theta), r(\theta)\sin(\theta)), \quad 0 \leq \theta \leq b.$$

Therefore, the arc length of $C$ is

$$\int_a^b \sqrt{\left(\frac{dx}{d\theta}\right)^2 + \left(\frac{dy}{d\theta}\right)^2}\, d\theta.$$

Let's simplify the expression under the radical sign:

$$\left(\frac{dx}{d\theta}\right)^2 + \left(\frac{dy}{d\theta}\right)^2 = \left(\frac{d}{d\theta}(r(\theta)\cos(\theta))\right)^2 + \left(\frac{d}{d\theta}(r(\theta)\sin(\theta))\right)^2$$

$$= \left(\frac{dr}{d\theta}\right)^2 \cos^2(\theta) + r^2(\theta)\sin^2(\theta)$$

$$+ \left(\frac{dr}{d\theta}\right)^2 \sin^2(\theta) + r^2(\theta)\cos^2(\theta)$$

$$= \left(\frac{dr}{d\theta}\right)^2 (\cos^2(\theta) + \sin^2(\theta)) + r^2(\theta)(\sin^2(\theta) + \cos^2(\theta))$$

$$= \left(\frac{dr}{d\theta}\right)^2 + r^2(\theta).$$

Therefore, the arc length of $C$ is

$$\int_a^b \sqrt{\left(\frac{dr}{d\theta}\right)^2 + r^2(\theta)}\, d\theta,$$

as claimed. ∎

**Example 3** Let $r = \sin(2\theta)$.
a) Express the arc length of one loop of the polar curve determined by $r = \sin(2\theta)$ as an integral.
b) Make use of your computational utility to approximate the integral of part a).

**Solution**

a) As we discussed in Example 7 of Section 10.2, the first loop of the curve is traversed as $\theta$ varies from 0 to $\pi/2$.

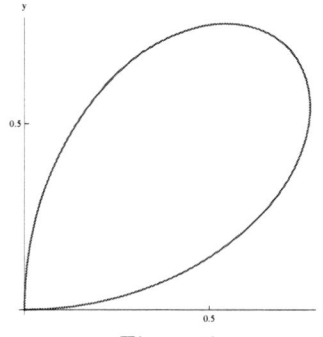

Figure 4

## 10.4. ARC LENGTH OF PARAMETRIZED CURVES

Therefore, the arc length of the loop is

$$\int_0^{\pi/2} \sqrt{\left(\frac{d}{d\theta}\sin(2\theta)\right)^2 + (\sin(2\theta))^2}\,d\theta = \int_0^{\pi/2} \sqrt{4\cos^2(2\theta) + \sin^2(2\theta)}\,d\theta.$$

b) The indefinite integral

$$\int \sqrt{4\cos^2(2\theta) + \sin^2(2\theta)}\,d\theta$$

cannot be expressed in terms of familiar special function. In any case the numerical integrator of your computational utility should give you the approximation 2.422 1, rounded to 6 significant digits. □

## Problems

In problems 1-4,
a) Set up the integral for the calculation of the length of the curve $C$ that is parametrized by the function. Simplify the integrand as much as possible but do not evaluate unless you recognize an antiderivative for the integrand readily.
b) [C] Make use of your computational utility in order to approximate the integral of part a).
c) [CAS] If you have access to a computer algebra system, try to determine an antiderivative for the integrand of part a). Do you recognize the special functions in the expression?

**1.**
$$\sigma(t) = (3\cos(t), 3\sin(t)),\ \pi/4 \le t \le \pi$$

($C$ is an arc of a circle).

**2.**
$$\sigma(t) = (3\cos(t), 2\sin(t)),\ t \in [0, 2\pi]$$

($C$ is an ellipse).

**3.**
$$\sigma(t) = (\cosh(t), \sinh(t)),\ -1 \le t \le 1$$

($C$ is part of a hyperbola).

**4.**
$$\sigma(t) = (t, t^2 + 4),\ 0 \le t \le 2$$

($C$ is part of a parabola).

In problems 5-8,
a) Set up the integral for the calculation of the length of the polar graph curve $C$ that is the graph of the given function $r(\theta)$. Simplify the integrand as much as possible but do not evaluate unless you recognize an antiderivative for the integrand readily.
b) [C] Make use of your computational utility in order to approximate the integral of part a).
c) [CAS] If you have access to a computer algebra system, try to determine an antiderivative for the integrand of part a). Do you recognize the special functions in the expression?

**5.**
$$r(\theta) = \cos(2\theta),\ 0 \le \theta \le \pi/4.$$

**6.**
$$r(\theta) = 1 - 2\sin(\theta),\ \frac{\pi}{6} \le \theta \le \frac{5\pi}{6}.$$

**7.**
$$r(\theta) = \cos(4\theta), \ -\frac{\pi}{8} \leq \theta \leq \frac{\pi}{8}.$$

**8.**
$$r(\theta) = \sin(3\theta), \ 0 \leq \theta \leq 2\pi.$$

## 10.5 Conic Sections

In this section we will characterize parabolas, ellipses and hyperbolas in terms of foci and directrixes (singular "directrix"). These curves are referred to as conic sections since they can be obtained as intersections of a plane and a right circular cone.

### Parabolas

Let $l$ be a line in the plane and assume that the point $F$ is a point in the plane that is not on $l$. **A parabola with focus $F$ and directrix $l$ consists of points $P$ in the plane such that the distance of $P$ from $F$ is equal to the distance of $P$ from the line $l$.** Let's denote the line segment that connects $P$ to $F$ as $PF$, the length of $PF$ as $|PF|$, and the foot of the perpendicular from $P$ to $l$ by $D$ so that the distance of $P$ from $l$ is $|PD|$. Thus, $P$ is on the parabola with focus $F$ and directrix $l$ if and only if $|PF| = |PD|$, as illustrated in Figure 1.

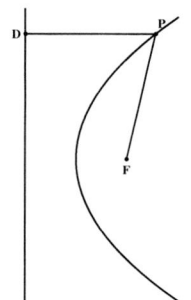

Figure 1: $|PF| = |PD|$

We can express the parabola as the graph of a simple equation if we place its focus and directrix appropriately. For example, in the Cartesian $xy$-plane, we can set $F = (p, 0)$ and have the directrix $l$ as the line $x = -p$, where $p > 0$, as in Figure 2.

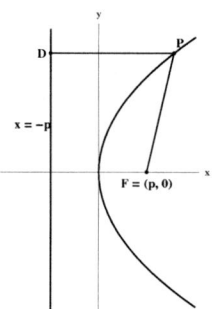

Figure 2: $y^2 = 4px$

Let $P = (x, y)$ be a point on the parabola so that $|PF| = |PD|$. Thus,

$$\sqrt{(x-p)^2 + y^2} = x + p.$$

We square both sides:

$$(x-p)^2 + y^2 = x^2 + 2px + p^2 \Rightarrow x^2 - 2px + p^2 + y^2 = x^2 + 2px + p^2$$
$$\Rightarrow y^2 = 4px$$

Conversely, if $y^2 = 4px$, the point $P = (x, y)$ is equidistant from $F$ and the line $l$ (check). Therefore, the parabola with focus at $(p, 0)$ and directrix $x = -p$ is the graph of the equation $y^2 = 4px$.

Similarly, if $p > 0$, the focus of the parabola is placed at $(-p, 0)$ and the directrix is the line $x = p$, the parabola is the graph of the equation $-y^2 = 4px$, as illustrated in Figure 3.

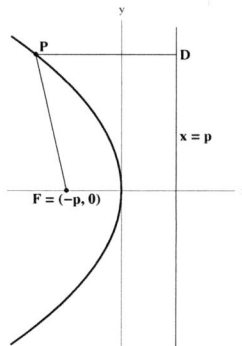

Figure 3: $y^2 = -4px$

The roles of $x$ and $y$ can be interchanged: If $p > 0$, the parabola with focus at $(0, p)$ and directrix $y = -p$ is the graph of the equation $x^2 = 4py$, as illustrated in Figure 4.

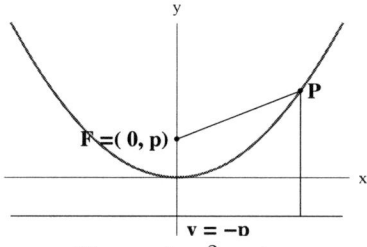

Figure 4: $x^2 = 4py$

If $p > 0$, the parabola with focus at $(0, -p)$ and directrix $y = p$ is the graph of the equation $-x^2 = 4py$, as illustrated in Figure 5.

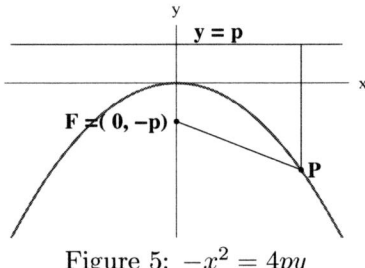

Figure 5: $-x^2 = 4py$

**Example 1** Identify the focus and directrix of the parabola that is the graph of the equation $y^2 = 8x$. Sketch the parabola.

**Solution**

The equation is in the form $y^2 = 4px$, with $p = 2$. Therefore, the focus of the parabola is $F = (2, 0)$ and the directrix is the line $x = -2$. Figure 6 shows the parabola.

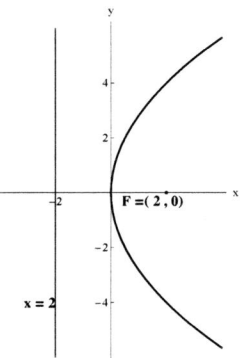

Figure 6: $y^2 = 8x$

## Ellipses

Let $F_1$ and $F_2$ be two points in the plane and let $C > 0$ be a constant. **An ellipse with foci $F_1$ and $F_2$ consists of all points $P$ in the plane such that $|PF_1| + |PF_2|$ is a constant.** If $F = F_1 = F_2$, an ellipse with the single focus $F$ is simply a circle. Thus, assume that $F_1 \neq F_2$.

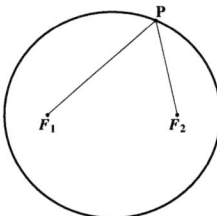

Figure 7: $|PF_1| + |PF_2|$ is a constant

## 10.5. CONIC SECTIONS

We can express an ellipse as the graph of a simple equation if we place the foci on the $x$-axis symmetrically with respect to the origin. Thus, assume that $F_1 = (-c, 0)$ and $F_2 = (c, 0)$, where $c > 0$, and assume that $P = (x, y)$ is on the ellipse if and only if $|PF_1| + |PF_2| = 2a$, where $a$ is a positive constant, as illustrated in Figure 8.

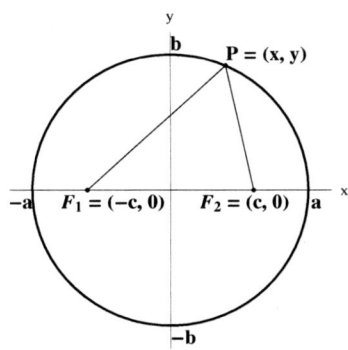

Figure 8

Thus,
$$\sqrt{(x+c)^2 + y^2} + \sqrt{(x-c)^2 + y^2} = 2a$$

so that
$$\sqrt{(x+c)^2 + y^2} = 2a - \sqrt{(x-c)^2 + y^2}.$$

Squaring both sides,
$$x^2 + 2cx + c^2 + y^2 = 4a^2 - 4a\sqrt{(x-c)^2 + y^2} + x^2 - 2cx + c^2 + y^2.$$

Therefore,
$$cx + a^2 = -a\sqrt{(x-c)^2 + y^2}.$$

Again, we square both sides of the equation:
$$c^2x^2 + 2a^2cx + a^4 = a^2x^2 - 2a^2xc + a^2c^2 + a^2y^2,$$

so that
$$a^2\left(a^2 - c^2\right) = \left(a^2 - c^2\right)x^2 + a^2y^2.$$

With reference to Figure 8, the side $F_1F_2$ of the triangle $F_1F_2P$ is less than the sum of the other two sides. Therefore,
$$2c < 2a \Rightarrow c < a.$$

We set $b = \sqrt{a^2 - c^2}$. Thus,
$$b^2x^2 + a^2y^2 = a^2b^2 \Rightarrow \frac{x^2}{a^2} + \frac{y^2}{b^2} = 1.$$

You can confirm that the sum of the distances of $(x, y)$ from $(\pm c, 0)$ is $2a$ if $x$ and $y$ satisfy the above equation. Note that $b = \sqrt{a^2 - c^2}$ so that $a^2 = b^2 + c^2$. Figure 9 illustrates the relationships between $a$, $b$ and $c$.

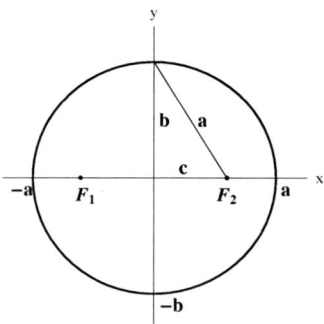

Figure 9: $b = \sqrt{a^2 - c^2}$

The points $(a, 0)$ and $(-a, 0)$ are on the ellipse, and are referred to as **the vertices** of the ellipse. The line segment that joins $(-a, 0)$ to $(a, 0)$ is **the major axis** of the ellipse. The points $(0, -b)$ and $(0, b)$ are also on the ellipse. The line segment that joins $(0, -b)$ to $(0, b)$ is **the minor axis** of the ellipse. We may consider the circle of radius $a$ centered at $(0, 0)$ as the special case where both foci $F_1$ and $F_2$ coincide with the origin, so that $c = 0$ and $b = a$.

In summary, if $a \geq b > 0$ and $c = \sqrt{a^2 - b^2}$, the graph of the equation

$$\frac{x^2}{a^2} + \frac{y^2}{b^2} = 1$$

is an ellipse with foci $(\pm c, 0)$ and vertices $(\pm a, 0)$. The major axis is along the $x$-axis.

**Example 2** Identify the foci and the major axis of the ellipse that is the graph of the equation

$$\frac{x^2}{25} + \frac{y^2}{16} = 1.$$

Sketch the ellipse.

**Solution**

With the notation of the above discussion, $a = 5$ and $b = 4$. Thus, the major axis of the ellipse is along the $x$-axis. We have

$$c = \sqrt{25 - 16} = \sqrt{9} = 3.$$

Therefore, the foci of the ellipse are $(-3, 0)$ and $(3, 0)$. Figure 10 shows the ellipse.

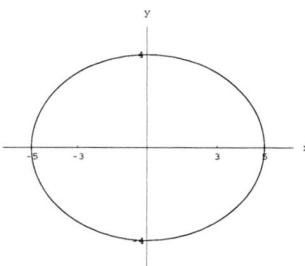

Figure 10

## 10.5. CONIC SECTIONS

**Remark** The roles of $x$ and $y$ may be interchanged. If the ellipse is the graph of the equation

$$\frac{y^2}{a^2} + \frac{x^2}{b^2} = 1,$$

where $a \geq b > 0$, $c = \sqrt{a^2 - b^2}$ then $F_1 = (0, -c)$ and $F_2 = (0, c)$ are the foci of the ellipse. ◊

It is possible to describe an ellipse as a parametrized curve as in the following example:

**Example 3** Let $\sigma(\theta) = (4\cos(\theta), 5\sin(\theta))$, where $0 \leq \theta \leq 2\pi$, and let $C$ be the curve that is parametrized by $\sigma$. Show that $C$ is an ellipse. Identify the foci of the ellipse.

**Solution**

If $(x, y)$ is on the curve $C$ then

$$\frac{y^2}{25} + \frac{x^2}{16} = \frac{25\sin^2(\theta)}{25} + \frac{16\cos^2(\theta)}{16} = \sin^2(\theta) + \cos^2(\theta) = 1.$$

Therefore $C$ is an ellipse with its major axis on the $y$-axis. The points $(0, \pm 5)$ and $(\pm 4, 0)$ are on the ellipse. The foci are the points $(0, \pm c)$ where

$$c = \sqrt{25 - 16} = \sqrt{9} = 3.$$

Thus, the foci are $(0, 3)$ and $(0, -3)$.

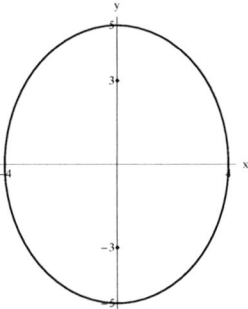

Figure 11

## Hyperbolas

Let $F_1$ and $F_2$ be two distinct points in the plane and let $C > 0$ be a constant. **A hyperbola with foci $F_1$ and $F_2$ consists of all points $P$ in the plane such that $||PF_1| - |PF_2||$ is a constant.**

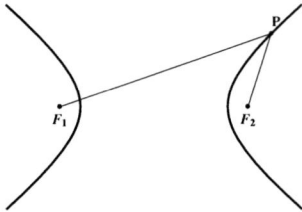

Figure 12

We can express a hyperbola as the graph of a simple equation if we place the foci on the $x$-axis symmetrically with respect to the origin. Thus, assume that $F_1 = (-c, 0)$ and $F_2 = (c, 0)$, where $c > 0$, and that $P = (x, y)$ is on the hyperbola if and only if $||PF_1| + |PF_2|| = 2a$, where $a$ is a positive constant. Thus,

$$\sqrt{(x+c)^2 + y^2} + \sqrt{(x-c)^2 + y^2} = \pm 2a$$

In a way that is similar to the calculations in the case of an ellipse, the above relationship leads to the equation

$$\frac{x^2}{a^2} - \frac{y^2}{b^2} = 1,$$

where $b = \sqrt{c^2 - a^2}$. The points $(a, 0)$ and $(-a, 0)$ are on the hyperbola, and are referred to as **the vertices**. Note that the hyperbola does not intersect the $y$-axis and has two disjoint branches on either side of the $y$-axis. Also note that a point $(x, y)$ on the hyperbola is close to a point on a line of the form

$$y = \pm \frac{b}{a} x$$

if $|x|$ is large.

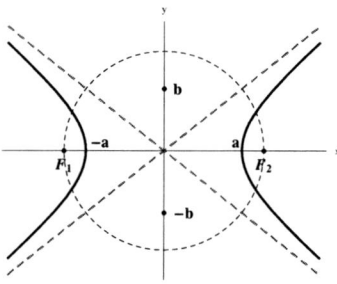

Figure 13

Indeed,

$$\frac{x^2}{a^2} - \frac{y^2}{b^2} = 1 \Rightarrow y^2 = \frac{b^2}{a^2} x^2 + b^2 \Rightarrow y = \pm \sqrt{\frac{b^2}{a^2} x^2 + b^2}$$

so that

$$y \cong \pm \sqrt{\frac{b^2}{a^2} x^2} = \pm \frac{b}{a} x$$

if $|x|$ is large. You can show that

$$\lim_{x \to \pm \infty} \left( y - \frac{b}{a} x \right) = 0$$

if $(x, y)$ is on the hyperbola and $y > 0$, and that

$$\lim_{x \to \pm \infty} \left( y - \left( -\frac{b}{a} x \right) \right) = 0$$

if $(x, y)$ is on the hyperbola and $y < 0$. The lines

$$y = \pm \frac{b}{a} x$$

## 10.5. CONIC SECTIONS

are **asymptotes** for the hyperbola
$$\frac{x^2}{a^2} - \frac{y^2}{b^2} = 1.$$

In summary,
If $a > 0$, $c > 0$, $b > 0$ and $c^2 = a^2 + b^2$, the graph of the equation
$$\frac{x^2}{a^2} - \frac{y^2}{b^2} = 1$$

is a hyperbola with foci $(\pm c, 0)$ and vertices $(\pm a, 0)$. The hyperbola has two branches on either side of the $y$-axis. The lines
$$y = \pm \frac{b}{a} x$$

as asymptotic to the hyperbola as $|x| \to \infty$.

**Example 4** Identify the foci of the hyperbola that is the graph of the equation
$$\frac{x^2}{25} - \frac{y^2}{16} = 1.$$

Sketch the graph of the hyperbola. Indicate the asymptotes.

**Solution**

Here we have $a = 5$ and $b = 4$. Therefore
$$c = \sqrt{a^2 + b^2} = \sqrt{5^2 + 4^2} = \sqrt{41} \cong 6.40312$$

Thus,
$$F_1 = (-c, 0) = \left(-\sqrt{41}, 0\right) \text{ and } F_2 = (c, 0) = \left(\sqrt{41}, 0\right)$$

are the foci of the hyperbola. The eccentricity of the hyperbola is
$$E = \frac{c}{a} = \frac{\sqrt{41}}{5} \cong 1.28062$$

The directrixes of the hyperbola are
$$x = -\frac{a}{E} = -\frac{5}{\frac{\sqrt{41}}{5}} = -\frac{25}{\sqrt{41}} \cong -3.90434$$

(corresponding to $F_1$) and
$$x = \frac{a}{E} = \frac{5}{\frac{\sqrt{41}}{5}} = \frac{25}{\sqrt{41}} \cong 3.90434$$

(corresponding to $F_2$).
The asymptotes are
$$y = \pm \frac{b}{a} x = \pm \frac{4}{5} x.$$

Figure 14 shows the hyperbola and the asymptotes. □

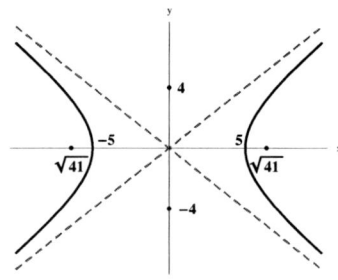

Figure 14

**Example 5** In Example 3 of Section 10.1 we saw that the curve $C$ that is parametrized by $\sigma(t) = (\cosh(t), \sinh(t))$, $t \in \mathbb{R}$, is the branch of the hyperbola $x^2 - y^2 = 1$ in the right half-plane.

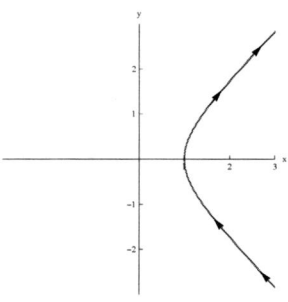

Figure 15

The focus that is on the right half-plane is

$$(\sqrt{1+1}, 0) = (\sqrt{2}, 0).$$

□

**Remark** The roles of the $x$ and $y$ axes may be interchanged. The graph of the equation

$$\frac{y^2}{a^2} - \frac{x^2}{b^2} = 1$$

is a hyperbola with foci $(0, \pm c)$ and vertices $(0, \pm a)$. The hyperbola has two branches, one above the $x$-axis and the other below the $x$-axis. The lines

$$y = \pm \frac{a}{b}x$$

as asymptotic to the hyperbola as $|x| \to \infty$. ◊

## Problems

In problems 1-6,
a) Identify the graph of the equation as a parabola, ellipse or hyperbola,
b) Determine the foci, directrices, major and minor axes and asymptotes of the conic section of part a), if applicable,

c) Sketch the conic section of part a). Indicate the foci, major and minor axes and asymptotes of the conic section, if applicable.

1.
$$y^2 = 16x$$

2.
$$x^2 = 8y$$

3.
$$\frac{x^2}{9} + \frac{y^2}{4} = 1$$

4.
$$\frac{y^2}{9} + \frac{x^2}{4} = 1$$

5.
$$\frac{y^2}{25} - \frac{x^2}{16} = 1$$

6.
$$\frac{y^2}{9} - \frac{x^2}{4} = 1$$

## 10.6 Conic Sections in Polar Coordinates

In some applications it is convenient to describe conic sections in polar coordinates. Let's begin with the case of a parabola. We will place the parabola in the $xy$-plane so that the focus $F$ is the origin and the directrix is the line $x = d$ where $d > 0$. Thus, the parabola is the set of points $P = (x, y)$ such that the distance of $P$ from the origin is the same as the distance of $P$ from the line $d$. Let $(r, \theta)$ be a pair of polar coordinates for $P$.
let $F = (0, 0)$ be a focus and assume that the directrix is the line $x = d$, where $d > 0$.

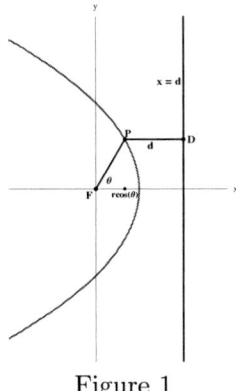

Figure 1

With reference to Figure 1,

$$r = d - r\cos(\theta) \Rightarrow (1 - \cos(\theta))\,r = d \Rightarrow r = \frac{d}{1 - \cos(\theta)}.$$

Ellipses and hyperbolas can be characterized in terms of foci and directrixes as well:

**Theorem 1** Let $F$ be a fixed point (a **focus**) and $l$ be a fixed line (a **directrix**) and $E$ be a positive number Assume that the curve $C$ is the set of points $P$ such that the ratio of the distance of $P$ from the focus $F$ and the distance of $P$ from the directrix $l$ is the eccentricity $E$. Then $C$ is an ellipse if $E < 1$, a parabola if $E = 1$ and a hyperbola if $E > 1$.

We will not prove the above theorem but confirm its statement in specific cases. It turns out that ellipses and hyperbolas have two focus-directrix pairs.

As in the case of the parabola, let's place a focus at the origin and the corresponding directrix as the line $x = d > 0$. Let $(r, \theta)$ be a pair of polar coordinates for the point $P = (x, y)$ on the curve. With reference to Figure 2, If the eccentricity of the conic section is $E$ and the foot of the perpendicular from $P$ to the line $x = d$ is $D$, then

$$\frac{|PF|}{|PD|} = E.$$

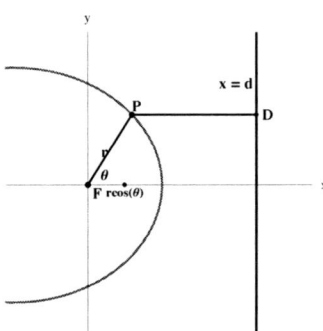

Figure 2

Thus,
$$\frac{r}{d - r\cos(\theta)} = E.$$

Therefore,
$$r = Ed - Er\cos(\theta) \Rightarrow (1 + E\cos(\theta))r = Ed \Rightarrow r = \frac{Ed}{1 + E\cos(\theta)}.$$

**Remark** Assume that the focus-directrix pair are placed so that $F = (0,0)$, as before, but the directrix is the line $x = -d$. Then, the expression

$$r = \frac{Ed}{1 - E\cos(\theta)}$$

replaces the expression

$$r = \frac{Ed}{1 + E\cos(\theta)}.$$

We can interchange the roles of $x$ and $y$. In that case, $F = (0,0)$ and the directrix is the graph of the equation $y = d$ or $y = -d$. In those cases, the conic section is the graph of an equation of the form

$$r = \frac{Ed}{1 \pm E\sin(\theta)}.$$

◊

## 10.6. CONIC SECTIONS IN POLAR COORDINATES

**Example 1** Let $C$ be the conic section that is the graph of the equation

$$r = \frac{4}{1 + \cos(\theta)}, \quad 0 \leq \theta \leq 2\pi,$$

in polar coordinates. Identify the conic section. Determine the relevant focus and directrix. Sketch $C$. Express the equation in Cartesian coordinates.

**Solution**

The equation is in the form

$$r = \frac{Ed}{1 + E\cos(\theta)},$$

where $E = 1$ and $d = 1$. Thus $C$ is a parabola with its focus at the origin and the directrix is the line $x = d$.

Figure 3 shows the graph of

$$r = \frac{4}{1 + \cos(\theta)}, \quad 0 \leq \theta \leq 2\pi,$$

in the $\theta r$-plane. Note that

$$1 + \cos(\theta) = 0 \text{ if } \theta = \pi$$

and the line $\theta = \pi$ is a vertical asymptote.

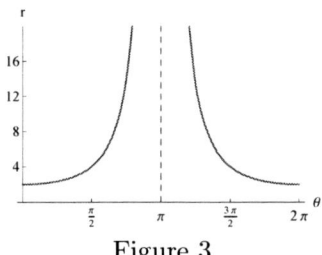

Figure 3

Figure 4 shows the parabola in the $xy$-plane. The focus is at the origin. The dashed line shows the directrix.

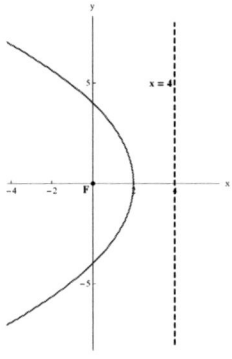

Figure 4

Now we will express the equation in Cartesian coordinates:

$$r = \frac{4}{1 + \cos(\theta)}$$

$\Rightarrow$
$$r + r\cos(\theta) = 4$$

$\Rightarrow$
$$r = 4 - r\cos(\theta)$$

$\Rightarrow$
$$\sqrt{x^2 + y^2} = 4 - x$$

$\Rightarrow$
$$x^2 + y^2 = (4-x)^2 = 16 - 8x + x^2$$

$\Rightarrow$
$$y^2 = 16 - 8x \Rightarrow x - 2 = -\frac{1}{8}y^2$$

This equation also confirms that $C$ is a parabola. The vertex of the parabola is at $(2, 0)$. $\square$

**Example 2** Let $C$ be the conic section that is the graph of the equation

$$r = \frac{4}{3 - \sin(\theta)}, \quad 0 \leq \theta \leq 2\pi,$$

in polar coordinates. Identify the conic section. Determine the relevant focus and directrix. Sketch $C$. Express the equation in Cartesian coordinates.

**Solution**

We have

$$r = \frac{4}{3 - \sin(\theta)} = \frac{4/3}{1 - \frac{1}{3}\sin(\theta)}.$$

Thus, the graph of the equation is an ellipse with eccentricity $1/3$. The focus is at the origin and the relevant directrix is the line $y = -4$.
Figure 5 shows the graph of

$$r = \frac{4}{3 - \sin(\theta)}, \quad 0 \leq \theta \leq 2\pi,$$

in the $\theta r$-plane.

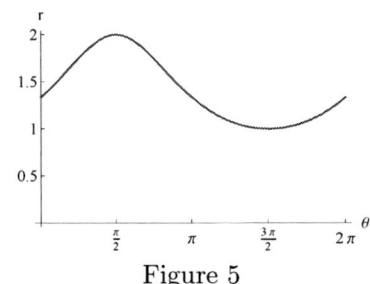

Figure 5

Figure 6 shows the ellipse in the $xy$-plane. The focus is at the origin. The dashed line shows the directrix.

## 10.6. CONIC SECTIONS IN POLAR COORDINATES

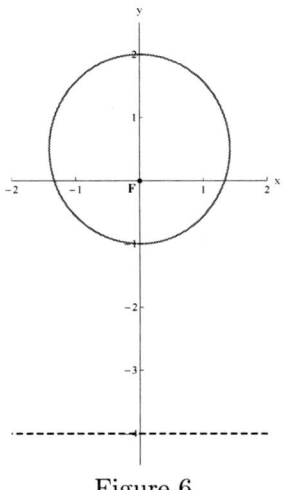

Figure 6

Now we will express the equation in Cartesian coordinates:

$$r = \frac{4}{3 - \sin(\theta)}$$

$\Rightarrow$
$$3r - r\sin(\theta) = 4$$

$\Rightarrow$
$$3r - y = 4$$

$\Rightarrow$
$$3r = y + 4$$

$\Rightarrow$
$$3\sqrt{x^2 + y^2} = y + 4$$

$\Rightarrow$
$$9x^2 + 9y^2 = y^2 + 8y + 16$$

$\Rightarrow$
$$9x^2 + 8y^2 - 8y = 16.$$

We will complete the square:

$$9x^2 + 8\left(y - \frac{1}{2}\right)^2 - 2 = 16 \Rightarrow 9x^2 + 8\left(y - \frac{1}{2}\right)^2 = 18.$$

This equation confirms that the curve is an ellipse. The ellipse is centered at $(0, 1/2)$. $\square$

**Example 3** Let $C$ be the conic section that is the graph of the equation

$$r = \frac{4}{1 - 2\cos(\theta)}, \quad 0 \leq \theta \leq 2\pi,$$

in polar coordinates. Identify the conic section. Determine the relevant focus and directrix. Sketch $C$. Express the equation in Cartesian coordinates.

# 330 CHAPTER 10. PARAMETRIZED CURVES AND POLAR COORDINATES

**Solution**

The eccentricity is $2 > 1$ so that the curve is a hyperbola. The focus is at the origin and the relevant directrix is the line $x = -2$.

Figure 7 shows the graph of
$$r = \frac{4}{1 - 2\cos(\theta)}, \quad 0 \leq \theta \leq 2\pi,$$
in the $\theta r$-plane. Note that
$$1 - 2\cos(\theta) = 0 \text{ if } \theta = \pi/3 \text{ and } \theta = 5\pi/3$$
and the lines $\theta = \pi/3$ and $\theta = 5\pi/3$ are vertical asymptotes.

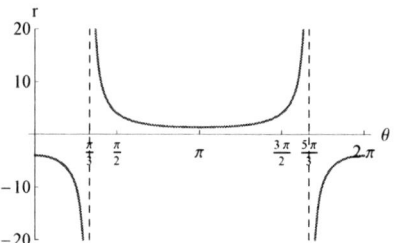

Figure 7

Figure 8 shows the hyperbola in the $xy$-plane. The focus is at the origin. The vertical dashed line shows the directrix. The other lines indicate the asymptotes to the curve at $\pm\infty$.

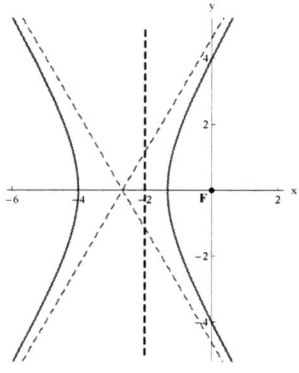

Figure 8

Now we will express the equation in Cartesian coordinates:
$$r = \frac{4}{1 - 2\cos(\theta)}$$

$\Rightarrow$
$$r - 2r\cos(\theta) = 4$$

$\Rightarrow$
$$r = 4 + 2x \Rightarrow \sqrt{x^2 + y^2} = 4 + 2x$$

$\Rightarrow$
$$x^2 + y^2 = 16 + 16x + 4x^2 \Rightarrow y^2 - 3x^2 - 16x = 16.$$

## 10.6. CONIC SECTIONS IN POLAR COORDINATES

We complete the square:

$$y^2 - 3x^2 - 16x = 16 \Rightarrow y^2 - 3\left(x + \frac{8}{3}\right)^2 + \frac{64}{3} = 16$$

$$\Rightarrow$$

$$y^2 - 3\left(x + \frac{8}{3}\right)^2 = -\frac{16}{3} \Rightarrow 3\left(x + \frac{8}{3}\right)^2 - y^2 = \frac{16}{3}.$$

This equation confirms that the curve is a hyperbola. The hyperbola is centered at $(-8/3, 0)$. Since

$$3\left(x + \frac{8}{3}\right)^2 = \frac{16}{3} + y^2 \Rightarrow \frac{y^2}{3\left(x + \frac{8}{3}\right)^2} = 1 - \frac{\frac{16}{3}}{3\left(x + \frac{8}{3}\right)^2},$$

we have

$$\lim_{x \to \pm\infty} \frac{y^2}{3\left(x + \frac{8}{3}\right)^2} = 1 \Rightarrow \lim_{x \to \pm\infty} \left(\pm \frac{y}{\sqrt{3}\left(x + \frac{8}{3}\right)}\right) = 1.$$

Therefore, the lines

$$y = \pm\sqrt{3}\left(x + \frac{8}{3}\right)$$

are asymptotic to the hyperbola at $\pm\infty$. Figure 8 is consistent with this observation. □

**Example 4** Let $C$ be the ellipse of Example 2 so that $C$ is the graph of the equation

$$r = \frac{4}{3 - \sin(\theta)}$$

where $r$ and $\theta$ are polar coordinates and $0 \le \theta \le 2\pi$.
a) Express the arc length of the ellipse as an integral.
b) Make use of your computational utility in order to approximate the integral of part a).

**Solution**

a) As we discussed in Section 10.4, we can express the arc length of a curve in polar coordinates as

$$\int_0^{2\pi} \sqrt{r^2(\theta) + \left(\frac{dr}{d\theta}\right)^2}\, d\theta.$$

We have

$$\frac{dr}{d\theta} = \frac{d}{d\theta}\left(\frac{4}{3 - \sin(\theta)}\right) = -4 \frac{\frac{d}{d\theta}(3 - \sin(\theta))}{(3 - \sin(\theta))^2} = \frac{4\cos(\theta)}{(3 - \sin(\theta))^2}.$$

Therefore the arc length of the ellipse is

$$\int_0^{2\pi} \sqrt{\left(\frac{4}{3 - \sin(\theta)}\right)^2 + \left(\frac{4\cos(\theta)}{(3 - \sin(\theta))^2}\right)^2}\, d\theta = \int_0^{2\pi} \sqrt{\frac{16}{(3 - \sin(\theta))^2} + \frac{16\cos^2(\theta)}{(3 - \sin(\theta))^4}}\, d\theta$$

b) With the help of a computational utility,

$$\int_0^{2\pi} \sqrt{\frac{16}{(3 - \sin(\theta))^2} + \frac{16\cos^2(\theta)}{(3 - \sin(\theta))^4}}\, d\theta \cong 9.15726.$$

□

**Example 5** Let $C$ be the hyperbola that is the graph of

$$r = \frac{4}{1 - 2\cos(\theta)}$$

as in Example 3.

a) Figure 9 shows the region $G$ that is swept by $(r, \theta)$ as $\theta$ varies from $\pi/2$ to $\pi$. Express the area of $G$ as an integral.
b) Make use of your computational utility in order to approximate the integral of part a).

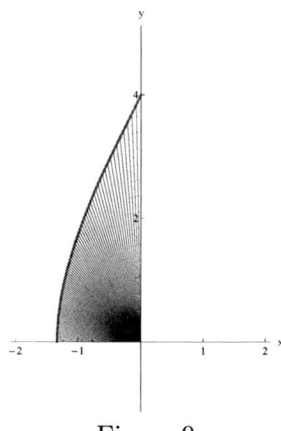

Figure 9

**Solution**
a) As we discussed in Section 10.3, we can express the area of $G$ as

$$\frac{1}{2}\int_{\pi/2}^{\pi} r^2(\theta) \, d\theta = \frac{1}{2}\int_{\pi/2}^{\pi} \left(\frac{4}{1 - 2\cos(\theta)}\right)^2 d\theta = 8\int_{\pi/2}^{\pi} \frac{1}{(1 - 2\cos(\theta))^2} d\theta$$

b) With the help of a computational utility,

$$8\int_{\pi/2}^{\pi} \frac{1}{(1 - 2\cos(\theta))^2} d\theta \cong 3.30574.$$

□

## Problems

In problems 1-6
a) Identify the conic section that is the graph of the given equation $r = f(\theta)$ in polar coordinates,
b) Sketch the graph of $r = f(\theta)$, where $0 \leq \theta \leq 2\pi$, in the $\theta r$-Cartesian coordinate plane. Indicate the points at which $f(\theta) = 0$, the points at which the graph has vertical asymptotes,
c) Sketch the graph of $r = f(\theta)$ in the $xy$-plane (the polar graph of $f$).

1.                                                    2.

$$r = \frac{4}{2 - \sin(\theta)} \qquad\qquad r = \frac{5}{3 - 2\cos(\theta)}$$

3. $$r = \frac{3}{1+\cos(\theta)}$$

4. $$r = \frac{2}{1-\sin(\theta)}$$

5. $$r = \frac{4}{\sqrt{2}-2\cos(\theta)}$$

6. $$r = \frac{4}{1-2\sin(\theta)}$$

# Appendix H

# Taylor's Formula for the Remainder

**Lemma 1 (The Generalized Mean Value Theorem for Integrals)** Assume that $f$ and $g$ are continuous on $[a,b]$ and $g(x) \geq 0$ for each $x \in [a,b]$ or $g(x) \leq 0$ for each $x \in [a,b]$. Then there exists $c \in [a,b]$ such that

$$\int_a^b f(x) g(x) \, dx = f(c) \int_a^b g(x) \, dx.$$

**Proof**

Since $f$ and $g$ are continuous on $[a,b]$, so is their product. Thus, $fg$ is integrable on $[a,b]$. If $g(x) = 0$ for each $x \in [a,b]$, then

$$\int_a^b g(x) \, dx = 0,$$

so that the equality

$$\int_a^b f(x) g(x) \, dx = f(c) \int_a^b g(x) \, dx$$

is satisfied for any choice of $c \in [a,b]$ ($0 = 0$).

Now assume that $g(x_0) > 0$ for some $x_0 \in [a,b]$. Then

$$\int_a^b g(x) \, dx > 0,$$

thanks to the continuity of $g$. We will provide the justification for $x_0 \in (a,b)$ (you can provide the modification of the argument if $x_0$ is an endpoint). There exists $\delta > 0$ such that

$$x \in (x_0 - \delta, x_0 + \delta) \Rightarrow |g(x) - g(x_0)| < \frac{1}{2} g(x_0)$$

$$\Rightarrow g(x) - g(x_0) > -\frac{1}{2} g(x_0)$$

$$\Rightarrow g(x) > \frac{1}{2} g(x_0).$$

Thus,

$$\int_a^b g(x) \, dx \geq \int_{x_0-\delta}^{x_0+\delta} g(x) \, dx > \int_{x_0-\delta}^{x_0+\delta} \frac{1}{2} g(x_0) \, dx = g(x_0) \delta > 0.$$

Let $M$ be the maximum of $f$ on $[a,b]$ and let $m$ be its minimum on $[a,b]$. We have

$$m\int_a^b g(x)\,dx \le \int_a^b f(x)g(x)\,dx \le M\int_a^b g(x)\,dx$$

since $g(x) \ge 0$ for each $x \in [a,b]$. Since $\int_a^b g(x)\,dx > 0$, we conclude that

$$m \le \frac{\int_a^b f(x)g(x)\,dx}{\int_a^b g(x)\,dx} \le M.$$

Since $f$ is continuous on $[a,b]$ there exists $c \in [a,b]$ such that

$$f(c) = \frac{\int_a^b f(x)g(x)\,dx}{\int_a^b g(x)\,dx},$$

thanks to the intermediate value theorem. Thus,

$$\int_a^b f(x)g(x)\,dx = f(c)\int_a^b g(x)\,dx.$$

■

**Theorem 2 (Taylor's formula for the remainder) Assume that $f$ has continuous derivatives up to order $n+1$ in an open interval $J$ containing the point $c$. If $x \in J$ there exists a point $c_n(x)$ between $x$ and $c$ such that**

$$f(x) = P_{c,n}(x) + R_{c,n}(x)$$
$$= f(c) + f'(c)(x-c) + \frac{1}{2!}f''(c)(x-c)^2 + \cdots + \frac{1}{n!}f^{(n)}(c)(x-c)^n + R_{c,n}(x),$$

where

$$R_{c,n}(x) = \frac{1}{(n+1)!}f^{(n+1)}(c_n(x))(x-c)^{n+1}.$$

The expression

$$P_{c,n}(x) = f(c) + f'(c)(x-c) + \frac{1}{2!}f''(c)(x-c)^2 + \cdots + \frac{1}{n!}f^{(n)}(c)(x-c)^n$$

is referred to as **the Taylor polynomial for $f$ based at $c$**. Thus, the Taylor polynomial for $f$ based at $c$ is the $(n+1)$st partial sum of the Taylor series for $f$ based at $c$.
The expression

$$R_{c,n}(x) = f(x) - P_{c,n}(x)$$

represents the error in the approximation of $f(x)$ by $P_{c,n}(x)$, and it is referred to as **the remainder** in such an approximation.
In almost all the examples, we will be interested in Maclaurin series, so that $c = 0$. In such a case we will use the notations,

$$P_n(x) = f(0) + f'(0)x + \frac{1}{2!}f''(0)x^2 + \cdots + \frac{1}{n!}f^{(n)}(0)x^n,$$
$$R_n(x) = f(x) - P_n(x),$$

**The proof of Theorem 2**

The starting point is the Fundamental Theorem of Calculus:

$$f(x) - f(c) = \int_c^x f'(t)dt.$$

Let us apply integration by parts to the integral by setting $u = f'(t)$ and $dv = dt$. Ordinarily, we would set $v = t$. In the present case, we will be somewhat devious, and set $v = t - x$. We still have

$$dv = \frac{dv}{dt}dt = \left(\frac{d}{dt}(t-x)\right)dt = dt.$$

Thus,

$$\begin{aligned}
f(x) - f(c) &= \int_c^x f'(t)dt \\
&= \int_c^x u\,dv \\
&= (uv|_c^x) - \int_c^x v\,du \\
&= (f'(x)(x-x) - f'(c)(c-x)) - \int_c^x (t-x)f''(t)\,dt \\
&= f'(c)(x-c) + \int_c^x f''(t)(x-t)\,dt.
\end{aligned}$$

Therefore,

$$\begin{aligned}
f(x) &= f(c) + f'(c)(x-c) + \int_c^x f''(t)(x-t)dt \\
&= P_{c,1}(x) + \int_c^x f''(t)(x-t)dt.
\end{aligned}$$

Let's focus our attention on the integral, and apply integration by parts, setting $u = f''(t)$ and $dv = (x-t)\,dt$. Then,

$$v = \int dv = \int (x-t)\,dt = -\frac{1}{2}(x-t)^2.$$

Thus,

$$\begin{aligned}
\int_c^x f''(t)(x-t)dt &= \int_c^x u\,dv \\
&= (uv|_c^x) - \int_c^x v\,du \\
&= \left(-\frac{1}{2}f''(t)(x-t)^2\bigg|_c^x\right) - \int_c^x -\frac{1}{2}(x-t)^2 f^{(3)}(t)dt \\
&= \left(-\frac{1}{2}f''(x)(x-x)^2 + \frac{1}{2}f''(c)(x-c)^2\right) + \int_c^x \frac{1}{2}(x-t)^2 f^{(3)}(t)dt \\
&= \frac{1}{2}f''(c)(x-c)^2 + \int_c^x \frac{1}{2}(x-t)^2 f^{(3)}(t)dt.
\end{aligned}$$

Therefore,

$$f(x) = f(c) + f'(c)(x-c) + \int_c^x f''(t)(x-t)dt$$
$$= f(c) + f'(c)(x-c) + \frac{1}{2}f''(c)(x-c)^2 + \int_c^x \frac{1}{2}(x-t)^2 f^{(3)}(t)dt$$
$$= P_{c,2}(x) + \int_c^x \frac{1}{2}(x-t)^2 f^{(3)}(t)dt.$$

Let us integrate by parts again, setting

$$u = f^{(3)}(t) \text{ and } dv = \frac{1}{2}(x-t)^2 dt.$$

Thus,

$$v = \int \frac{1}{2}(x-t)^2 dt = -\frac{1}{(2)(3)}(x-t)^3 = -\frac{1}{3!}(x-t)^3.$$

Therefore,

$$\int_c^x f^{(3)}(t)\frac{1}{2}(x-t)^2 dt = \int_c^x u\,dv$$
$$= \left(uv\Big|_c^x\right) - \int_c^x v\,du$$
$$= \left(-\frac{1}{3!}f^{(3)}(t)(x-t)^3\Big|_c^x\right) - \int_c^x -\frac{1}{3!}(x-t)^3 f^{(4)}(t)dt$$
$$= \frac{1}{3!}f^{(3)}(c)(x-c)^3 + \int_c^x \frac{1}{3!}(x-t)^3 f^{(4)}(t)dt.$$

Thus,

$$f(x) = f(c) + f'(c)(x-c) + \frac{1}{2}f''(c)(x-c)^2 + \int_c^x \frac{1}{2}(x-t)^2 f^{(3)}(t)dt$$
$$= f(c) + f'(c)(x-c) + \frac{1}{2}f''(c)(x-c)^2 + \frac{1}{3!}f^{(3)}(c)(x-c)^3$$
$$+ \int_c^x \frac{1}{3!}(x-t)^3 f^{(4)}(t)dt$$
$$= P_{c,3}(x) + \int_c^x \frac{1}{3!}(x-t)^3 f^{(4)}(t)dt.$$

The general pattern is emerging:

$$f(x) = f(c) + f'(c)(x-c) + \cdots + \frac{1}{n!}f^{(n)}(c)(x-c)^n + R_{c,n}(x)$$
$$= P_{c,n}(x) + R_{c,n}(x),$$

where

$$R_{c,n}(x) = \int_c^x \frac{1}{n!}f^{(n+1)}(t)(x-t)^n dt.$$

It is not difficult to supply an inductive proof.:Let us assume that the statement is valid for $n$. We set

$$u = f^{(n+1)}(t) \text{ and } dv = \frac{1}{n!}(x-t)^n dt.$$

Then,
$$v = \int \frac{1}{n!}(x-t)^n \, dt = -\frac{1}{n!(n+1)}(x-t)^{n+1} = -\frac{1}{(n+1)!}(x-t)^{n+1},$$
so that
$$\int_c^x f^{(n+1)}(t) \frac{1}{n!}(x-t)^n \, dt = \int_c^x u \, dv$$
$$= (uv|_c^x) - \int_c^x v \, du$$
$$= \left(-\frac{1}{(n+1)!} f^{(n+1)}(t)(x-t)^{n+1}\bigg|_c^x\right)$$
$$- \int_c^x -\frac{1}{(n+1)!}(x-t)^{n+1} f^{(n+2)}(t) \, dt$$
$$= \frac{1}{(n+1)!} f^{(n+1)}(c)(x-c)^{n+1} + \int_c^x \frac{1}{(n+1)!}(x-t)^{n+1} f^{(n+2)}(t) \, dt.$$

Therefore,
$$f(x) = f(c) + f'(c)(x-c) + \cdots + \frac{1}{n!} f^{(n)}(c)(x-c)^n$$
$$+ \int_c^x f^{(n+1)}(t) \frac{1}{n!}(x-t)^n \, dt$$
$$= f(c) + f'(c)(x-c) + \cdots + \frac{1}{n!} f^{(n)}(c)(x-c)^n$$
$$+ \frac{1}{(n+1)!} f^{(n+1)}(c)(x-c)^{n+1} + \int_c^x \frac{1}{(n+1)!}(x-t)^{n+1} f^{(n+2)}(t) \, dt$$
$$= P_{c,n+1}(x) + R_{c,n+1}(x),$$

where
$$R_{c,n+1}(x) = \int_c^x \frac{1}{(n+1)!}(x-t)^{n+1} f^{(n+2)}(t) \, dt.$$

This completes the induction.

We have shown that
$$f(x) = P_{c,n}(x) + R_{c,n}(x),$$
where
$$R_{c,n}(x) = \int_c^x \frac{1}{n!} f^{(n+1)}(t)(x-t)^n \, dt.$$

The above expression for the remainder in the approximation of $f(x)$ by the Taylor polynomial of order $n$ based at $c$ for $f$ is referred to as the integral form of the remainder. We can obtain the form of the remainder as announced in the statement of the theorem by appealing to the generalized mean value theorem for integrals. If $x > c$,
$$(x-t)^n \geq 0 \text{ for } t \in [c, x].$$

Therefore, there exists $c_x \in [c, x]$ such that
$$\int_c^x \frac{1}{n!} f^{(n+1)}(t)(x-t)^n \, dt = \frac{1}{n!} f^{(n+1)}(c_x) \int_c^x (x-t)^n \, dt$$
$$= \frac{1}{n!} f^{(n+1)}(c_x) \left(\frac{1}{n+1}(x-c)^{n+1}\right)$$
$$= \frac{1}{(n+1)!} f^{(n+1)}(c_x)(x-c)^{n+1}.$$

Similarly, if $x < c$, $(x-t)^n = (-1)^n (t-x)^n$ does not change sign on the interval $[x, c]$, so that

$$\int_c^x \frac{1}{n!} f^{(n+1)}(t) (x-t)^n \, dt = -\int_x^c \frac{1}{n!} f^{(n+1)}(t) (x-t)^n \, dt$$
$$= -\frac{1}{n!} f^{(n+1)}(c_x) \int_x^c (x-t)^n \, dt$$
$$= -\frac{1}{n!} f^{(n+1)}(c_x) \left( -\frac{1}{n+1} (x-c)^{n+1} \right)$$
$$= \frac{1}{(n+1)!} f^{(n+1)}(c_x) (x-c)^{n+1}$$

for some $c_x$ in $[x, c]$. ∎

# Appendix I

# Answers to Some Problems

**Answers of Some Problems of Section 6.1**

**Remark: An arbitrary constant can be added to an antiderivative.**

**1.**
$$\int xe^{2x}\,dx = \frac{1}{2}xe^{2x} - \frac{1}{4}e^{2x}$$

**3.**
$$\int x\cos\left(\frac{1}{2}x\right)dx = 2x\sin(x/2) + 4\cos(x/2)$$

**5.**
$$\int x\sinh(x)\,dx = x\cosh(x) - \sinh(x)$$

**7.**
$$\int \arccos(x)\,dx = x\arccos(x) + \int \frac{x}{\sqrt{1-x^2}}dx$$

**9.**
$$\int \ln(x)\,x^2\,dx = \frac{1}{3}x^3\ln(x) - \frac{1}{9}x^3$$

**11.**
$$\int e^{-x}\sin(x)\,dx = -e^{-x}\cos(x) - \int e^{-x}\cos(x)\,dx$$

**13.**
$$\int e^{-x}\cos\left(\frac{x}{2}\right)dx = \frac{2}{5}e^{-x}\sin\left(\frac{x}{2}\right) - \frac{4}{5}e^{-x}\cos\left(\frac{x}{2}\right)$$

**15.**
$$\int_e^{e^2} x^2\ln^2(x)\,dx = \frac{26}{27}e^6 - \frac{5}{27}e^3$$

**17.**
$$\int_{1/4}^{1/2} x\sin(\pi x)\,dx = \frac{\sqrt{2}}{8\pi} + \frac{1}{\pi^2} - \frac{\sqrt{2}}{2\pi^2}$$

**Answers of Some Problems of Section 6.2**

**1.**
$$\int \frac{1}{4x^2 + 8x + 5}dx = \frac{1}{2}\arctan(2x+2)$$

**3.**
$$\int \frac{1}{16x^2 - 96x + 153}dx = \frac{1}{12}\arctan\left(\frac{4}{3}x - 4\right)$$

**5.**

a)
$$\frac{-3x+16}{x^2+x-12} = -\frac{4}{x+4} + \frac{1}{x-3}$$

b)
$$\int \frac{-3x+16}{x^2+x-12} dx = -4\ln(|x+4|) + \ln(|x-3|).$$

**7.**
a)
$$\frac{x+5}{x^2+5x+6} = -\frac{2}{x+3} + \frac{3}{x+2}$$

b)
$$\int \frac{x+5}{x^2+5x+6} dx = -2\ln(|x+3|) + 3\ln(|x+2|)$$

**9.**
a)
$$\frac{2x^3-5x^2-24x+13}{x^2-x-12} = 2x-3 + \frac{2}{x+3} - \frac{5}{x-4}$$

b)
$$\int \frac{2x^3-5x^2-24x+13}{x^2-x-12} dx = x^2 - 3x + 2\ln(|x+3|) - 5\ln(|x-4|)$$

**11.**
a)
$$\frac{x-5}{(x-2)^2} = \frac{1}{x-2} - \frac{3}{(x-2)^2}$$

b)
$$\int \frac{x-5}{(x-2)^2} dx = \ln|x-2| - 3\left(-\frac{1}{x-2}\right) = \ln|x-2| + \frac{3}{x-2}$$

**13.**
a) The expression
$$\frac{x-1}{x^2+4}$$
is its own partial fraction decomposition.

b)
$$\int \frac{x-1}{x^2+4} dx = \frac{1}{2}\ln(x^2+4) - \frac{1}{2}\arctan\left(\frac{x}{2}\right).$$

**15.**
a)
$$\frac{4x^2+3x+5}{(x+2)(x^2+1)} = \frac{3}{x+2} + \frac{x+1}{x^2+1}$$

b)
$$\int \frac{4x^2+3x+5}{(x+2)(x^2+1)} dx = 3\ln(|x+2|) + \frac{1}{2}\ln(x^2+1) + \arctan(x)$$

**17.**
a)
$$\frac{6}{(x+4)(x-2)} = -\frac{1}{x+4} + \frac{1}{x-2}$$

b)
$$\int \frac{6}{(x+4)(x-2)} dx = -\ln(|x+4|) + \ln(|x-2|).$$

c)
$$\int_{-3}^{1} \frac{6}{(x+4)(x-2)} dx = -2\ln(5)$$

## Answers of Some Problems of Section 6.3

**1.**
$$\int \cos^3(x) \sin^2(x) \, dx = \frac{1}{3} \sin^3(x) - \frac{1}{5} \sin^5(x)$$

**7.**
$$\int \cos^2(x) \, dx = \frac{1}{2} \cos(x) \sin(x) + \frac{1}{2} x$$

**3.**
$$\int \cosh^3(x) \sinh^2(x) \, dx = \frac{1}{5} \sinh^5(x) + \frac{1}{3} \sinh^3(x)$$

**9.**
$$\int \sinh^2(x) = \frac{1}{2} \cosh(x) \sinh(x) - \frac{1}{2} x.$$

**5.**
$$\int \cos^2(x) \, dx = \frac{x}{2} - \frac{1}{4} \sin(2x)$$

**11.**
$$\sin^6(x) \, dx = -\frac{1}{6} \sin^5(x) \cos(x) + \frac{5}{6} \int \sin^4(x) \, dx$$

**13.**

a)
$$\int \csc(x) \, dx = \ln(|\tan(x/2)|)$$

b) According to Mathematica 7,
$$\int \csc(x) \, dx = -\ln(2\cos(x/2)) + \ln(2\sin(x/2)).$$

According to Maple 12,
$$\int \csc(x) \, dx = -\ln(\csc(x) + \cot(x))$$

**15.**

a)
$$\int \text{sech}(x) \, dx = 2 \arctan(e^x)$$

b) According to Mathematica 7,
$$\int \text{sech}(x) \, dx = 2 \arctan(\tanh(x/2))$$

According to Maple 12,
$$\int \text{sech}(x) \, dx = \arctan(\sinh(x)).$$

## Answers to Some Problems of Section 6.4

**1.**
a) We have
$$\int \sqrt{9-4x^2}\,dx = \frac{9}{4}\arcsin\left(\frac{2}{3}x\right) + \frac{1}{2}x\sqrt{9-4x^2}$$

b)
$$\int_0^{3/4} \sqrt{9-4x^2}\,dx = \frac{3}{8}\pi + \frac{9}{16}\sqrt{3}$$

**3.** The area of the ellipse is $6\pi$.

**5.**
a)
$$\int \sqrt{9x^2+16}\,dx = \frac{1}{2}x\sqrt{9x^2+16} + \frac{8}{3}\operatorname{arcsinh}\left(\frac{3}{4}x\right)$$

or
$$\int \sqrt{9x^2+16}\,dx = \frac{1}{2}x\sqrt{9x^2+16} + \frac{8}{3}\ln\left(3x + \sqrt{9x^2+16}\right)$$

**7.**
a)
$$\int_{2\sinh(2)}^{2\sinh(3)} \sqrt{x^2+4}\,dx = \int_2^3 4\cosh^2(u)\,du.$$

b)
$$\int_2^3 4\cosh^2(u)\,du = \int_2^3 4\cosh^2(u)\,du.$$

$$\begin{aligned}
\int_{2\sinh(2)}^{2\sinh(3)} \sqrt{x^2+4}\,dx &= \int_2^3 4\cosh^2(u)\,du \\
&= 4\left(\frac{1}{2}\cosh(u)\sinh(u) + \frac{1}{2}u\Big|_2^3\right) \\
&= 2(\cosh(3)\sinh(3) + 3) - 2(\cosh(2)\sinh(2) + 2) \\
&= 2(\cosh(3)\sinh(3) - \cosh(2)\sinh(2)) + 2.
\end{aligned}$$

c)
$$\int_{2\sinh(2)}^{2\sinh(3)} \sqrt{x^2+4}\,dx \cong 176.423$$

**9.**
$$\int \sqrt{-x^2+2x+3}\,dx = 2\arcsin\left(\frac{x-1}{2}\right) + \frac{(x-1)}{2}\sqrt{-x^2+2x+3}$$

**11.**
a) The area is
$$2\sqrt{3} - \ln\left(2+\sqrt{3}\right)$$

b) $2.14714$

# Answers of Some Problems of Section 6.5

**1.**

| $n$ | $l_n$ | $\|l_n - \pi/4\|$ |
|-----|-------|-------------------|
| 100 | 0.793252 | $8 \times 10^{-3}$ |
| 200 | 0.789325 | $4 \times 10^{-3}$ |
| 300 | 0.788016 | $2.6 \times 10^{-3}$ |
| 400 | 0.787362 | $2 \times 10^{-3}$ |

**3.**

| $n$ | $m_n$ | $\|m_n - (e^{-1} - e^{-2})\|$ |
|-----|-------|-------------------------------|
| 4 | 0.231940 | $6 \times 10^{-4}$ |
| 8 | 0.23239 | $1.5 \times 10^{-4}$ |
| 16 | 0.232506 | $3.8 \times 10^{-5}$ |
| 32 | 0.232535 | $9.5 \times 10^{-6}$ |

**5.**

| $n$ | $T_n$ | $\|T_n - \frac{\pi}{6}\|$ |
|-----|-------|---------------------------|
| 4 | 0.524596 | $10^{-3}$ |
| 8 | 0.523849 | $2.5 \times 10^{-4}$ |
| 16 | 0.523661 | $6.2 \times 10^{-5}$ |
| 32 | 0.523614 | $1.5 \times 10^{-5}$ |

**7.**

| $n$ | $S_n$ | $\|S_n - \int_2^4 \sqrt{1+x^2}\,dx\|$ |
|-----|-------|---------------------------------------|
| 2 | 6.33571 | $3.3 \times 10^{-5}$ |
| 4 | 6.33568 | $3.9 \times 10^{-6}$ |

# Answers to Some Problems of Sections 6.6

**1.**
$$\int_0^\infty \frac{x}{(x^2+4)^2}\,dx = \frac{1}{8}.$$

**3.**
$$\int_6^\infty \frac{1}{(x-4)^2}\,dx = \frac{1}{2}.$$

**5.**
$$\int_{\sqrt{3}/2}^\infty \frac{1}{4x^2+9}\,dx = \frac{\pi}{18}.$$

**7.**
$$\int_0^\infty x^2 e^{-x}\,dx = 2.$$

**9.** The given improper integral diverges.

**11.** The given improper integral diverges.

**13.**
$$\int_{-\infty}^\infty \frac{1}{x^2-8x+17}\,dx = \pi.$$

**15.**

a) The given integral is an improper integral since
$$\lim_{x\to 2^-} \frac{1}{\sqrt{2-x}} = +\infty.$$

b)
$$\int_0^2 \frac{1}{\sqrt{2-x}}\,dx = 2\sqrt{2}.$$

**17.**

a) The given integral is an improper integral since
$$\lim_{x\to 1^-} \frac{x}{1-x^2} = +\infty.$$

b) The given improper integral diverges.

**19.**

a) The given integral is an improper integral since
$$\lim_{x\to 2^+} \frac{1}{(2-x)^{1/3}} = -\infty \quad \text{and} \quad \lim_{x\to 2^-} \frac{1}{(2-x)^{1/3}} = +\infty.$$

b)
$$\int_0^3 \frac{1}{(2-x)^{1/3}} dx = \frac{3}{2}\left(2^{2/3} - 1\right).$$

**21.**
a) The given integral is improper since
$$\lim_{x \to 1} \frac{1}{(x-1)^{4/5}} = +\infty,$$

and $1 \in [-1, 2]$.
b)
$$\int_{-1}^2 \frac{1}{(x-1)^{4/5}} dx = 5\left(2^{1/5} + 1\right).$$

## Answers to Some Problems of Section 6.7

**1.** The improper integral converges.

**3.** The improper integral diverges.

**5.** The improper integral converges.

**7.** The integral is improper since
$$\lim_{x \to 4^-} \frac{x^{1/3}}{(4-x)^2} = +\infty.$$

The improper integral diverges.

## Answers to Some Problems of Section 7.1

**1.**
$$\frac{1}{3}\pi r^2 h$$

**3.**
$$\pi\left(\frac{\pi}{2} + 1\right)$$

**5.**
$$\frac{\pi^2}{2}$$

**7.**
$$8\pi\left(\frac{1}{3}\pi - \frac{1}{4}\sqrt{3}\right)$$

**9.**
$$\frac{\pi}{2}$$

**11.**
$$\pi\left(e^{-1} - e^{-4}\right)$$

**13.**
$$2\pi\left(4 - \sqrt{7}\right)$$

**15.**
$$2\pi(6\ln(2) - 2)$$

**17.**
$$2\pi\left(\frac{\sqrt{2}}{4}\pi - 1\right)$$

## Answers to Some Problems of Section 7.2

**1.**
a) The length is
$$\int_3^4 \sqrt{\frac{2x^2 - 4}{x^2 - 4}} dx$$

b)
$$\int_3^4 \sqrt{\frac{2x^2 - 4}{x^2 - 4}} dx \cong 1.58403$$

c) According to Mathematica 7,
$$\int \sqrt{\frac{2x^2 - 4}{x^2 - 4}} dx = 2\text{EllipticE}\left(\arcsin\left(\frac{x}{2}\right), 2\right).$$

The special function EllipticE is beyond the scope of this course. Maple's response is more complicated and involves such elliptic functions.

**3.**

a) The length is
$$\int_0^{\pi/2} \sqrt{1+\cos^2(x)}\,dx$$

b)
$$\int_0^{\pi/2} \sqrt{1+\cos^2(x)}\,dx \cong 1.9101$$

c) According to Mathematica 7,
$$\int \sqrt{1+\cos^2(x)}\,dx = \sqrt{2}\text{EllipticE}(x, 1/2).$$

Maple's response is more complicated and involves such elliptic functions.

**5.** The length is
$$\int_8^{27} \frac{1}{3x^{1/3}} \sqrt{9x^{2/3}+4}\,dx = \frac{1}{27}\left(85^{3/2} - 40^{3/2}\right)$$

**7.** The required area is
$$2\pi \int_1^2 (x^2+4)\sqrt{1+4x^2}\,dx$$

**9.** The area is
$$\int_1^{25} 2\pi\sqrt{5+4x}\,dx = \frac{\pi}{3}\left(105^{3/2}-27\right)$$

**11.** The required area is
$$2\pi \int_1^2 x\sqrt{1+4x^2}\,dx$$

**13.** The area is
$$\frac{2\pi}{3}\int_1^2 x^{1/3}\sqrt{9x^{4/3}+1}\,dx = \frac{\pi}{27}\left(9\left(2^{4/3}\right)+1\right)^{3/2} - \frac{\pi}{27}\left(10^{3/2}\right)$$

## Answers to Some Problems of Section 7.3

**1.** The mass of the rod is
$$\frac{\pi}{4} \cong 0.785\,398$$

Tthe center of mass is at
$$\frac{1}{4\pi}\left(12\pi + \pi^2 - 4\right) \cong 3.467\,09$$

**3.** The work done by the force is $-2$.

**5.** The work done against gravity is 367.5 joules.

## Answers to Some Problems of Section 7.4

**1.**

a) The density function is

$$f(x) = \begin{cases} 0 & \text{if } x < 0, \\ \frac{\ln(10/9)}{2} e^{-x \ln(10/9)/2} & \text{if } x \geq 0, \end{cases}$$

b)
$$0.853\,815 \cong 85\%$$

**3.**

a) The density function is
$$\frac{1}{16\sqrt{2\pi}} e^{-(x-105)^2/512}.$$

b)
$$0.174\,251 \cong 17\%$$

## Answers to Some Problems of Section 8.1

**1.**

a)
$$y(t) = Ce^{-t/4}; \quad y(t) = 6e^{-t/4}; \quad \lim_{t\to\infty} y(t) = 0.$$

b)

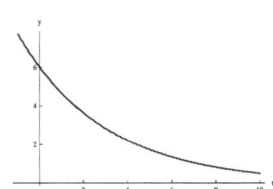

**3.**

a) $y = 50$ is the steady-state solution of the differential equation.

b)
$$y(t) = 50 + Ce^{-t/10}; \quad \lim_{t\to\infty} y(t) = 50.$$

c) The solutions corresponding to $y_0 = 20, 40, 60, 80$ are

$$50 - 30e^{-t//10}, \ 50 - 10e^{-t//10}, \ 50 + 10^{-t//10}, \ 50 + 30^{-t//10}.$$

d)

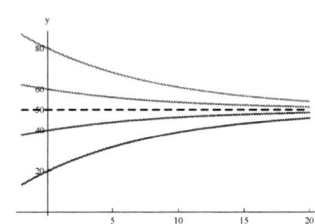

**5.**

a)
$$y(t) = -5t - 25 + Ce^{t/5}$$
b)
$$y(t) = -5t - 25 + 28e^{t/5}$$

**7.**
a)
$$y(t) = 1 + Ce^{t^2/2}$$
b)
$$y(t) = 1 + 3e^{t^2/2}$$

**9.**
a)
$$y(t) = \frac{1}{2}t^3 + Ct$$
b)
$$y(t) = \frac{1}{2}t^3 + \frac{7}{2}t$$

**11.**
a)
$$y(t) = \frac{4}{257}\sin(4t) - \frac{64}{257}\cos(4t) + \frac{578}{257}e^{-t/4}$$
b)
$$y_p(t) = \frac{4}{257}\sin(4t) - \frac{64}{257}\cos(4t) \text{ and } y_{tran}(t) = \frac{578}{257}e^{-t/4},$$
c)

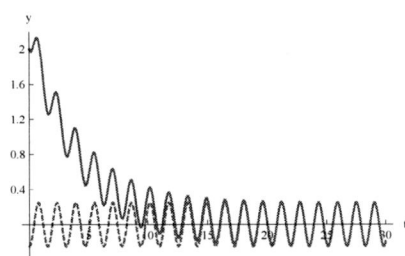

The dashed is the graph of $y_p$. We can hardly distinguish between the graph of the solution and the graph of $y_p$ past $t = 15$. This is consistent with the fact that

$$\lim_{t \to \infty}(y(t) - y_p(t)) = \lim_{t \to \infty} y_{tran}(t) = 0.$$

**13.**
$$y(t) = e^{-t/3}\int_1^t \frac{1}{\sqrt{u^2+1}}e^{u/3}\,du + 10e^{1/3}e^{-t/3}$$

**15.**
a)
$$y(t) = e^{-t^2/2}\int_0^t \sin(u)e^{u^2/2}\,du + 3e^{-t^2/2}$$

b)
$$y(1) \cong 2.17876$$

c)

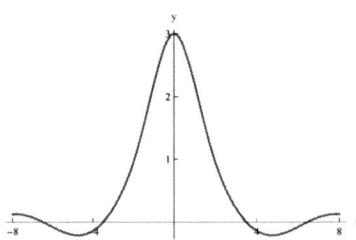

## Answers to Some Problems of Section 8.2

**1.**
a)
$$v(t) = 0.98\left(1 - e^{-10t}\right) \text{ (meters/second)}$$
The terminal velocity is 0.98 (meters/second).

b)

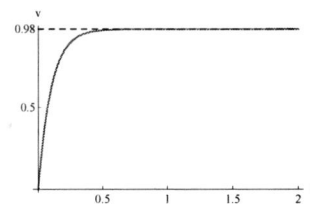

**3.**
a)
$$y(t) = 1500\left(1 - e^{-18\left(10^{-3}\right)t}\right), \quad \lim_{t \to \infty} y(t) = 1500 \text{ grams}$$

b)

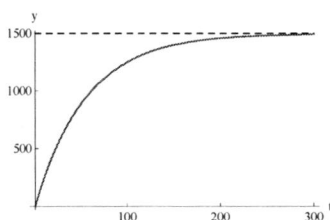

**5.**
a)
$$I(t) = \frac{150}{22\,501}\left(-60\cos(60t) + \frac{4}{10}\sin(60t)\right) + \frac{9000}{22\,501}e^{-4t/10}$$

b) The sinusoidal part of the current is
$$\frac{150}{22\,501}\left(-60\cos(60t) - \frac{4}{10}\sin(60t)\right),$$

and the transient part is

$$\frac{9000}{22\,501}e^{-4t/10}; \quad \lim_{t\to\infty}\frac{9000}{22\,501}e^{-4t/10} = 0$$

c)

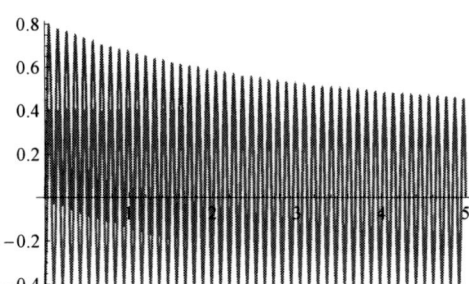

## Answers to Some Problems of Section 8.3

**1.**
a)
$$y(x) = \frac{1}{x+C}$$
b)
$$y(x) = \frac{1}{x+1}$$

**3.**
a)
$$y(x) = \sinh(x+C)$$
b)
$$y(x) = \sinh(x+\operatorname{arcsinh}(2))$$

**5.**
a)
$$y(x) = \tan(\arctan(x)+C)$$
b)
$$y(x) = \tan\left(\arctan(x)+\frac{\pi}{4}\right).$$

**7.**
a)
$$y(x) = -\frac{1}{\ln(1+x^2)+C}$$
b)
$$y(x) = -\frac{1}{\ln(1+x^2)+\frac{1}{2}}$$

**9.**
a)
$$y(t) = \frac{1}{\cos(t)-C}$$

b)
$$y(t) = \frac{1}{\cos(t) + \frac{5}{4}}.$$

**11.**
a) The steady-state solutions are $\pm 20$.
b) The solution $f_1$ of the initial value problem corresponding to $y(0) = 0$ is

$$f_1(t) = 20\left(\frac{e^{2t/5} - 1}{e^{2t/5} + 1}\right) = 20\left(\frac{1 - e^{-2t/5}}{1 + e^{-2t/5}}\right)$$

The domain of $f_1$ is the entire number line $\mathbb{R}$ since $e^{2t/5} + 1 > 0$ for all $t$.

$$\lim_{t \to +\infty} f_1(t) = 20, \quad \lim_{t \to -\infty} f_1(t) = -20$$

The line $y = 20$ is a horizontal asymptote for the graph of $f_1$ at $+\infty$ and the line $y = -20$ is a horizontal asymptote for the graph of $f_1$ at $-\infty$.
The solution $f_2$ of the initial value problem corresponding to $y(0) = 40$ is

$$f_2(t) = 20\left(\frac{3e^{2t/5} + 1}{3e^{2t/5} - 1}\right) = 20\left(\frac{3 + e^{-2t/5}}{3 - e^{-2t/5}}\right).$$

The domain of the solution is $\{t \in \mathbb{R} : t \neq -\frac{5}{2}\ln(3)\}$. The line $t = -5\ln(3)/2$ is a vertical asymptote for the graph of $f_2$. The line $y = 20$ is a horizontal asymptote for the graph of $f_2$ at $+\infty$ and the line $y = -20$ is a horizontal asymptote for the graph of $f_2$ at $-\infty$.
c) The graph of $f_1$ is shown below:

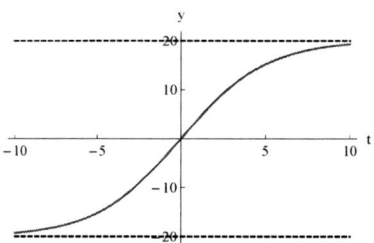

The picture is consistent with the facts that $\lim_{t \to \infty} f_1(t) = 20$ and $\lim_{t \to -\infty} f_1(t) = 20$.
The graph of $f_2$ is shown below:

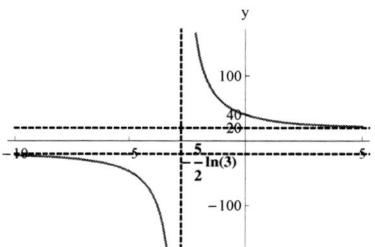

The picture is consistent with our observations about the vertical and horizontal asymptotes.

## Answers to Some Problems of Section 8.4

**1.**

a)
$$v(t) = \sqrt{\frac{9.8}{0.02}}\left(\frac{1-e^{-2\sqrt{9.8(0.02)}t}}{1+e^{-2\sqrt{9.8(0.02)}t}}\right) \cong 22.1359\left(\frac{1-e^{-0.885\,438\,t}}{1+e^{-0.885\,438\,t}}\right).$$

The terminal velocity is approximately 22.1 meters per second.

b)

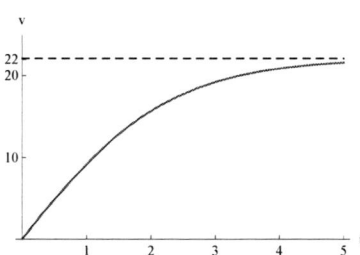

**3.**

a)
$$y(t) = \frac{4\,(10^7)}{1+3e^{-4(10^{-2})t}}.$$

The growth rate of the population peaks at $t_m$ when
$$y(t_m) = \frac{c}{2b} = \frac{4\,(10^{-2})}{2\,(10^{-9})} = 20 \times 10^6.$$

We have
$$t_m \cong 27.5 \text{ years}.$$

b) The smallest upper limit for $y(t)$ is
$$\frac{c}{b} = \frac{4\,(10^{-2})}{10^{-9}} = 40 \times 10^6.$$

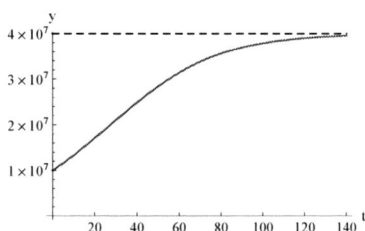

## Answers to Some Problems of Section 8.5

**1.**

a)
$$Y_{j+1} = \left(1 - \frac{1}{2}\Delta t\right)Y_j + \Delta t.$$

b) If $\Delta t = 0.1$,

$$Y_{j+1} = \left(1 - \frac{0.1}{2}\right) Y_j + 0.1 = 0.95 Y_j + 0.1, \quad j = 0, 1, 2, \ldots,$$

where $Y_0 = 6$.

The picture shows the points $(j\Delta t, Y_j)$, where $\Delta t = 0.1$ and $j = 0, 1, 2, \ldots, 40$, and the graph of $y(t) = 2 + 4e^{-t/2}$ on the interval $[0, 4]$

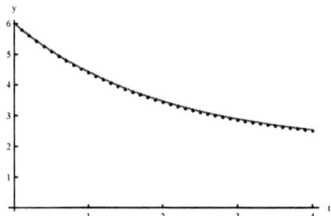

3.

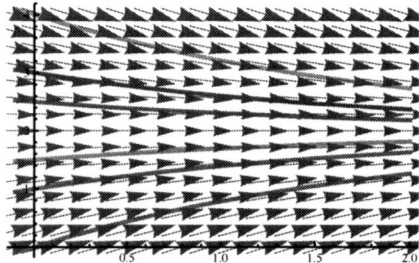

The picture of the slope field give some indication of the behavior of the solutions, even though it is more useful to plot several solutions with the help of our computational/graphing utility.

## Answers to Some Problems of Section 9.1

**1.**
a)

$$\begin{aligned} P_2(x) &= 1 - \frac{1}{2}x^2, \\ P_4(x) &= 1 - \frac{1}{2}x^2 + \frac{1}{4!}x^4, \\ P_6(x) &= 1 - \frac{1}{2}x^2 + \frac{1}{4!}x^4 - \frac{1}{6!}x^6. \end{aligned}$$

b)

$$P_{2n+1}(x) = P_{2n}(x) = \sum_{k=0}^{n} (-1)^k \frac{1}{(2k)!} x^{2k}, \quad n = 0, 1, 2, 3, \ldots$$

c)

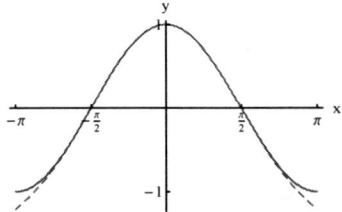

In the picture, the dashed curve is the graph of $P_6$. Since we can hardly distinguish between the graphs of cosine and $P_6$ on the interval $[-\pi/2, \pi/2]$, the picture indicates that the approximation of $\cos(x)$ by $P_6(x)$ is very accurate if $x$ is close to the basepoint 0.

**3.**

a)

$$P_1(x) = x$$
$$P_2(x) = x - \frac{1}{2}x^2$$
$$P_3(x) = x - \frac{1}{2}x^2 + \frac{1}{3}x^3$$
$$P_4(x) = x - \frac{1}{2}x^2 + \frac{1}{3}x^3 - \frac{1}{4}x^4$$

b)

$$P_n(x) = \sum_{k=1}^{n} \frac{(-1)^{k+1}}{k} x^k$$

**5.**

a)

$$P_3(x) = x - \frac{1}{3}x^3.$$

b)

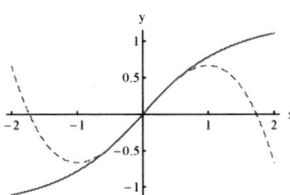

In the picture, the dashed curve is the graph of $P_3$. Since we can hardly distinguish between the graphs of arctangent and $P_3$ on the interval $[-0.5, 0.5]$, the picture indicates that the approximation of $\arctan(x)$ by $P_3(x)$ is very accurate if $x$ is close to the basepoint 0.

**7.**

$$P_3(x) = x + \frac{1}{6}x^3.$$

## Answers to Some Problems of Section 9.2

**1.** The required integer is 6.

**3.** The required integer is 2, with the corresponding polynomial $P_5(x)$.

# APPENDIX I. ANSWERS TO SOME PROBLEMS

**5.**

$$\lim_{n\to\infty}\left(1+\left(-\frac{2}{3}\right)+\left(-\frac{2}{3}\right)^2+\cdots+\left(-\frac{2}{3}\right)^n\right)=\lim_{n\to\infty}\frac{3}{5}\left(1-\left(-\frac{2}{3}\right)^{n+1}\right)=\frac{3}{5}.$$

**7.**

| n | $P_{2n}(\pi/4)$ | $|\cos(\pi/4) - P_{2n}(\pi/4)|$ |
|---|---|---|
| 2 | 0.70743 | $3.2243 \times 10^{-4}$ |
| 3 | 0.7071 | $3.5664 \times 10^{-6}$ |
| 4 | 0.70711 | $2.4497 \times 10^{-8}$ |
| 5 | 0.70711 | $1.1462 \times 10^{-10}$ |

The numbers indicate that $\lim_{n\to\infty} P_{2n}(\pi/4) = \cos(\pi/4)$. Note that the decimal representations of $P_4(\pi/4)$ and $P_5(\pi/4)$ are the same as the decimal representation of $\cos(\pi/4)$, rounded to 6 significant digits, since the absolute error is less than $10^{-6}$.

## Answers to Some Problems of Section 9.3

**1.**
a)
$$S_n = \frac{4}{3}\left(1 - \frac{1}{4^n}\right)$$

b)
$$S = \lim_{n\to\infty}\frac{4}{3}\left(1 - \frac{1}{4^n}\right) = \frac{4}{3}$$

**3.**
a)
$$S_n = 4\left(1 - \left(\frac{3}{4}\right)^{n+1}\right) - 1$$

b)
$$S = \lim_{n\to\infty}\left[4\left(1 - \left(\frac{3}{4}\right)^{n+1}\right) - 1\right]$$

**5.**
a)
$$S_n = 1 - \frac{1}{n+1}$$

b)
$$S = \lim_{n\to\infty}\left(1 - \frac{1}{n+1}\right) = 1.$$

**7.**

| n | $S_n$ | $|2 - S_n|$ |
|---|---|---|
| 4 | 1.625 | 0.375 |
| 8 | 1.9609 | $3.9063 \times 10^{-2}$ |
| 16 | 1.9997 | $2.7466 \times 10^{-4}$ |
| 32 | 2.0000 | $7.9162 \times 10^{-9}$ |

The numbers support the claim that the sum of the series is 2. The decimal representation of $S_{32}$ is 2, rounded to 6 significant digits since the error $|S - S_{32}| < 10^{-6}$.

**9.** Since
$$\lim_{n\to\infty} \left(\frac{4}{3}\right)^n = \infty \neq 0$$
the series diverges.

**11.** By L'Hôpital's rule,
$$\lim_{n\to\infty} \frac{2^n}{n^2} = \infty \neq 0$$
so that the series diverges.

**13.** The sequence of partial sums is
$$1, 1, 0, 0, 1, 1, 0, 0, 1, 1, 0, 0, \ldots$$

This sequence does not converge to 0. Therefore, the given series diverges.

## Answers to Some Problems of Section 9.4

**1.** The series converges

**3.** The series diverges

**5.** The series converges absolutely

**7.** The series converges absolutely.

**9.** The series diverges.

**11.** The ratio test is inconclusive

**13.** The series converges absolutely

**15.** The series converges absolutely

**17.** The ratio test is inconclusive.

**19.**
b) We have
$$\sum_{n=1}^{\infty} \frac{n}{2^n} = 2.$$
The following table displays the required data:

| $n$ | $S_n$ | $|S_n - 2|$ |
|---|---|---|
| 2 | 1 | 1 |
| 4 | 1.625 | 0.375 |
| 8 | 1.9609 | $3.9063 \times 10^{-2}$ |
| 16 | 1.9997 | $2.7466 \times 10^{-4}$ |
| 32 | 2.0000 | $7.9162 \times 10^{-9}$ |

The numbers support the claim of the fast convergence of the sequence of partial sums to the sum of the series. Note that the decimal representation of $S_{32}$ is 2, rounded to 6 significant digits since $|S_{32} - 2| < 10^{-8}$.

**21.**
b) The following table displays the required data:

| $n$ | $\dfrac{n!}{4^n}$ |
|---|---|
| 2 | 0.125 |
| 4 | 0.09375 |
| 8 | 0.61523 |
| 16 | 4871.5 |
| 32 | $1.4264 \times 10^{16}$ |

Even though $a_n$ is small for $n = 2, 4$ and $8$, $a_n$ increases to $3.7289 \times 10^{50}$ when $n = 32$. Thus, the numbers definitely support the claim that $|a_n|$ grows rapidly as $n$ increases.

**23.** The series converges absolutely

**25.** The root test is inconclusive.

**27.** The series converges absolutely (note that the series is the geometric series with $x = -1/3$).

**29.** The series converges absolutely.

**31.**
b) We have
$$\sum_{n=1}^{\infty} \frac{n}{3^n} = \frac{3}{4} = 0.75.$$

The following table displays the required data:

| $n$ | $S_n$ | $\left|S_n - \frac{3}{4}\right|$ |
|---|---|---|
| 2 | 0.55556 | 0.19444 |
| 4 | 0.71605 | $3.3951 \times 10^{-2}$ |
| 8 | 0.74928 | $7.2398 \times 10^{-4}$ |
| 16 | 0.75000 | $2.0327 \times 10^{-7}$ |
| 32 | 0.75000 | $9.0393 \times 10^{-15}$ |

The numbers support the claim of the fast convergence of the sequence of partial sums to the sum of the series. Note that the decimal representations of $S_{16}$ and $S_{32}$ are 0.75, rounded to 6 significant digits since the absolute errors are less than $10^{-6}$.

**33.**
b) The following table displays the required data:

| $n$ | $|a_n|$ |
|---|---|
| 2 | 4.5 |
| 4 | 20.25 |
| 8 | 820.13 |
| 16 | $2.6904 \times 10^6$ |
| 32 | $5.7907 \times 10^{13}$ |

The numbers support the claim that $|a_n|$ grows rapidly as $n$ increases.

## Answers to Problems of Section 9.5

1. Radius of convergence is $1/2$ and the open interval of convergence is $(-1/2, 1/2)$.

3. The radius of convergence is $1/4$ and the open interval of convergence is $(-1/4, 1/4)$.

5. The radius of convergence is $e$ and the open interval of convergence is $(0, 2e)$.

7. The power series converges absolutely on the entire number line. Its radius of convergence is $\infty$.

9.
$$\frac{1}{1+x^2} = 1 - x^2 + x^4 - x^6 + \cdots + (-1)^n x^{2n} + \cdots = \sum_{n=0}^{\infty} (-1)^n x^{2n}.$$

11.
$$e^{-x^2} = 1 - x^2 + \frac{1}{2}x^4 - \frac{1}{3!}x^6 + \cdots + \frac{(-1)^n}{n!}x^{2n} + \cdots = \sum_{n=0}^{\infty} \frac{(-1)^n}{n!}x^{2n}.$$

13.
$$\cosh(x) = 1 + \frac{1}{2}x^2 + \frac{1}{4!}x^4 + \frac{1}{6!}x^6 + \cdots = \sum_{n=0}^{\infty} \frac{1}{(2n)!}x^{2n}.$$

15.
$$\frac{\sin(x) - x + \dfrac{x^3}{6}}{x^5} = \frac{1}{5!} - \frac{1}{7!}x^2 + \frac{1}{9!}x^4 - \frac{1}{11!}x^6 + \cdots + \frac{(-1)^n}{(2n+1)!}x^{2n-4} + \cdots$$
$$= \sum_{n=2}^{\infty} \frac{(-1)^n}{(2n+1)!}x^{2n-4}$$

17.
$$\lim_{x \to 0} \frac{\cos(x) - 1}{x^2} = \lim_{x \to 0} \left(-\frac{1}{2} + \frac{1}{4!}x^2 - \frac{1}{6!}x^4 + \cdots\right) = -\frac{1}{2}$$

19.
$$\lim_{x \to 0} \frac{\sin(x) - x + \dfrac{x^3}{6}}{x^5} = \lim_{x \to 0} \left(\frac{1}{5!} - \frac{1}{7!}x^2 + \cdots\right) = \frac{1}{5!} = \frac{1}{120}$$

## Answers to Some Problems of Section 9.6

1.
$$F(x) = x - \frac{1}{12}x^3 + \frac{1}{160}x^5 - \frac{1}{2688}x^7 + \frac{(-1)^n}{n! 4^n (2n+1)}x^{2n+1} + \cdots$$

The expansion is valid for each $x \in \mathbb{R}$.

3.
$$\operatorname{Si}(x) = x - \frac{1}{3!(3)}x^3 + \frac{1}{5!(5)}x^5 - \frac{1}{7!(7)}x^7 + \cdots + \frac{(-1)^n}{(2n+1)!(2n+1)}x^{2n+1} + \cdots$$

The expansion is valid for each $x \in \mathbb{R}$.

5.
$$F(x) = \frac{1}{3}x^3 - \frac{1}{5}x^5 + \frac{1}{2(7)}x^7 - \frac{1}{3!(9)}x^9 + \cdots + \frac{(-1)^n}{n!(2n+3)}x^{2n+3} + \cdots$$

The expansion is valid for each $x \in \mathbb{R}$.

7. An antiderivative for $e^x / (1 + x^2)$ is
$$x + \frac{1}{2}x^2 - \frac{1}{6}x^3 - \frac{5}{24}x^4 + \frac{13}{120}x^5 + \cdots$$

9.
$$\frac{x^2}{1 + x^2} = x^2 - x^4 + x^6 - x^8 + x^{10} + \cdots$$

## Answers to Some Problems of Section 9.7

1. The series converges absolutely.   3. The series converges absolutely.

5.
$$S_{1001} \cong 0.65941$$

and
$$S_{1001} \leq S \leq S_{1000} + 10^{-3} \Rightarrow 0.6594 < S < 0.6604$$

The decimal representation of the exact value of $S$ is 0.6604, rounded to 6 significant digits.

7. The series converges absolutely.   11. The series diverges.

9. The series converges absolutely.   13. The series diverges.

## Answers to Some Problems of Section 9.8

5. The given series converges conditionally.

7. The series diverges.

9. The series converges conditionally.

11. The series converges absolutely.

13. The series converges absolutely.

15. The open interval of convergence is $(-1, 1)$. The series converges conditionally at $x = -1$ and diverges at $x = 1$.

17. The open interval of convergence of the series is $(1, 3)$. The series diverges at $\pm 1$.

## Answers to Problems of Section 9.9

1.
a) The Fourier series of the function is

$$\frac{2}{3}\pi^2 + 4\sum_{j=1}^{\infty}\left(\frac{1}{(2j-1)^2}\cos((2j-1)x) - \frac{1}{(2j)^2}\cos(2jx)\right)$$

c) We have

$$F_n(\pi/2) = \frac{2}{3}\pi^2 + \sum_{l=1}^{n/4}\left(\frac{1}{(2l-1)^2} - \frac{1}{4l^2}\right),$$

and

$$F_n(\pi) = \frac{2}{3}\pi^2 - 4\sum_{j=1}^{j=n/2}\left(\frac{1}{(2j-1)^2} + \frac{1}{4j^2}\right)$$

The following tables show $F_n(\pi/2)$, $|F_n(\pi/2) - f(\pi/2)|$ and $F_n(\pi)$ for $n = 4, 8, 16, 32, 64$ (We have $f(\pi/2) = \frac{3}{4}\pi^2 \cong 7.4022$ and $f(\pi) = 0$):

| $n$ | $F_n(\pi/2)$ | $|F_n(\pi/2) - f(\pi/2)|$ |
|---|---|---|
| 4 | 7.3297 | $7.2467 \times 10^{-2}$ |
| 8 | 7.3783 | $2.3856 \times 10^{-2}$ |
| 16 | 7.3954 | $6.8505 \times 10^{-3}$ |
| 32 | 7.4004 | $1.8315 \times 10^{-3}$ |
| 64 | 7.4017 | $4.7304 \times 10^{-4}$ |

The numbers indicate that $\lim_{n\to\infty} F_n(\pi/2) = f(\pi/2)$.

| $n$ | $F_n(\pi)$ |
|---|---|
| 4 | 6.5797 |
| 8 | 0.47005 |
| 16 | 0.24235 |
| 32 | 0.12307 |
| 64 | $6.2014 \times 10^{-2}$ |

The numbers indicate that $\lim_{n\to\infty} F_n(\pi) = f(\pi) = 0$.

d) We have

$$F_8(x) = \frac{2}{3}\pi^2 + 4\sum_{j=1}^{4}\left(\frac{1}{(2j-1)^2}\cos((2j-1)x) - \frac{1}{(2j)^2}\cos(2jx)\right)$$

The figure shows the graphs of $f$ and $F_8$. The graphs are not distinguishable from each other. This indicates that $F_8(x)$ approximates $f(x)$ well for each $x \in [-3\pi, 3\pi]$, as to be expected since $f$ does not have any discontinuities.

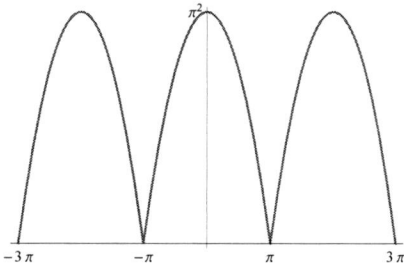

3.
a) The Fourier series for $f$ is

$$\frac{4}{\pi}\left(\sin(\pi x) + \frac{1}{3}\sin(3\pi x) + \frac{1}{5}\sin(5\pi x) + \cdots + \frac{1}{(2j-1)}\sin((2j-1)\pi x) + \cdots\right)$$
$$= \frac{4}{\pi}\sum_{j=1}^{\infty}\frac{1}{(2j-1)}\sin((2j-1)\pi x).$$

c) We have

$$F_n(1/2) = \frac{4}{\pi}\sum_{l=1}^{n/4}\frac{2}{16l^2 - 16l + 3}$$

The following table shows $F_n(1/2)$ and $|F_n(1/2) - f(1/2)| = |F_n(1/2) - 1|$ for $n = 4, 8, 16, 32, 64$:

| $n$ | $F_n(1/2)$ | $|F_n(1/2) - f(1/2)|$ |
|---|---|---|
| 4 | 0.84883 | 0.15117 |
| 8 | 0.92158 | $7.8417 \times 10^{-2}$ |
| 16 | 0.96036 | $3.9636 \times 10^{-2}$ |
| 32 | 0.98012 | $1.9875 \times 10^{-2}$ |
| 64 | 0.99006 | $9.9448 \times 10^{-3}$ |

The numbers indicate that $\lim_{n\to\infty} F_n(1/2) = f(1/2)$.

d)

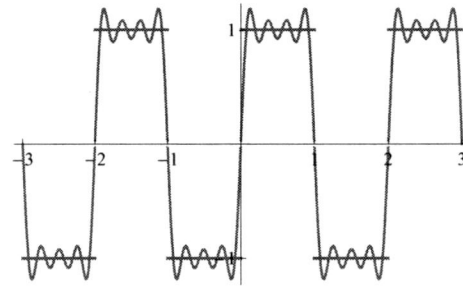

The picture suggests that $F_8(x)$ approximates $f(x)$ if $f$ is not discontinuous at $x$. The average of the right and left limits of $f$ at the jump discontinuities is 0 and $F_8$ at these points has value 0. This is consistent with the fact that the limit of the sequence $\{F_n\}_{n=1}^{\infty}$ is predicted to converge to 0 by the general theory.

## Answers to Some Problems of Section 10.1

**1.**
a)

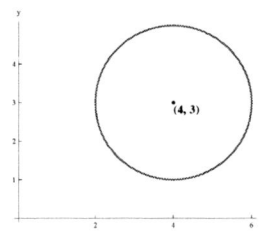

b) $C$ is a circle of radius 2 centered at $(4, 3)$.

**3.**

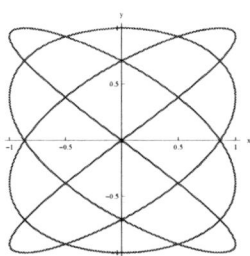

**5.**
a)

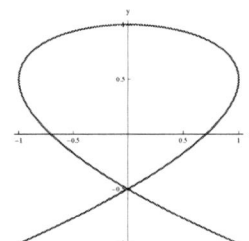

The picture shows the curve that is parametrized both by $\sigma_1$ and $\sigma_2$.

b)
$$\sigma_1\left(\frac{\pi}{2}\right) = \left(\sin\left(\frac{3\pi}{2}\right), \cos(\pi)\right) = (-1, -1),$$

whereas
$$\sigma_2\left(\frac{\pi}{2}\right) = (\sin(3\pi), \cos(2\pi)) = (0, 1).$$

**7.**
$$\sigma(x) = (x, \cos(x)), \quad -2\pi \leq x \leq 2\pi.$$

**9.**
a) Let's set
$$x(t) = \sin(3t) \text{ and } y(t) = \cos(2t).$$

We have
$$\frac{dx}{dt} = 3\cos(3t) \text{ and } \frac{dy}{dt} = -2\sin(2t).$$

Thus,
$$\frac{dx}{dt}\left(\frac{\pi}{4}\right) = 3\cos\left(\frac{3\pi}{4}\right) = 3\left(-\frac{\sqrt{2}}{2}\right) = -\frac{3\sqrt{2}}{2} \neq 0.$$

Therefore, there exists an open interval $J$ that contains $\pi/4$ such that the part of $C$ corresponding to $t \in J$ is the graph of a function $y(x)$.

b) The tangent line to $C$ at
$$\sigma(\pi/4) = (\sin(3\pi/4), \cos(\pi/2)) = \left(\frac{\sqrt{2}}{2}, 0\right)$$

is the graph of the equation
$$y = \frac{2\sqrt{2}}{3}\left(x - \frac{\sqrt{2}}{2}\right).$$

c)

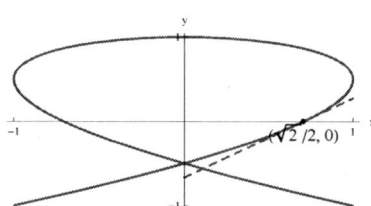

The picture suggests that we have determined the required tangent line.

**11.**
a)

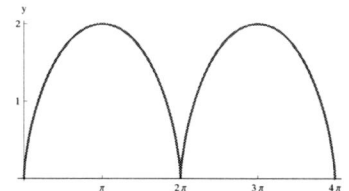

364               APPENDIX I. ANSWERS TO SOME PROBLEMS

The picture indicates that $C$ has a cusp at $(x(2\pi), y(2\pi)) = (2\pi, 0)$. In particular, $C$ does not appear to be the graph of a function of $x$ that is differentiable at $2\pi$.

c) The picture of part a) supports our response.

## Answers to Some Problems of Section 10.2

1. 
$$x = \sqrt{3}, \ y = 1.$$

3. 
$$x = -\frac{3}{2}, \ y = \frac{3\sqrt{3}}{2}.$$

5. 
$$x = \sqrt{3}, \ y = -1.$$

7. 
$$r = 4, \ \theta = 0.$$

9. 
$$r = 2, \ \theta = -\frac{\pi}{2}.$$

11. 
$$r = 4, \ \theta = \frac{5\pi}{6}.$$

13.

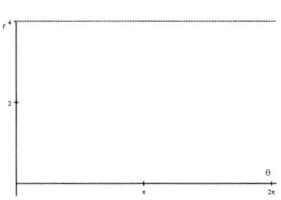

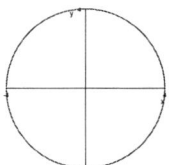

15.

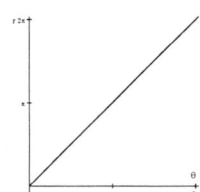

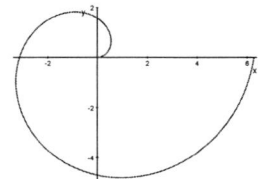

17.

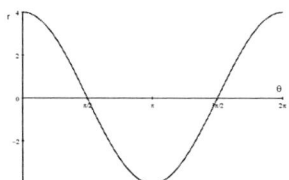

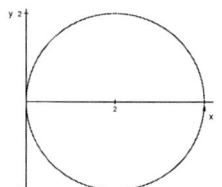

19.

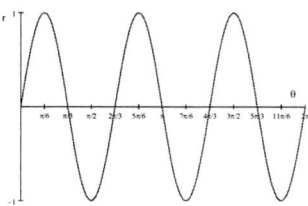

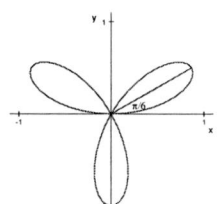

21.

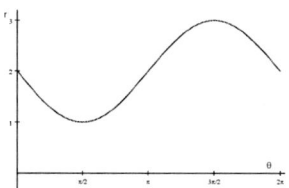

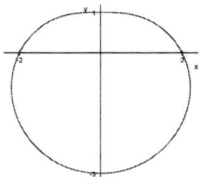

## Answers to Some Problems of Section 10.3

**1.**

a) The required tangent line is the graph of

$$y = \frac{3}{2} - \sqrt{3}\left(x - \frac{\sqrt{3}}{2}\right).$$

b)

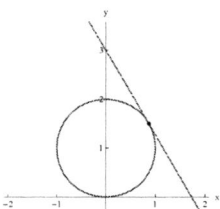

**3.**

a) The required tangent line is the graph of

$$\begin{aligned} y &= y(3\pi/4) + \left(\frac{31}{28} - \frac{3}{14}\sqrt{2}\right)(x - x(3\pi/4)) \\ &= 2\sqrt{2} + \frac{1}{2} + \left(\frac{31}{28} - \frac{3}{14}\sqrt{2}\right)\left(x + 2\sqrt{2} + \frac{1}{2}\right) \end{aligned}$$

b)

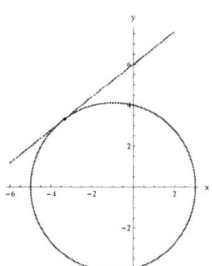

**5.**

a)

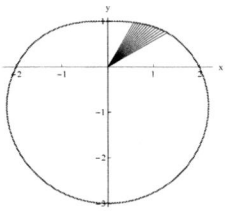

$G$ is inside the graph, between the rays $\theta = \pi/6$ and $\theta = \pi/3$.

b)
$$Area = \frac{1}{2}\int_{\pi/6}^{\pi/3} (2 - \sin(\theta))^2 \, d\theta = \frac{3}{8}\pi - \sqrt{3} + 1$$

**7.**

a)

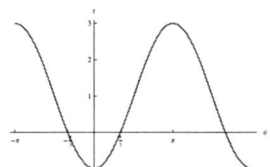

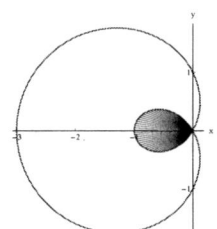

b) The area of the region in the inner loop is
$$\frac{1}{2}\int_{-\pi/3}^{\pi/3} (1 - \cos(\theta))^2 \, d\theta = \frac{1}{2}\pi - \frac{7}{8}\sqrt{3}$$

## Answers to Some Problems of Section 10.4

**1.**

a) The length of $C$ is
$$\int_{\pi/4}^{\pi} 3 \, dt = 3\left(\pi - \frac{\pi}{4}\right) = \frac{9\pi}{4}.$$

b)
$$\int_{\pi/4}^{\pi} 3 \, dt = 7.068\,6$$

c) There is no need for a CAS.

**3.**

a) The length of $C$ is
$$\int_{-1}^{1} \sqrt{\sinh^2(t) + \cosh^2(t)}\, dt$$

b)
$$\int_{-1}^{1} \sqrt{\sinh^2(t) + \cosh^2(t)}\, dt \cong 2.634\,38$$

c) According to Mathematica 7,
$$\int \sqrt{\sinh^2(t) + \cosh^2(t)}\, dt = -i\text{EllipticE}\,(it, 2)$$

($i = \sqrt{-1}$). Even though the expression involves the imaginary number $i$, the evaluation of such an integral by Mathematica yields the correct real number. For example, according to Mathematica,

$$\int_{-1}^{1} \sqrt{\sinh^2(t) + \cosh^2(t)}\, dt = -2i\text{EllipticE}\,(i, 2) = 2.63438 - 5.55112 \times 10^{-16} i.$$

It appears that Mathematica evaluates the expression $-2i\text{EllipticE}(i, 2)$ in a way that involves $i$ with a coefficient of order $10^{-16}$ that should be ignored.

**5.**
a)
$$\int_0^{\pi/4} \sqrt{4\sin^2(2\theta) + \cos^2(2\theta)}\, d\theta$$

b)
$$\int_0^{\pi/4} \sqrt{4\sin^2(2\theta) + \cos^2(2\theta)}\, d\theta \cong 1.211\,06$$

c) According to Mathematica 7,
$$\int \sqrt{4\sin^2(2\theta) + \cos^2(2\theta)}\, d\theta = \frac{1}{2}\text{EllipticE}\,(2\theta, -3).$$

The special function EllipticE is not in our portfolio of special functions.

**7.**
a)
$$\int_{-\pi/8}^{\pi/8} \sqrt{16\sin^2(4\theta) + \cos^2(4\theta)}\, d\theta$$

b)
$$\int_{-\pi/8}^{\pi/8} \sqrt{16\sin^2(4\theta) + \cos^2(4\theta)}\, d\theta \cong 2.144\,61$$

c) According to Mathematica 7,
$$\int \sqrt{16\sin^2(4\theta) + \cos^2(4\theta)}\, d\theta = \frac{1}{4}\text{EllipticE}\,(4\theta, -15).$$

The special function EllipticE is not in our portfolio of special functions.

## Answers to Some Problems of Section 10.5

**1.**
a) The graph is a parabola. The focus is at $(4,0)$ and the directrix is the line $x = -4$.
b)

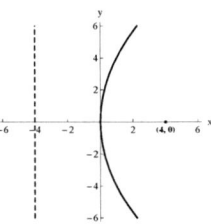

**3.**
a) The graph is an ellipse with its major axis along the $x$-axis. The foci are at $(\pm\sqrt{5}, 0)$.
b)

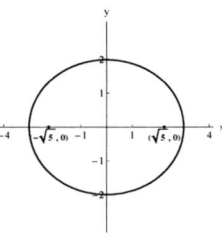

**5.**
a) The graph is a hyperbola that does not intersect the $y$-axis. The foci are at $(\pm\sqrt{13}, 0)$. The hyperbola intersects the $x$-axis at $(\pm 3, 0)$. The lines

$$y = \pm\frac{2}{3}x$$

are the asymptotes.
b)

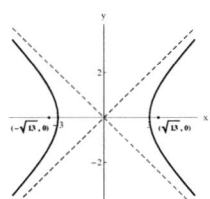

## Answers to Some Problems of Section 10.6

**1.**
a) The conic section is an ellipse with eccentricity $1/2$.

b)

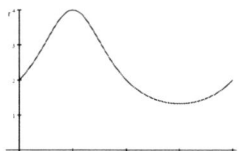

c)

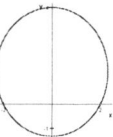

3.
a) The conic section is a parabola.
b) The graph has a vertical asymptote at $\theta = \pi$.

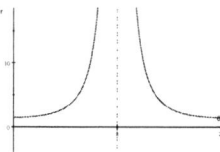

c)

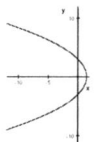

5.
a) The conic section is a hyperbola with eccentricity $\sqrt{2} \cong 1.41421$.
b) The graph has vertical asymptotes at $\pi/4$ and $7\pi/4$.

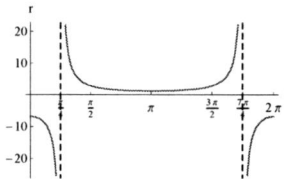

c)

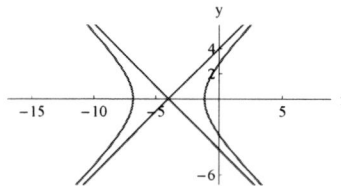

(The lines are asymptotic to the graph).

# Appendix J

# Basic Differentiation and Integration formulas

### Basic Differentiation Formulas

1. $\dfrac{d}{dx} x^r = r x^{r-1}$

2. $\dfrac{d}{dx} \sin(x) = \cos(x)$

3. $\dfrac{d}{dx} \cos(x) = -\sin(x)$

4. $\dfrac{d}{dx} \sinh(x) = \cosh(x)$

5. $\dfrac{d}{dx} \cosh(x) = \sinh(x)$

6. $\dfrac{d}{dx} \tan(x) = \dfrac{1}{\cos^2(x)}$

7. $\dfrac{d}{dx} a^x = \ln(a) a^x$

8. $\dfrac{d}{dx} \log_a(x) = \dfrac{1}{x \ln(a)}$

9. $\dfrac{d}{dx} \arcsin(x) = \dfrac{1}{\sqrt{1-x^2}}$

10. $\dfrac{d}{dx} \arccos(x) = -\dfrac{1}{\sqrt{1-x^2}}$

11. $\dfrac{d}{dx} \arctan(x) = \dfrac{1}{1+x^2}$

### Basic Antidifferentiation Formulas

$C$ denotes an arbitrary constant.

1. $\int x^r \, dx = \dfrac{1}{r+1} x^{r+1} + C \ (r \neq -1)$

2. $\int \dfrac{1}{x} \, dx = \ln(|x|) + C$

3. $\int \sin(x) \, dx = -\cos(x) + C$

4. $\int \cos(x) \, dx = \sin(x) + C$

5. $\int \sinh(x) \, dx = \cosh(x) + C$

6. $\int \cosh(x) \, dx = \sinh(x) + C$

7. $\int e^x \, dx = e^x + C$

8. $\int a^x \, dx = \dfrac{1}{\ln(a)} a^x + C \ (a > 0)$

9. $\int \dfrac{1}{1+x^2} \, dx = \arctan(x) + C$

10. $\int \dfrac{1}{\sqrt{1-x^2}} \, dx = \arcsin(x) + C$

# Index

Area of a Surface of Revolution, 106

Conic Sections
    Conic sections in polar coordinates, 328
        focus and directrix, 328
    Ellipses, 321
        Foci, 321
    Hyperbolas, 324
        Foci, 324
    Parabolas, 319
        Directrix, 319
        Focus, 319

Differential Equations
    equilibrium solutions, 139
        stable, 139
        unstable, 139
    Euler's method, 181
    Linear differential equations, 135
        integrating factor, 136, 144
    Logistic equation, 176
    Newton's law of cooling, 153
    Newtonian damping, 173
    Separable differential equations, 159
    Slope fields, 185
    steady-state solutions, 139
    Viscous damping, 151

Gaussian Elimination, 21

Improper Integrals, 66
    Comparison theorems, 81
Infinite Series
    Absolute convergence, 219, 220
    Alternating series, 266
    Comparison test, 261
    Concept of an infinite series, 209
    Conditional convergence, 220, 266
    Fourier series, 274
    Geometric series, 212
    Harmonic series, 217
    Integral test, 255
    Limit comparison test, 263
    Ratio test, 219, 221
    Root test, 219, 223
Integration Techniques
    Integration by parts, 1
        for definite integrals, 10
        for indefinite integrals, 1
    Integration of hyperbolic functions, 28
    Integration of rational functions, 14
    Integration of trigonometric functions, 28
    Numerical integration, 55
        midpoint rule, 56
        Simpson's rule, 60
        trapezoid rule, 58
    Trigonometric and hyperbolic substitutions, 44

Length of a Graph, 101

Maclaurin Polynomials, 199
Maclaurin's Polynomials, 191
Mass and Density, 113
Monotone Convergence Principle, 219

Parametrized Curves, 287
    Arc length, 313
    Tangents to parametrized curves, 290
Partial Fraction Decomposition, 18
Polar Coordinates, 296
    Arc length in polar coordinates, 316
    Area in polar coordinates, 310
    Tangents to curves in polar coordinates, 308
Power Series, 230
    Binomial series, 251
    Differentiation of power series, 234
    Integration of power series, 243
    Interval of convergence, 232
    Maclaurin series, 231, 243
    Radius of convergence, 232
    Taylor series, 230, 243
Probability, 123
    Random variable, 124
        distribution function, 127
        mean, 129
        normal distribution, 130

# INDEX

    probability density function, 124
    variance, 129

Taylor Polynomials, 189, 199
    Taylor's formula for the remainder, 201, 337

Volumes by Cylindrical Shells, 98
Volumes by Disks and Washers, 94
Volumes by Slices, 91

Work, 116

CPSIA information can be obtained at www.ICGtesting.com
Printed in the USA
LVOW030744100812

293759LV00003B/22/P